Innovation in aeronautics

Related titles:

Welding and joining of aerospace materials
(ISBN 978-1-84569-532-3)
Welding and joining techniques play an essential role in both the manufacture and in-service repair of aerospace vehicles and components. This important book provides in-depth information on different techniques for joining metallic and non-metallic aerospace materials and their applications. Part I reviews different types of welding such as inertia friction, laser and hybrid laser-arc welding and issues relating to their use. The second part focuses on other joining techniques such as riveting, bonding and brazing and methods of assessing their quality and effectiveness. Finally, an important appendix to this book covers linear friction welding in the aerospace industry.

Introduction to aerospace materials
(ISBN 978-1-85573-946-8)
The structural materials used in airframe and propulsion systems influence the cost, performance and safety of aircraft. This essential new book is intended for undergraduate students studying aerospace and aeronautical engineering, and is also a valuable resource for postgraduate students and practising aerospace engineers. The first three chapters introduce the reader to aerospace materials. The next group of chapters go on to discuss the properties, production and metallurgy of materials for aerospace structures, followed by chapters on performance issues and recycling. The final chapter covers materials selection for aerospace structures and engines.

Leveraging information technology for optimal aircraft maintenance, repair and overhaul (MRO)
(ISBN 978-1-84569-982-6)
Aimed at professionals in the aviation industry and students of aircraft maintenance, this book covers the use and management of technology in the aviation maintenance, repair, and overhaul (MRO) context. The book provides a backdrop of current trends in the industry where airlines are tending to retain their aircraft longer on the one hand, and rapidly introducing a new genre of aircraft such as the A380 and B787 into service. This book provides industry professionals and students of aviation MRO (at all levels) with the necessary principles, approaches and tools to respond effectively and efficiently to the constant development of new technologies – both in general and within the aviation MRO profession.

Details of these and other Woodhead Publishing materials books can be obtained by:

- visiting our web site at www.woodheadpublishing.com
- contacting Customer Services (e-mail: sales@woodheadpublishing.com; fax: +44 (0) 1223 832819; tel.: +44 (0) 1223 499140 ext. 130; address: Woodhead Publishing Limited, 80 High Street, Sawston, Cambridge CB22 3HJ, UK)
- in North America, contacting our US office (e-mail: usmarketing@ woodheadpublishing.com; tel.: (215) 928 9112; address: Woodhead Publishing, 1518 Walnut Street, Suite 1100, Philadelphia, PA 19102-3406, USA)

If you would like e-versions of our content, please visit our online platform: www. woodheadpublishingonline.com. Please recommend it to your librarian so that everyone in your institution can benefit from the wealth of content on the site.

Innovation in aeronautics

Edited by
Trevor M. Young and Mike Hirst

WP

**WOODHEAD
PUBLISHING**

Oxford   Cambridge   Philadelphia   New Delhi

Published by Woodhead Publishing Limited,
80 High Street, Sawston, Cambridge CB22 3HJ, UK
www.woodheadpublishing.com
www.woodheadpublishingonline.com

Woodhead Publishing, 1518 Walnut Street, Suite 1100, Philadelphia,
PA 19102-3406, USA

Woodhead Publishing India Private Limited, G-2, Vardaan House,
7/28 Ansari Road, Daryaganj, New Delhi – 110002, India
www.woodheadpublishingindia.com

First published 2012, Woodhead Publishing Limited
© Woodhead Publishing Limited, 2012 except Chapter 6 © Ian Poll, 2012 and Chapter 15
© E. Murman, 2012.
The authors have asserted their moral rights.

British Library Cataloguing in Publication Data
A catalogue record for this book is available from the British Library.

Library of Congress Control Number: 2012936686

ISBN 978-1-84569-550-7 (print)
ISBN 978-0-85709-609-8 (online)

The publisher's policy is to use permanent paper from mills that operate a sustainable forestry policy, and which has been manufactured from pulp which is processed using acid-free and elemental chlorine-free practices. Furthermore, the publisher ensures that the text paper and cover board used have met acceptable environmental accreditation standards.

Typeset by RefineCatch Limited, Bungay, Suffolk
Printed by Lightning Source

Contents

D. HARRIS, HFI Solutions Ltd, UK

H. SMITH, Cranfield University, UK

M. HENSHAW, Loughborough University, UK

Contributor contact details

(* = main contact)

Editors and Chapters 1, 5, 12 and 16

Dr Trevor M. Young
Department of Mechanical and
    Aeronautical Engineering
University of Limerick
Ireland

Email: trevor.young@ul.ie

Mike Hirst
Independent consultant
UK

Email: mhirst883@btinternet.com

Chapter 2

Y. Bar-Cohen
Jet Propulsion Laboratory (JPL)
California Institute of Technology
Pasadena
CA
USA

Email: yosi@jpl.nasa.gov

Chapter 3

R. W. Guiler, PhD
Principal Research Scientist
Physical Sciences, Inc.
20 New England Business Center
Andover
Massachusetts 01810

Email: guiler@psicorp.com

S. D. Hamburg
WVU Research Corporation
Aerospace Engineer
B-47 Engineering Sciences Bldg
Morgantown
West Virginia 26506

Email: Shanti.Hamburg@mail.wvu.edu

Professor W. W. Huebsch*
Department of Mechanical and
    Aerospace Engineering
West Virginia University
331 Engineering Sciences Bldg
Morgantown
West Virginia 26506

Email: Wade.Huebsch@mail.wvu.edu

Chapter 4

R. Singh,* G. Ameyugo and Dr F. Noppel
Cranfield University
Cranfield, Bedfordshire MK43 OAL
UK

Email: r.singh@cranfield.ac.uk

Chapter 6

D. I. A. Poll
Cranfield University
Cranfield, Bedfordshire MK43 OAL
UK

Email: D.I.A.Poll@cranfield.ac.uk

Chapter 7

Don Harris
HFI Solutions Ltd
26 Bradgate Road
Bedford MK40 3DE
UK

Email: don.harris@hfisolutions.co.uk

Chapter 8

H. Smith
Cranfield University
Cranfield, Bedfordshire MK43 OAL
UK

Email: howard.smith@cranfield.ac.uk

Chapter 9

Professor Michael J. de C. Henshaw
Loughborough University
Leicestershire LE11 3TU
UK

Email: M.J.d.Henshaw@lboro.ac.uk

Chapter 10

R. Henke
RWTH Aachen University
Templergraben 55
Germany

Email: rolf.henke@dlr.de

Chapter 11

J. R. Wilson (freelance writer)
Las Vegas, NV
USA

Email: Scribe@TheFreelancer.com

Chapter 13

D. A. McCarville (Boeing)
The Boeing Company
16918 223rd Ave E
Orting
Washington 98360
USA

Email: douglas.a.mccarville@boeing.com

Chapter 14

T. Browning
Texas Christian University
Fort Worth Texas 76129
USA

Email: t.browning@tcu.edu

Chapter 15

E. M. Murman
MIT
325 Lincoln St
Port Townsend
WA 98368
USA

Email: murman@mit.edu

1

Introduction to innovation in aeronautics

T.M. YOUNG, University of Limerick, Ireland, and M. HIRST,
Independent consultant, UK

1.1 Introduction

Innovation is a natural talent of humanity. Over several recorded millennia, curiosity has spurred discovery and the creation of artefacts that utilised natural substances or phenomena to serve human needs. The goal has been, often, to achieve something more quickly or with less effort or, in some cases, to extend curiosity by setting new goals.

Aeronautics has been one of the most recent and most significant of the latter human pursuits, at first simply answering a need expressed by curiosity-struck philosophers, who had the desire to view or traverse the earth as a bird does. Kites and balloons, the innovators' first 'flight' inventions, were significant flying machines, but in practical terms they had limited capability. Consequently, human conquest of the air is largely attributed to the attainment of manned flight using a wing-borne machine.

From the 1600s onwards, philosophers and spirited individuals conducted experiments that proved that man could be carried aloft on wings, but it was the Wright brothers who first melded wing-borne and controllable, powered flight together. In essence, the essential elements of aeronautical engineering had materialised much earlier than 17 December 1903, but on that date, like a toddler now taking its first steps, the child began to walk and to reveal how profound and significant were the visions of these extraordinary brothers. The world, at that time, had no idea as to where these visions might lead; within little over half a century their invention had opened global air travel, created weapon systems of devastating capability, and carried humans to the fringes of the earth's atmosphere.

Many of the readers of this book will be doyens in the aeronautics profession. They will have witnessed, and indeed been part of, this remarkable journey. What was achieved by visionaries who joined the business 50 years ago has been astonishing – from every superlative-glossed viewpoint. In civil aviation, aircraft have become faster and bigger, and the world has shrunk. Indeed, from certain hub airports, such as those located in the Middle East, modern long-range airliners can travel to every other major airport in the world. In military aviation, the ability to

1

wage war from the air has been through several metamorphoses: bombers led the development of large, high-speed, high-altitude aircraft, but today large aircraft are relegated, on the whole, to transport roles, and there is an emphasis on more on-board multi-role capability to deliver ordnance more effectively. Today, unmanned air vehicles prosper, and they vie with helicopters and novel vertical take-off and landing (VTOL) platforms to re-invent the answer to military role requirements. The scene is as dynamic as ever before. Major civil and military projects, however, have become gargantuan, and the number of nations that have the resources to lead them has declined. Multi-national alliances and collaborations are now commonplace, with supply chains that stretch across the globe. The situation is, however, ever-changing. Wars, global financial upheavals and large-scale socio-political events continuously alter this landscape. Opportunities shine ever more brightly for the 'emerging' economies; the so-called BRIC nations (Brazil, Russia, India and China) leading the way, with several Asian nations being only a half-step behind. Innovation metrics, such as the number of patents registered, PhDs conferred and scientific papers published, support anecdotal evidence of the ever-increasing scientific and manufacturing capabilities of these countries. But to leap ahead they need to do more than emulate the trend-setters; they need to become more innovative and to take risks by adopting novel, unconventional strategies.

There is a tacit belief that what has been exciting and innovative for aeronautical engineers of the past would remain true for subsequent generations, but times have changed, and it is worth considering what lies ahead for today's graduates. It is self-evident that, apart from the advent of stealth as a state-of-the-art technology, there have been no truly profound technical developments – say, on the scale of the jet engine – for several decades now. Yes, the engine has progressed from being a relatively simple machine to a highly-optimised, exceedingly complex machine, but the thermodynamic design is essentially unchanged, and the likelihood of us witnessing many more profoundly innovative developments of this magnitude now seems increasingly remote. Similarly, the aerodynamic shape of airliners has seemed to atrophy, with the swept-wing, tube-shaped fuselage configuration, as exemplified by the Boeing 707 (which first flew in 1957), likely to dominate civil aviation for, at least, another 40 years. Military aircraft types have adopted recognisable role-related shapes, with relatively few variations. Even new space vehicle variations have been slow to emerge. That does not mean, of course, that innovation in the industry has stopped; just that its nature has changed as the industry has matured. Novel, incremental developments take place all the time, with each new generation of aircraft containing lighter, stiffer, stronger materials, more capable navigation and control systems and quieter, more fuel-efficient engines. But – and this is central to what is explored in this book – we ask: Is this enough? Will these evolutionary developments deliver the step-change improvements needed to address future challenges?

There is less need to address issues related to the development of faster, bigger or more powerful aeroplanes. But aeronautics does face challenges, and it does still cry for innovation, and that is because there are new frontiers on which developments are essential. Leading amongst these are ever-growing environmental issues related to the ability of our planet to cope with anthropogenic environmental changes. The key issues now concern sustainability – for example: the continued ability of the earth's natural resource stocks to supply raw materials for the manufacture of new aircraft; the diminishing crude oil reservoirs, which are needed to satisfy the ever-increasing demand for aviation fuel; the increasing levels of carbon dioxide and other aviation-related emissions in the atmosphere and in the oceans; and the limited ability of the natural world to adjust to these changes. The design 'drivers' of new aircraft are evolving in response to new priorities. It is not just about more fuel-efficient engines, lower drag airframes and quieter engines. There are also profound social question marks over whether the prolific use of aircraft for business communication, sight-seeing and pleasure activities can be justified at the scale that is evident in certain societies today. Furthermore, the true 'cost' of transporting commodity items, in terms of its environmental impact, is not properly accounted for – for example, there is no environmental damage 'cost' added to the price of fresh fruit flown half-way around the world. Social changes are inevitable, and they will bring about new design priorities.

Capacity limits are another concern. With over 20000 jet airliners in service world-wide, the skies are getting so crowded in some regions as to require energetic operational changes. Many of the major international airports operate at or near capacity levels, most of the time. This factor supports the development of very large aircraft types, as slot constraints limit the number of daily operations at international hub airports. The political agenda is beginning to reflect the fact that this is an auspicious time for civil aviation, and governments find nothing enterprising when research offers just more of the same, only a little better – they want solutions that are radical and wide-ranging. Meanwhile, global market surveys from the suppliers to this domain remain confident that, unchecked, demand will grow and the skies will be more crowded than ever. Recent Boeing and Airbus forecasts suggest that there will be a 4–5% annual growth in global air travel, which will lead to a near doubling of traffic over the next 10–15 years.

Military policies, in most developed countries, are ham-strung by budgetary constraints, and already, in most nuclear-capable societies, the manned aircraft has abrogated the right to deliver deterrence to the long-range ballistic missile and the submarine. The roles of manned aircraft have been further curtailed by cruise missiles and newer-generation unmanned air vehicles, which have not just assumed the role of delivering deterrence, but have undertaken many of the traditional air-to-air and air-to-ground fighter and attack tasks. Air force personnel vie with their navy and army counterparts to justify their slice of the ever-decreasing financial cake, and the arena cries out for innovation to do more with less.

Getting new ideas from concept to fruition is not easy, however, so there has to be innovation, too, in the way that ideas are generated, nurtured and applied. The aerospace industry, by necessity, is dominated by large companies with strong vertically and laterally integrated capabilities. Such large entities, with their bureaucratic structures and legacy-design traditions, tend to stifle innovation as there is a tendency to adopt low-risk, incremental design developments over more radical alternatives. Creating a corporate climate in which radical ideas will not only be tolerated, but encouraged, is vital for the development of new technologies that can produce step-changes in capability or performance. Novelty often relies on under-developed methodologies and incomplete knowledge bases; it employs technologies that are not always fully understood. Consequently, there are risks that radical concepts will not yield the promised performance improvements or that there will be undesirable side-effects. Establishing such a corporate climate requires a tolerance and an understanding of the fact that many novel, radical concepts will fail. As there is a general lack of understanding of how humans innovate, it is difficult – and, for most managers, uncomfortable – to plan and manage development projects based on radical ideas that fly in the face of convention. It is possible that our poor understanding of the process that transforms initial, tentative ideas into useful products is the most limiting aspect of all.

Aeronautical products operate on a world-wide stage, and it is because the world is a changing place that there is reason to wish for new paradigms. In the latter part of the twentieth century, for example, the Cold War was a prime driver for military solutions, some of which would endure – stealth is one such technology. It is thought that this period of our history is over, but there are still political hot-spots that will have to be addressed, and it is in such spots that the past lingers. In other areas, there are hot-spots of a very different nature: national security is one of them, especially when facing the threat of guerrilla tactics with suicide 'troops' (unidentifiable from the civilian public) or when military targets are protected by 'human shields'. Will we see fleets of minute aircraft that can pass muster as townscape birds and pry to counter security threats? This is where innovation hits boundaries that have never been traversed before, and it is in such realms that new blood is likely to be innovative.

It is not easy to draw together these issues in an ordered manner, but in this book some of the world's leading specialists in evolutionary and revolutionary technologies, which could be the essential springboards for further development, present a series of papers that will serve to remind all, young and old, that there are legacies to carry and that new opportunities beckon. The papers have been grouped into three sections: Concepts, Change and Challenges.

1.2   Concepts

1.2.1   Biological inspiration

The book opens with a thought-provoking chapter on biologically-inspired technologies, compiled by Dr Yoseph Bar-Cohen, of the Jet Propulsion

Laboratories (JPL), California Institute of Technology, Pasadena, CA. His review considers the insect world, looking at macro- and micro-scale aerospace applications. His viewpoints stress how innovative aeronautical engineers can be, their vision roaming across conventional air vehicles and unmanned robotic systems for space exploration.

Bar-Cohen reviews simple analogies: wings are good for flying and fins for swimming, but he wonders if legs will replace the wheel – giving the phrase 're-inventing the wheel' a new meaning. New materials offer the possibility of 'muscular' components to provide mechanical motion, for steerage/control or even propulsion. His sights focus on many intriguing subjects, considering whether planetary exploration with novel systems, such as 'tumbleweed' robotic explorers, is sensible, and asks whether smart structures will break conventional boundaries. He considers the possibility of sensors being so integrated as to be classified as a part of the vehicle's structure.

1.2.2   Aircraft morphing

The potential to do something new, in terms of the shape of air vehicles, in the future is addressed by Professor Wade Huebsch, from West Virginia University, WV, and Dr R. W. Guiler (Andover, MA). This is a hard-engineering view of the arena that Poll (see Chapter 6) has described as rife for change. It looks at the engineering acumen needed to manage the physical attributes of the aerodynamic components of an aircraft to maximise its performance in forever demanding operational situations. Their views embrace many historical facets, starting from the Wright brothers' use of wing warping, and consider many ways of morphing, or changing, the aerodynamic shape of an aircraft's wing in the course of its mission.

Huebsch and Guiler hold attention with their charting of aircraft morphing experiments across numerous invaluable, albeit sometimes obscure or even forgotten, research aircraft. They remind us too of the numerous variable-geometry wing aircraft that have been in production over many decades, many still in front-line military service, and this leads into the current scene, where the latest and most promising new technologies are the stepping-stones for some innovators. In particular, they investigate alternative actuator technologies, ranging from small-scale rotary and actuation devices to novel shape-changing techniques that use solid-state electronic technology within structural components.

1.2.3   Jet engines

Professor Riti Singh, with colleagues at the School of Engineering, Cranfield University, UK, introduces the concept of disruptive technology: innovations that change the rules of the game. Starting with considerations of the most topical subject in this domain, the openrotor, the paper puts current thinking amongst engine specialists into perspective, in terms of efficiencies and operational attributes. This sets landmarks by illustrating the trends already experienced in gas-turbine development, and it outlines the expectations in the short term. There are considerations of biofuels and other alternative fuel technologies, such as fuel cells and the use of hydrogen. Issues that arise from such developments, including arable land usage, micro-algae production and the freezing properties of biofuels are discussed.

Innovative concepts such as smaller and more numerous engines within the airframe are shown to have a dichotomy of attributes that technical innovation could change, and the ever-present environmental impact considerations are introduced as essential future system design drivers. Innovative heat exchangers and pulse detonation technology, even the possible resuscitation of interest in concepts of the 1950s such as the compound Nomad engine, are given space.

1.2.4   Avionics

Co-editor Mike Hirst looks at the evolution of avionic systems through technology (analogue to digital) and integration, affecting the architecture of avionic systems, and he shows how these systems are evolving through wider synthesis. This can be on the aircraft, or it could reside anywhere else within the systems that support an aircraft throughout its operational life.

There is consideration of how the human factors aspect is being integrated such that aircraft crews in the future will have a persona-less 'associate' whose guidance will be, ideally, pure wisdom, but the implementation is difficult to perceive at this stage of development. The use of neural network systems architectures, which incorporate fuzzy logic processes to speed up decision-making, is believed to be the way in which research will proceed at the current time, and there is a challenge to develop perspectives on the systems' properties that have to be tracked and exchanged and managed through the wider system. This will involve not only the avionics on board the aircraft, but all the systems it interacts with through ports that communicate data, decisions and ideas.

1.2.5   Environment

Professor Ian Poll, of Cranfield University, UK, considers the environment as the key design driver that is currently shaping the challenges that face the civil aviation business. Rather than considering what is possible, he considers what will be allowed and how it will affect the industry. Poll looks at how the airliner manufacturing business can live up to the challenges that society will place upon it. He does not expect the airliner to remain so cosily-wed to the configurations and technologies it has used for several decades, and throws light on how the constraints we face in current technologies will have to be addressed by innovative thinkers in the future.

His method is to build theoretical models and to analyse their attributes through a series of theorems that deliver rational guidelines on where, why and how the airliner design community will have to change its paradigm. The chapter provides a clearer view than what has been apparent from current political debate on how to address aviation-related environmental concerns, and it suggests waypoints for visionary innovators. Poll considers the potential of new technologies (e.g. advanced materials, alternative fuels, novel structures, advanced aerodynamics and new propulsion technologies) to reduce global aircraft emissions, and he speculates on the likely magnitude of these improvements in the short, medium and long terms.

1.2.6   Human factors

Dr Don Harris, formerly with the School of Engineering at Cranfield, UK, follows with a chapter that reminds all engineers that the human element is still as critical as ever, perhaps even more so as innovative ideas consider sidelining the operator from some front-office decision-making in the modern air vehicle.

He provides a clear overview of the milestones in crew–aircraft development, presenting to the reader a clear picture of the foundations of the science of human factors. It leaves no doubt that there is more to be done, and examines in a structured way the mechanisms by which we might expect to see developments. He examines too the qualities that might be expected of future commercial passenger air vehicles without the pilot on-board. It is hard to say, definitively, that this development will not occur, and the reasons for holding out are perhaps more visceral than subjective; with this chapter casting light on the processes that innovators will need to address in moving forward with what could be the ultimate down-sizing of crew complement.

1.2.7   Supersonic passenger air travel

In looking ahead, Dr Howard Smith, Cranfield University, UK, looks at supersonic business aircraft proposals. His work shows why large supersonic transport (SST) aircraft must be accompanied by a sonic boom reduction to make it acceptable, and reviews published data on shock-wave reduction, largely through appropriate aircraft configuration development. He presents information on the structural, aerodynamic and propulsion aspects of supersonic aircraft and concludes that the most likely next-generation SST will be small, to alleviate boom nuisance, and will have a small payload to accommodate the fuel requirement associated with increased range demands. These considerations lead him to consider that the SST business jet, as proposed by several design teams, already has merit, and he provides a clear view of the attributes of those designs on which there is sufficient data. For those people who see innovation as being solely about 'bigger, faster and better' he leaves no doubt that these are not necessarily the best criteria to apply in the future.

1.3   Change

1.3.1   The process of innovation

A non-technical, but technically cognizant, view of innovation is presented in the chapter prepared by Dr Michael Henshaw, Loughborough University, UK, in which he seeks to define the very essence of innovation, where it fits into the overall scene, and what is involved in what is termed an 'innovative process'. This is a mind-provoking review of whether successful innovation can be defined simply or whether it is a more prosaic pursuit that defies commonly-acknowledged lines and slips through simple alliterations. He roots his debate in the qualities that are required in modern project teams, and reminds us that new ideas have to be linked to the real worlds of science and business, and that, whilst these might not constrain thinking, they may restrain implementation.

Ideally, new ideas should not impose any restraint, and his consideration of processes already used, which are often model-based, are rounded off by the interpretation of how systemic thinking can be integrated with evaluations of risk. The use of a Technology Readiness Level (TRL) matrix, with levels of TRL associated with the stage of maturity of a project, is an example of how risk is managed within complex projects today. The review considers the past, and the outcomes of the processes he assimilates define his view of what we might expect to frame innovative enterprise in the future. The framework he sets out is invaluable for the mind-set of the reader who faces uncertainty with disdain.

1.3.2   Managing innovation

Professor Rolf Henke, from RWTH Aachen, Germany, investigates a dilemma that besets many technologists, who, while working in one discipline, discover that their contribution to a design will impact other disciplines, and that these consequences have to be investigated or their own innovative contribution will be put at risk. The Technology Assessment (TA) approach is introduced as an early design stage review process that will allow the implications across the disciplines to be assessed, not just in terms of performance, but throughout the product lifecycle, by considering the impact on both the manufacturer and the operator.

He shows how this is a decision-support process and stresses that it is not the quantitative basis for decisions to be taken. There are examples of two technical developments involving post-service entry introduction, and he uses these to illustrate the justifications that overrule or mitigate the technical advantage that a designer will see.

1.3.3 Mining the 'far side' of technology

It is difficult to find a reliable source on an issue where the responsible organisation prefers not to discuss its programmes publicly, and certainly not until they have matured to production (or even been in covert service). The Defense Advanced Research Projects Agency of the USA – widely known as DARPA – is such an organisation. Intriguing lessons regarding innovation can be learnt from this organisation. What makes this relatively small organisation so successful? How does it pick winning ideas, that is, intelligently mining the 'far side', while ignoring the duds? How does it progress with concepts that are at odds with widely-accepted engineering thinking and conventions? So much of revolutionary innovation depends on the selection of people who can operate on the far side of technology, people who with imagination can realise that which is conventionally deemed impossible. A related issue is the establishment of a corporate culture that tolerates failures from reaching too far.

Mr J. R. Wilson is a freelance writer who had interviewed senior DARPA representatives, and received permission to disclose the contents of his interviews. In this chapter, he looks at DARPA's unique approach to innovation – an organisation that he has described as one of the world's most consistent and pre-eminent contributors to revolutionary advances in aerospace. Using DARPA's impressive list of radical aircraft as evidence, the chapter reviews this management approach for innovative projects. One interesting aspect is the way that DARPA has so few long-term staff – people join, do their unique research, then disappear, usually back into industry. The corporate culture, as well as the funding and working premises, is discussed.

1.3.4 Future air transport

Co-editor Mike Hirst has taken a long-term look at the air-transport scene, opening with consideration of air-vehicle developments and proceeding into the perceivable trends in transportation overall, and the way that aviation will have a role within that bigger picture.

A framework within which to assess processes is presented. This is based on financial, compliancy, efficiency and effectiveness targets, and their trade-offs. The issue of understanding what is required in the future is considered, with guidance on the way that requirements can be viewed, from the perspectives of consistency, completeness and clarity, before considering whether a requirement is sufficiently objective. The need grasps the nettle of environmental impact, whereby long-term goals are set, and incremental improvement of service quantity is regarded not alone, but alongside overall usage of air transport.

The purposes for which air travel might be used could change, and society might even impose that change. A property-led systemic debate is encouraged, using the framework postulated with the notes to provide some perspectives on the way that thinking will need to develop.

1.4   Challenges

1.4.1   Intellectual property

Dr Douglas McCarville, from The Boeing Company, Orting, WA, investigates the innovation process that occurs within industrial systems. McCarville looks at how business processes stimulate new ideas, and investigates the extent to which an organisation should, or can, shield or distribute new knowledge. He analyses the Boeing 787 project from the aircraft manufacturer's point of view, considering the legacy in the processes that the prime company needs to access, and in which its suppliers are already stakeholders, and presents an understanding of the techniques – there are many – that can be used not just to protect the benefits of innovation, but to assist in their exploitation and stimulate the implementation of more intellectual content in future programmes.

1.4.2   Cost, schedule and technical performance risk management

Professor Tyson Browning, from the Texas Christian University, Fort Worth, TX, provides a chapter that is an overview of adaptive decision-making in practice. He reviews techniques, such as the risk value method (RVM), and concentrates on how risk is managed within an innovative project.

It is clear that risks, with three perspectives – based on cost (cheaper), lead time (faster) and technical performance (better) – are difficult properties of products to enumerate, and, when considering a new working regime, they get even more difficult. Browning surmises that an automated RVM process evaluation, through software, would be invaluable, but concedes that managing the holistic model that will accommodate the depth of the evaluations in an uncharted regime is especially challenging. He provides a clear example of the difficulties that were encountered in a UCAV (un-manned combat air vehicle) programme, and this becomes a case study and a beacon for the reader.

1.4.3   Lean engineering

Professor Earll Murman, of the Aeronautics and Astronautics Department at Massachusetts Institute of Technology, Cambridge, MA, sees process innovation as a key component for the success of new aerospace products. He presents a framework for engineers adapted from the Lean Thinking approach, which is based on the same fundamental principles that led to Total Quality Management (TQM), Six Sigma and other similar quality processes – all of which have their roots in the post-World War II Japanese automotive industry. Three case studies – based on the HondaJet, Citation X and Engine Testing at AEDC – are presented, illustrating how the framework can apply to different engineering tasks.

Murman dismisses the notion that Lean Engineering is less engineering, explaining that lean describes the right amount of engineering, at the right time to meet the right objectives. The advantages of improved processes can be measured, not only by the development of superior products, but also by the effect that such improved processes can have on the project team. Murman makes an interesting observation, which is so often ignored, that process improvements lead to greater job satisfaction and happier engineers. He describes the attributes of fluid, transitional and specific stages in design, correlating with cerebral, supplier-driven and purer engineering concepts at each stage. Some homilies emerge that reinforce his initial observation: that engineers don't need changes in job description but benefit from job rotation, as they are enabled to learn new skills. He encourages innovative employee programmes and cites the Technical Fellow concept used in several US companies and Boeing's Welliver programme as examples of good practice. The latter example involves integration across the company's boundaries and into associated industries and educational establishments.

The editors would have welcomed even more papers, but in this distillation there is evidence of the immense scope of opportunity that exists for innovation in aeronautics businesses. There are many ways in which practising engineers and researchers believe that change can be achieved and managed, as suggested in the chapters that are summarised above. Bringing new concepts to fruition requires a team effort, a multi-disciplinary team, with complementary skill-sets, working in a culture of openness with a sharing of ideas. Above all, there is a need to embrace radical concepts – concepts that have the potential for quantum improvements. The associated risks, with technical, commercial or societal facets, which may in the past have stifled the take-up of new opportunities, need to be quantified, understood and managed. For current and future generations of aeronautical engineers, there is a tremendous opportunity to utilise the ever-increasing capabilities of computer simulation tools, not just in a manner that relies on the 'coding' of traditional design and manufacturing practices, but in a holistic, concurrent approach. Through the debates they present, these chapters will assist in helping those who wish to see change for the better implemented in due course. Innovation is an international and cross-enterprise activity, where some individuals, firms or people do better than others. By stimulating honest discussion and disseminating information on best-practice techniques, a brighter future for all may be assured.

The editors commend these chapters as guidelines that will serve curious innovators well.

Part I
Concepts

2

Biologically inspired technologies for aeronautics

Y. BAR-COHEN, Jet Propulsion Laboratory
(JPL)/California Institute of Technology, USA

Abstract: Through evolution and employing principles of all the science and engineering fields, Nature addressed its challenges by trial and error and came up with inventions that work well and last. Technology inspired by Nature is known as biomimetics and offers enormous potential for many exciting capabilities. Biomimetics can be as simple as copying fins for swimming, and is providing numerous benefits, including the development of prosthetics that closely mimic real limbs. The focus of this chapter is on the aerospace-related innovations that were inspired by Nature, and it covers various examples, the potentials and the challenges.

Key words: aerospace, robotics, biomimetics, electroactive polymer, biologically inspired technology.

2.1 Introduction

Nature is a self-maintained experimental laboratory that is addressing its changing challenges through the trial-and-error process of evolution. In performing its experiments every field of science and engineering nature is involved, with testing the principles of physics, chemistry, mechanical engineering, materials science, mobility, control, sensors, and many others. Also, the process involves scaling from the nano and macro scales, as in the case of bacteria and viruses, to the macro and mega scales, including the scale of our lives and that of the whales. The extinction of the dinosaurs may suggest that mega-scale land-living animals are an unsustainable form of life as opposed to mega-size sea creatures such as the whales. Observing and studying the capabilities of living creatures suggest numerous possibilities that can be adapted to solve and support human needs. Nature has always served as a model for mimicking and an inspiration to humans in their efforts to improve the way we live. The subject of copying, imitating, and learning from biology is also known as biomimetics and it represents the studies, imitation and inspiration of Nature's, methods, designs and processes (Bar-Cohen, 2005; Benyus, 1998; Schmitt, 1969; Vincent, 2001; Vogel, 2003). Some of the capabilities were copied from Nature, while for others it served as an inspiring model. Flying was inspired by insects and birds using human-developed capabilities, whereas the design and function of fins, which divers use, were copied from the legs of water creatures such as the seal, goose and frog. Scientific approaches are helping humans understand Nature's capabilities and the associated principles, resulting in the development of effective tools, algorithms, approaches and other capabilities to benefit mankind. The ultimate goal of biomimetics may be the development of life-like robots that appear and function like humans. Efforts are currently underway to develop such robots, and impressive capabilities have already been reported where human-like robots can conduct conversation with limited vocabulary and respond to body and facial expressions, as well as avoid obstacles while walking and other capabilities (Bar-Cohen and Hanson, 2009). The focus of this chapter is on biologically inspired innovation in aerospace.

In general, Nature's materials and processes are far superior to man-made ones. The bodies of biological creatures are laboratories that process chemicals acquired from the surroundings and produce energy, construction materials, multifunctional structures and waste (Mann, 1995; Nemat-Nasser *et al.*, 2004). Some of the capabilities of Nature's materials include self-replication, reconfigurability, self-healing, and balancing the content of various chemicals, including the pH of fluids, as well as temperature and pressure. Recognizing the advantages of these materials, for thousands of years humans have used them as sources of food, clothing, comfort, construction and many other applications. These materials include fur, leather, honey, wax, milk and silk (Carlson *et al.*, 2005). The need to make these materials in any desired quantity led to developing approaches for enhancing their production from the related creatures as well as producing imitations. Many man-made materials are processed by heating and pressurizing, and this is in contrast to Nature, which uses ambient conditions. Materials such as bone, collagen or silk are made inside the organism's body using nature-friendly processes with minimal waste, and the resulting strong materials are biodegradable and recyclable by Nature.

Besides the multifunctional structures that make up biological creatures, they also have the capability to produce structures using materials that they make or pick up from the surroundings. The skeletons of animals' bodies are quite marvellous – they are able to support enormous physical actions even though they are not rigid structures. Also, the produced structures (such as the nest, cocoon's shell and underground tunnels that gophers and rats build) are quite robust and support the structure's required function over the duration that it is needed. Often the size of a structure can be significantly larger than the species that builds it, as in the case of the spider's web. An example of a creature that has a highly impressive engineering skill is the beaver, which constructs dams as its habitat on water streams. The honeycomb is also an inspiring structure, and it provides the bees with a highly efficient packing configuration (Gordon, 1976). Using the same configuration, the honeycomb is used to create aircraft structures benefiting from the low weight and high strength that are obtained. Even plants offer inspiration, where mimicking the adherence of seeds to animals' fur led to the invention of Velcro and to numerous applications including clothing and electric wire strapping.

The development of biomimetic systems and devices is supported by a growing number of biologically inspired technologies, including artificial intelligence, which mimic the control of biological systems (Amaral *et al.*, 2004; Hecht-Nielsen, 2005; Serruya *et al.*, 2002). The invention of the wheel made the most profound impact on human life, allowing the traverse of enormous distances and performance of tasks that otherwise would have been impossible to perform within the lifetime of a single person. While the wheel enabled enormous capabilities, it has significant limitations when used for mobility in complex terrains that have obstacles. Obviously, legged creatures can operate in many such conditions and in ways far superior to an automobile. Legged robots are increasingly becoming an objective for the developers of robotic machines, and these include even human-like ones (Bar-Cohen and Breazeal, 2003; Bar-Cohen and Hanson, 2009). Generally, the mobility of legged mechanisms currently is enabled via motors. While motors have numerous advantages, since they require gears they are relatively heavy, structurally complex and have many potential points of failure. Advances in electroactive polymers (EAP), also known as artificial muscles, are expected to enable new possibilities for legged robotics, with the potential of turning science fiction ideas into engineering applications (Bar-Cohen, 2004).

As a model for inspiration, it is important to remember that Nature's solutions are driven by survivability of the fittest, and these solutions are not necessarily optimal for technical performance. Effectively, all organisms need to do is to survive long enough to reproduce. Living systems archive the evolved and accumulated information by coding it into the species' genes and passing the information from generation to generation through self-replication.

There are great benefits to better understanding how Nature's marvels work and how to adapt them in human-made mechanisms. These include such capabilities as:

- The dragonfly's flight performance, its ability to fly backwards, as well as stopping and starting using its relatively small body (Huang and Sun, 2007).
- The toughness of spider silk and the ability of the spider to produce silk at room temperature and pressure conditions (Trotter *et al.*, 2000).
- The navigational capability of the Monarch butterfly, migrating over great distances and reaching targeted locations to which, as an individual, it has never before been. This information is coded into the genes of its small body (Sauman *et al.*, 2005).
- The strength of seashells, which is quite enormous even though they are made of calcium carbonate, which is, effectively, a soft material also known as chalk (Yang, 1995).
- Our ability to identify people whom we have not seen for many years and who have changed appearance enormously.

The above list is only a small number of examples, and covering them all can be an enormous task, and the challenge to adapt them can be much more complex. This chapter examines various examples that are relevant to aerospace.

2.2   Biologically inspired or independent human innovation

Nature and its biological systems were on Earth for many Millions of years before humans reached the level of intelligence that was enough to start making their own tools. In the effort of humans to become domesticated and to minimize their dependence on luck and their harvesting of the surrounding resources, they started seeking to improve the way in which they lived. Observing Nature was part of their daily life and it inspired them with ideas of how to acquire and handle food, how to protect themselves and their resources as well as many other things that were essential to their way of life.

One may wonder how inspiring for human innovation have been the various creatures that lived in their neighborhoods. The presence of spiders in human habitats should have had some inspiring role in making such things as wires, ropes, nets, sieves, screens and woven fabrics. One cannot avoid seeing the similarity of a spider's web and the fishing net or the screen in screen-doors or even the kitchen sieve, as shown in Fig. 2.1. In addition to the sieve, another inspired tool for application to food handling is the tong, which was probably inspired by the beak of birds, as shown in Fig. 2.2.

From another angle on the subject of biomimetics, one may wonder if all the inventions and tools that are commonly used by humans were inspired by Nature or, perhaps, it was just a coincidence that the solutions have similar features. An example might be the honeycomb, which looks very much the same in the natural and technological versions. In the case of bees, the honeycomb serves as a highly efficient packing container for their offspring and for food storage when their eggs

(a)

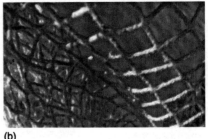

(b)

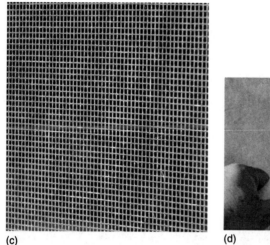

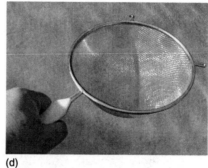

(c)   (d)

2.1 The spider web has probably been the inspiration for numerous human made tools. (a) Spider web; (b) Net; (c) Screen door; (d) Kitchen sieve.

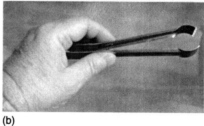

(a)   (b)

2.2 The tong may have been inspired by the beak of birds and the way they grab food.

hatch. On the other hand, the honeycomb in aircraft structures provides a highly efficient space filler for parts that is lightweight and has great strength. It is also interesting to note that many mammals are four-legged creatures and that most of our furniture, such as chairs and tables, is supported by four legs (see an illustration of this point in Fig. 2.3). It is hard to believe that all human-made solutions were pure inventions of individuals who ignored what they had seen in Nature and came up with them on their own.

(a)   (b)

2.3 One may wonder if the four legs of most animals have been an inspiring model for furniture such as chairs and tables.

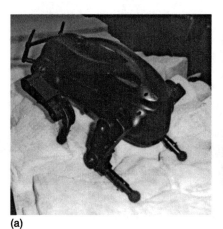

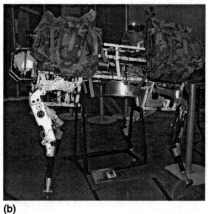

(a)   (b)

2.4 Military application of legged robotics.

Roboticists are well aware that making robots operate with legs significantly increases their capability and mobility in complex terrains. In recent years, there have been increasing efforts to create legged robots for various applications including space and the military, and examples are shown in Fig. 2.4 and 2.5.

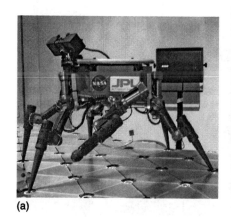

(a)   (b)

2.5 Examples of legged robots that were developed at JPL for space application. Future robots may climb mountains and perform lifelike functions. (Source: courtesy of JPL/Caltech/NASA.)

2.3 Nature as a source of innovation in aerospace

Inspired by insects' and birds' ability to fly, the field of aerospace as we know it today uses human-developed technology (Fig. 2.6). The enormous number of species capable of flying suggests that Nature has extensively 'experimented' with aerodynamics and has been quite successful. Birds are able to maneuver in flight with quite amazing capabilities, as well as flying while carrying prey that can be quite large and heavy compared with their bodies. They can even catch prey while flying, for example a running rabbit or a swimming fish; they are able to predict their path's intersection with the hunted creature. This capability to hunt while the hunter and the hunted creature are both moving fast (running, flying, or swimming) is increasingly within the capabilities of military weapons, for example allowing a tank to destroy a moving vehicle while they are both moving quickly. As another example, missiles are used to hit enemy fighter aircraft or other missiles by tracking the moving target and either adjusting the direction in flight or aiming at the moment of launch.

Another form of biologically inspired flight for potential future NASA missions is under consideration at the Ohio Aerospace Institute (Fig. 2.7). This study takes into consideration that flying on Mars is much more difficult than on Earth due to the lower air pressure and therefore it is necessary to operate within a very low Reynolds number regime. In addition to this restriction, one needs to take into

2.6 Inspiration by nature and aerodynamic principles led to the flying capabilities of aircraft such as the supersonic passenger plane, the Concorde. (Source: photographed by the author at the Boeing Aerospace Museum, Seattle, WA.)

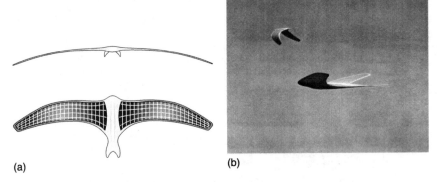

(a)   (b)

2.7 A flying mechanism that emulates the bird was proposed for planetary exploration missions. (Source: courtesy of Anthony Colozza, Ohio Aerospace Institute.)

account the practical size limitations of a vehicle that can be deployed from Earth. An entomopter vehicle was recently proposed that uses biomimetic configuration (http://www.niac.usra.edu/files/studies/abstracts/448Colozza.pdf) and circulation control techniques to achieve substantially higher lift. The concept is based on the use of a micro-scale vortex at the wing's leading edge as determined in 1994 by Charles Ellington of the University of Cambridge (Scott, 1999). Taking advantage of the lower gravity on Mars, one may be able to develop an insect-inspired flying machine with a size in the range of a meter. For power the wing will be covered with flexible solar cells throughout the structure (Fig. 2.7a). Under a DARPA sponsored study, researchers at the Georgia Tech Research Institute have preliminarily confirmed that this concept may be feasible for operation on Mars, with the vehicle able to take off, fly slowly or hover, and land.

Wagging the tail is the leading form of propulsion in water, and many sea creatures are able to develop significant swimming speeds. Inspired by this propulsion method and using balloon designs, with helium for operation in air, researchers at EMPA, Duebendorf, Switzerland, in collaboration with the Institute of Mechanical Systems of ETH, Zürich, Switzerland, are currently developing such a flying vehicle (Michel *et al.*, 2007). The project objective is to use electroactive polymers emulating muscles to produce a lighter-than-air vehicle (Fig. 2.8). In the first phase, a blimp was developed that has its fins bent to the left and right by EAP-based actuators, allowing the blimp to be steered. The goal is to develop a novel bionic-propelled blimp that is operated like a fish with tail-wagging capability (Fig. 2.9). For this purpose, fluid dynamics, structural mechanics, and flight performance are explored with systematic experimental studies. The commercial application of this technology is the development of larger blimps for use in transportation, observation and reconnaissance, as well as stratospheric platforms.

The dragonfly is an incredible flying insect that can maneuver in air at relatively high speeds. Its capability has been under study for many years in an effort to adapt or inspire aeronautic innovation and solution to existing problems (Huang and Sun, 2007). The dragonfly adjusts the effects of high gravity on its body during its flight and rapidly maneuvers using liquid-filled sacs that surround

2.8 Photographic view of the EAP activated blimp (the black strips are dielectric elastomer EAP). (Source: photographed by the author at the SPIE 2008 EAPAD Conference.)

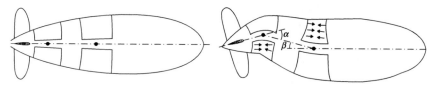

2.9 A graphic view of the envisioned EAP-activated blimp that is propelled like a fish. (Source: courtesy of Silvain Michel, EMPA – Materials Science & Technology, Duebendorf, Switzerland.)

its cardiac system. This method has inspired the Swiss company Life Support Systems to develop an anti-G suit that allows pilots to fly at high Mach speeds with significantly lower effects on their ability to stay coherent. The developed liquid-filled suit is called Libelle, which means dragonfly in German (http://www.lssag.ch/website%2003%2014.html). Tests of the Libelle suit have shown promise as far as the advantages over pneumatic (compressed air) anti-G suits and they are being tested by several air forces.

Plants use many methods of dispersing their seeds, including being blown in the wind and being shaped in an aerodynamic configuration to enable the largest distance to be traveled. Thus, plant species reduce the danger of crowding a specific type of plant into the same local area, may cause competition over the same resources, as well as being subjected to the same environmental risks that possibly endanger their survival. There are various aerodynamic configurations of seeds, and an example is shown in Fig. 2.10, where the seed of the tree *Tipuana tipu* (about 6.5 cm long) has a wing that propels it in the wind. It is also interesting to mention the tropical Asian climbing gourd *Alsomitra macrocarpa*, a tree with a relatively large seed having a 13 cm wingspan. The flight of this seed resembles that of a boomerang, and it is capable of gliding in wide circles through the rain forest. One may see quite a similarity between these seeds and helicopter blades, and it is most likely to have been an inspiration for the design of many aerodynamic parts of aircraft and other human-made flying machines.

Another aerospace-related area that is benefiting from biomimetics is the design and development of potential alternatives for planetary landing of rovers and landers on planets with atmospheres (such as Mars and Venus). Adapting such designs may offer better alternatives to the use of a parachute on Mars and possibly provide a better ability to steer the landing hardware toward selected sites. Some of the issues that are being studied include the determination of the appropriate vehicle size, mass distribution and platform shape to ensure stable autorotation and scalability from operation on Earth to performance on Mars.

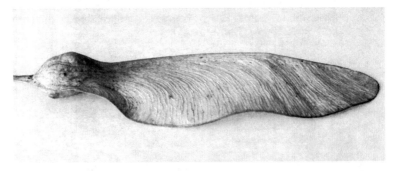

2.10 Seed of the *Tipuana tipu*, which has an aerodynamic shape for dispersion by the wind.

(a)   (b)

2.11 Tumbleweed (a) offered an inspiration for a futuristic design of a Mars rover (b). (Source: (b) courtesy of NASA. http://smartmachines. blogspot.com/2007/04/nasas-tumbleweed-inspired-rovers-for.html.)

The tumbleweed is another plant that offered an inspiring design for planetary mobility that is powered by wind (Wilson *et al.*, 2006). Generally, winds blow throughout Mars and they provide an attractive source for mobilizing a rover by mimicking tumbleweed. As shown in Fig. 2.11, the tumbleweed has inspired a futuristic lander that could one day be used as a vehicle for mobility on Mars for traversing great distances with minimal use of power. At NASA Langley, using three-dimensional dynamic modeling and simulations, Southard *et al.* (2007) have shown that dispersion and exploration of Mars with tumbleweed rovers is feasible. A likely mission scenario involves an organized search for geologically interesting features using a group of rovers with heterogeneous sensor packages. A tumbleweed rover can potentially travel longer distances and gain access to areas such as valleys and chasms that previously were inaccessible. Varying the location of the mass imbalance is one of the methods currently under consideration for controlling the motion of a wind-blown tumbleweed-like rover.

2.4   Biologically inspired mechanisms and systems

The manufacture of aerospace structures would benefit greatly if they could be made of materials that have Nature's characteristics of self-healing, self-replication, reconfigurability, chemical balance, durability and multi-functionality. The advantages of biological and botanical materials were well recognized by humanity, and were used for many applications (Carlson *et al.*, 2005). Learning how to process biologically inspired materials can make our choices greater and improve our ability to create recyclable materials that can better protect the environment. Mimicking natural materials will also benefit humans in many other ways, including the development of more life-like prosthetics, where increasingly artificial parts such as hips, teeth, structural support of bones and others are being produced. There are also many mechanisms that were biologically inspired. Some are discussed below.

2.4.1   Ground penetration inspired by gophers and crabs

Since 1997, the author, members of his group at JPL, and engineers from Cybersonics, Inc. have been involved with research and development of sampling techniques for future *in situ* exploration of planets in the Universe. The developed techniques are mostly driven by piezoelectric actuators and the Ultrasonic/Sonic Driller/Corer (USDC) in particular (Bao *et al.*, 2003; Bar-Cohen *et al.*, 2005a, 2005b). The general configuration of the USDC allows it to penetrate sub-surfaces to a depth that is no longer than the length of the bit, since the other parts are larger in diameter. In order to reach greater depth with less restriction on the depth, two models of deep drills were conceived that were inspired by the gopher and sand-crab (Bar-Cohen *et al.*, 2005b). A piezoelectric actuator induces vibrations that impact the medium with which it is in contact; and the mechanism consists of a bit with a diameter that is the same as or larger than the actuator. The device that emulates the biological gopher is lowered into the produced borehole, cores the medium, breaks and holds the core, and finally the core is extracted onto the surface. This ultrasonic/sonic device can be lowered and raised from the ground surface via cable as shown in Fig. 2.12. This device was called the Ultrasonic/ Sonic Gopher and it was designed analogously to the biological gopher that digs into the ground. It removes the loose soil out of the underground tunnel that it forms, bringing it to the surface, and resumes the process to reach great depths. The Ultrasonic/Sonic Gopher was developed to the level of a prototype and demonstrated at Mount Hood and in Antarctica to perform its intended function. Further, the Ultrasonic/Sonic Crab design emulates the sand crab, which shakes its body to penetrate sand on beaches. This device uses mechanical vibrations on the front surface of the end-effector to penetrate media that consist of loose soil, sand, or particulates. The Ultrasonic/Sonic Crab has not yet been produced; however, its implementation is not expected to pose major technical challenges.

2.4.2   Pumping mechanisms

Nature uses many pumping mechanisms that have inspired human-made mechanisms. The most common pumps operate by peristaltic pumping, where liquids are squeezed in the required direction. The lungs pump air in a tidal process using the diaphragm, which allows us to breathe. Pumping via valves and chambers that change volume is found in human and animal hearts, where the chambers expand and contract to allow the flow of the blood. Just as in mechanical

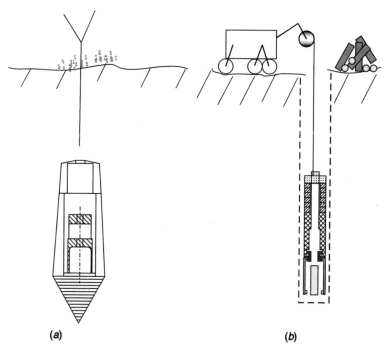

2.12 Biologically inspired ground penetrators. (a) Ultrasonic/Sonic Crab; (b) Ultrasonic/Sonic Gopher.

pumps, the flow of the blood is critically dependent on the action of the valves in the heart.

2.4.3 Artificial muscles

Muscles, which are both compliant and linear in behavior (Full and Meijer, 2004), are the actuators of biological systems, allowing all our physical movements. Emulating the characteristics of muscles is important, allowing us to make robots that function with life-like performance. The actuators that are closest to emulating natural muscles are the EAPs, which have emerged in recent years and gained the name 'artificial muscles' (Bar-Cohen, 2004). There are many types of EAP materials known today, and most of them emerged in the 1990s. Unfortunately, they are still not generating sufficient forces to perform significant tasks such as lifting heavy objects. In order to help advance the field rapidly, the author initiated and organized in March 1999 the first annual international EAP Actuators and Devices (EAPAD) Conference (Bar-Cohen, 1999). This conference is held annually by the International Society for Optics and Photonics (SPIE) as part of its Smart Structures and Materials Symposium. At the opening of the first conference, he posed a challenge to scientists and engineers worldwide to develop a robotic arm that is actuated by artificial muscles to win an arm-wrestling match against a human opponent (http://ndeaa.jpl.nasa.gov/nasa-nde/lommas/eap/EAP-armwrestling.htm). The icon of the challenge can be seen in Fig. 2.13, illustrating the wrestling of human with robotic arm driven by artificial muscles.

On 7 March 2005, the author organized the first arm-wrestling match with a human (17-year-old high school female student) as part of the EAP-in-Action Session of the SPIE's EAPAD Conference. In this contest, three EAP-actuated robotic arms participated and the girl won against all three (Fig. 2.14). Following this match in the second contest, rather than wrestling with a human opponent, the contest consisted of measuring the arm's performance and comparing the results. A measuring fixture was used to gauge the speed and pulling force. To establish a baseline for comparison, the capability of the above student was measured first and then three participating robotic arms were tested. The second Artificial Muscles Armwrestling Contest was held on 27 February 2006, and the results showed two orders of magnitude lower performance of the arms compared with the student. In a future conference, once advances in developing EAP-actuated arms lead to sufficiently high force, a professional wrestler will be invited for another human/machine wrestling match.

2.13 The icon of the armwrestling challenge for artificial muscles match against human.

2.14 The robotic arm driven by artificial muscles, made by Virginia Tech students, is being prepared for the 2005 match against the human opponent.

2.4.4   Inchworm motors

Another form of actuation that was biologically inspired is the movement of the inchworm, a caterpillar of a group of moths called *Geomeridae*. Emulating the mobility mechanism of this larva or caterpillar led to the development of high-precision motors and linear actuators that are known as inchworms. The forces that are generated by the commercial types of inchworms can reach over 30 N with zero-backlash and high stability. As opposed to biological muscles, the piezoelectric-actuated inchworms have zero-power dissipation when holding position. Inchworm mechanisms have many configurations, with the basic principle using two brakes and an extender. These motors perform cyclical steps, where the first brake clamps onto the shaft and the extender pushes the second brake forward. Clamping the second brake and stretching the extender makes the first step, which is then repeated as many times as needed. Inchworm motors were used already in the Telesat (NASA mission) in the mid-1980s, allowing high-precision articulation in the nanometer range.

2.4.5   Bio-sensors

It is well recognized that sensors are a critical part of any system, allowing it to monitor its functions and respond to the operation conditions as needed. Sensors emulate the senses in biological creatures, which provide inputs to the central nervous system about the environment around and within their body, and the muscles are then commanded to act after analysis of the received information (Hughes, 1999). Biological sensory systems are extremely sensitive and limited only by quantum effects (Bialek, 1987; Bar-Cohen, 2005). Sensors are widely used and it is not possible to imagine effective operation of any system without them. Pressure, temperature, optical and acoustical sensors are widely in use and continuously being improved in terms of their sensing capability, while reducing their size and consumed power. The eye is emulated by the camera, the whiskers of rodents are emulated with collision avoidance sensors, and acoustic detectors imitate the sonar in bats. Similarly to the ability of our body to monitor the temperature and keep it within healthy, acceptable limits, our homes, offices, and other enclosed areas have environmental controls that allow us to operate at comfortable temperature levels. One of the recent studies related to applications in aerospace includes the development of an artificial fly unmanned aircraft system with combined hearing and vision for navigation to inaccessible locations. This on-going research at the University of Maryland is funded by the US Air Force Office of Scientific Research (AFOSR) (http://www.af.mil/news/story. asp?id=123125017). In this study, the capability of a pair of mechanically-coupled ears that are separated by only 500 microns is being investigated while seeking to incorporate advances in microelectronics and other system-on-a-chip capabilities. This study is focused on the understanding, modeling and emulation of the ability of flies to combine hearing and vision at micro-scale levels as means of rapid flight and response.

2.4.6   Artificial intelligence (AI)

Controlling the operation of systems in an automatic way can be limited if simple software with known answers to any of the possibilities is used. Increasingly, systems are being made 'smart' using artificial intelligence, where the control algorithms are emulating Nature (Musallam *et al.*, 2004; Mussa-Ivaldi, 2000). The field of AI is providing important tools for making automatic and robotic mechanisms with capabilities such as knowledge capture, representation and reasoning, reasoning under uncertainty, planning, vision, face and feature tracking, language processing, mapping and navigation, natural language processing, and machine learning (Bar-Cohen and Breazeal, 2003; Kurzweil, 1999; Luger, 2001). Generally, AI is a branch of computer science that studies the computational requirements for such tasks as perception, reasoning, and learning, to allow the development of systems that perform these capabilities (Russell and Norvig, 2003). Through improvement of the understanding of human cognition (Hecht-Nielsen, 2005) scientists are able to understand the requirements for intelligence in general, and develop artifacts such as intelligent devices, autonomous agents, and systems that cooperate with humans to enhance their abilities. AI researchers are using models that are inspired by the computational capability of the brain and explaining them in terms of higher-level psychological constructs such as plans and goals.

2.5   Robotics as beneficiary of biomimetic technologies

Creating robots that mimic the shape and performance of biological creatures has always been a highly desirable engineering objective (Bar-Cohen and Breazeal, 2003; Bar-Cohen and Hanson, 2009). The term 'robot' refers to a biomimetic machine with human-like features and functions that consist of electro-mechanical mechanisms. Also, it suggests a machine that is capable of manipulating objects and sensing its environment as well as being equipped with a certain degree of intelligence. Searching the Internet for the word robot brings numerous links related to research and development projects that are involved with robots. Manipulator arms that are fixed to a single position and perform such tasks as painting and assembly are part of many production lines, including the manufacture of cars. Rovers that have locomotion with wheels or legs are already being made autonomous, and are able to perform quite sophisticated tasks. These include the Mars Rovers, which have been operating on Mars since 2003 on terrains that are unknown, and they are capable of avoiding obstacles while conducting various tasks in support of the exploration of this planet.

The entertainment and toy industries have greatly benefited from advancement in this technology. Toys that emulate the appearance and movement of such creatures as frogs, fish, dogs and even babies are now supplied by many stores. The higher-end robots and toys are becoming increasingly sophisticated, allowing them to walk and even appear to converse with humans using a limited vocabulary at the level of hundreds of words. Some of these robots can be operated autonomously or can be remotely reprogrammed to change the characteristic behavior. An example of a robot that expresses and reacts to human expressions facially and verbally is Kismet, which was developed at MIT (Bar-Cohen and Breazeal, 2003; Breazeal, 2004). As this technology evolves it is becoming more likely that, in the future, human-like robots may be part of our daily life, operating at our homes and offices and doing work that currently is done by humans. Beside the benefits of this technology there is a need for awareness of the potential risks that these robots may pose due to errors or even malicious intent.

Industry has increasingly benefited from advances in robotics and automation that are biologically inspired (Bar-Cohen, 2000; Bar-Cohen and Breazeal, 2003). Crawlers with the equivalent to legs as well as various manipulation devices are increasingly being used to perform a variety of nondestructive evaluation (NDE) tasks. At JPL, a multifunctional automated crawling system (MACS) was developed to allow rapid scanning of aircraft structures in field conditions (Fig. 2.15). MACS

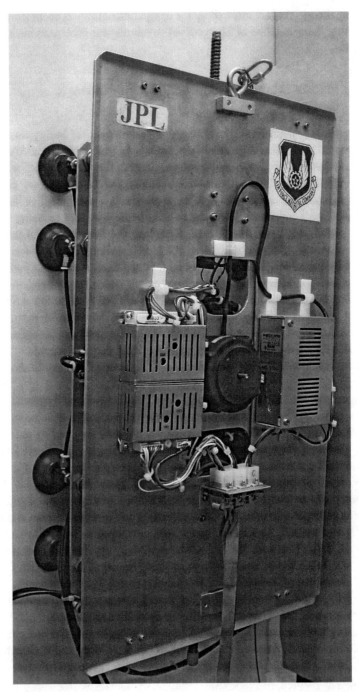

2.15 MACS crawling on a wall using suction cups on two simulated legs.

consists of two legs for mobility on structures, with one of the legs designed also to rotate. This crawler performs scanning by 'walking' on aircraft fuselages while adhering to the surface via suction cups, and is capable of walking upside down on such structures. Mobility on structures is critically dependent on the capability of the legs to have controlled adherence, and alternative forms that were reported include the use of magnetic wheels and electrostatic fields. The author and his co-investigator (Bar-Cohen and Joffe, 1997) conceived a rover that can operate on ships and submarines using magnetic wheels. Another legged robot is JPL's STAR, which has four legs and can perform multiple functions (Fig. 2.5), including grabbing objects as well as climbing rocks with the aid of the ultrasonic/sonic anchor on each of the legs (Bar-Cohen, 1999; Bar-Cohen and Sherrit, 2007; Kennedy *et al.*, 2006). This anchor provides the ability to 'hang onto' rocks via a mechanism that requires a relatively low axial force to drill into the rocks and then extract the bit. JPL's legged robots are developed for potential operation on future planetary missions, where a Lemur class robot will be able to autonomously negotiate its way through unknown terrain that is filled with obstacles (Fig. 2.16).

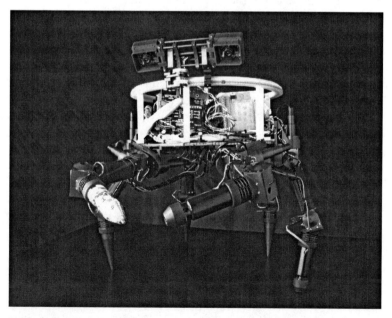

2.16 JPL's Lemur, a six-legged robot, in a staged operation. (Source: courtesy of Brett Kennedy, JPL/NASA.)

2.6   Conclusion: challenges and potential development

The evolution of Nature over billions of years led to highly effective and reasonably power-efficient biological mechanisms, which are appropriate for the intended tasks and that last (Petr, 1996). Ongoing evolution eliminates failed solutions and often leads to the extinction of specific species that do not survive the changing conditions. As it evolves, Nature archives its solutions in the genes of the creatures that make up the terrestrial life around us. Imitating Nature's mechanisms offers enormous potential for the improvement of our life and the tools we use. With the capability of today's science and technology we are significantly more capable of employing, extracting, copying and adapting Nature's inventions.

Nature offers a model for us as humans in our efforts to address our needs as well as a source for inspiring many human-made devices, processes and mechanisms. By studying Nature from the angle of seeking ideas for biologically inspired technologies many applications can result, including stronger fibers, multifunctional materials, improved drugs, superior robots, and many others. Preventing the loss of Nature's solutions that managed to survive, at least until we understand them well, is an important aspect of biomimetics, where we need to be assured that species are not extinct since they may harbor inventions that we have not yet appreciated. We can learn manufacturing techniques from animals and plants, such as the use of sunlight and simple production of compounds with no pollution, biodegradable fibers, ceramics, plastics, and various chemicals. One can envisage the emergence of extremely strong fibers that are woven as the spider does, and ceramics that are shatterproof, emulating the pearl, or possibly seashells. Besides providing models, Nature can serve as a guide to determine the appropriateness of our innovations in terms of durability, performance, and compatibility.

The inspiration of Nature on aerospace is expected to continue growing and to enable technological improvements with impacts on every aspect of our lives. Miniature flying devices that are as small as or smaller than a fly, with the speed and performance of a dragonfly, are still a challenge to mimic. However, some of the inspired future capabilities may be considered science fiction in today's terms, but as we improve our understanding of Nature and develop better capabilities this may become a reality that is closer than we think.

2.7   Acknowledgement

Some of the research reported in this chapter was conducted at the Jet Propulsion Laboratory (JPL), California Institute of Technology, under a contract with National Aeronautics and Space Administration (NASA).

2.8   References

Amaral J.F.M., Amaral J.L.M., Santini C., Tanscheiot R., Vellasco M. *et al.* (2004). 'Towards evolvable analog artificial neural networks controllers'. In *Proceedings of the 2004 NASA/DoD Conference on Evolvable Hardware* , pp. 46–52. Seattle, WA, 24–26 June 2004.

Bao X., Bar-Cohen Y., Chang Z., Dolgin B.P., Sherrit S. *et al.* (2003). 'Modeling and computer simulation of ultrasonic/sonic driller/corer (USDC).' *IEEE Transaction on Ultrasonics, Ferroelectrics and Frequency Control (UFFC)*, **50**: 9, 1147–60.

Bar-Cohen Y. (ed.) (2000). 'Automation, miniature robotics and sensors for nondestructive evaluation and testing.' *Topics on NDE (TONE) Series*, Vol. 4, pp.1–481. American Society for Nondestructive Testing, Columbus, OH.

Bar-Cohen Y. (2004). *Electroactive Polymer (EAP) Actuators as Artificial Muscles – Reality, Potential and Challenges*, 2nd ed., Vol. PM136. SPIE Press, Bellingham, Washington.

Bar-Cohen Y. (ed.) (1999). *Proceedings of the first SPIE's Electroactive Polymer Actuators and Devices (EAPAD) Conf., Smart Structures and Materials Symposium*, Volume 3669. SPIE Press, Bellingham, Washington.

Bar-Cohen Y. (ed.) (2005). *Biomimetics – Biologically Inspired Technologies*, CRC Press, Boca Raton, FL.

Bar-Cohen Y. and Breazeal C. (eds) (2003). *Biologically-Inspired Intelligent Robots*, Vol. PM122. SPIE Press, Bellingham, Washington.

Bar-Cohen Y. and Hanson D. (with the graphic artist A. Marom) (2009). *The Coming Robot Revolution – Expectations and Fears About Emerging Intelligent, Humanlike Machines*. Springer, New York.

Bar-Cohen Y. and B. Joffe (1997). *Magnetically Attached Multifunction Maintenance Rover (MAGMER)*, NTR, Docket 20229, Item No. 9854, 6 February 1997.

Bar-Cohen Y. and Sherrit S. (2007). '*Self-Mountable and Extractable Ultrasonic/Sonic Anchor.*' US Patent No. 7,156,189, 2 January 2007.

Bar-Cohen Y., Sherrit S., Dolgin B., Peterson T., Pal D. *et al.* (2005a). '*Smart-Ultrasonic/Sonic Driller/Corer.*' US Patent No. 6,863,136, 8 March 2005.

Bar-Cohen Y., Sherrit S., Dolgin B., Bao X. and Askins S. (2005b). '*Ultrasonic/Sonic Mechanism of Deep Drilling (USMOD).*' US Patent No. 6,968,910, 29 November 2005.

Bialek W. (1987). 'Physical limits to sensation and perception.' *Annual Review of Biophysics, Biophysics Chemistry*, **16**: 455–78.

Breazeal C.L. (2004). *Designing Sociable Robots*. MIT Press, Cambridge, MA.

Carlson J., Ghaey S., Moran S., Anh Tran C. and Kaplan D.L. (2005), 'Biological materials in engineering mechanisms.' In *Biomimetics – Biologically Inspired Technologies*, pp. 365–80. CRC Press, Boca Raton, FL.

Full R.J. and Meijir K. (2004). 'Metrics of Natural Muscle Function.' In *Electroactive Polymer (EAP) Actuators as Artificial Muscles – Reality, Potential and Challenges*, pp. 73–89. 2nd ed. Vol. PM136. SPIE Press, Bellingham, Washington.

Gordon J.E. (1976). *The New Science of Strong Materials, or Why You Don't Fall Through the Floor*, 2nd ed. Pelican-Penguin, London.

Hecht-Nielsen R. (2005). 'Mechanization of cognition.' In *Biomimetics – Biologically Inspired Technologies*, pp. 57–128. CRC Press, Boca Raton, FL.

Huang H. and Sun M. (2007). 'Dragonfly forewing-hindwing interaction at various flight speeds and wing phasing.' *AIAA J*, **45**: 508–11.

Hughes H.C. (1999). *Sensory Exotica: A World Beyond Human Experience*, MIT Press, Cambridge, MA.

Kennedy B., Okon A., Aghazarian H., Badescu M., Bao X. *et al.* (2006). 'Lemur IIb: A robotic system for steep terrain access.' *Industrial Robot*, **4**: 4, 265–69.

Kurzweil R. (1999). *The Age of Spiritual Machines: When Computers Exceed Human Intelligence*. Penguin Books Ltd, Harmondsworth.

Luger G.F. (2001). *Artificial Intelligence: Structures and Strategies for Complex Problem Solving*. Pearson Education Publishers, Harlow.

Mann S. (ed.), *Biomimetic Materials Chemistry*. John Wiley & Sons, Chichester.

Michel S., Dürager C., Zobel M. and Fink E. (2007). 'Electroactive polymers as a novel actuator technology for lighter-than-air vehicles.' In Y. Bar-Cohen (ed.), *Proceedings of the SPIE Electroactive Polymer Actuators and Devices (EAPAD) 2007*, Vol. 6524. SPIE Press, Bellingham, Washington.

Musallam S., Corneil B.D., Greger B., Scherberger H. and Andersen R.A. (2004), 'Cognitive control signals for neural prosthetics.' *Science*, **305**: 258–62.

Mussa-Ivaldi S. (2000). 'Real brains for real robots.' *Nature*, **408**: 305–6.

Nemat-Nasser S. and Thomas C. (2004). 'Ionic Polymer-Metal Composite (IPMC).' In Y. Bar-Cohen (ed.), *Electroactive Polymer (EAP) Actuators as Artificial Muscles – Reality, Potential and Challenges*, pp. 171–230. 2nd ed. Vol. PM136. SPIE Press, Bellingham, Washington.

Petr V. (1996). 'Animal extinctions in the fossil record: a developmental paradigm.' *Bulletin of the Czech Geological Survey*, **71**(4): 351–65.

Russell S.J. and Norvig P. (2003). *Artificial Intelligence: A Modern Approach*. Pearson Education, Harlow.

Sauman I., Briscoe A., Zhu H., Shi D., Froy O. *et al.* (2005). 'Connecting the navigational clock to sun compass input in monarch butterfly brain.' *Neuron*, **46**(3): 457–67.

Schmitt O.H. (1969). 'Some interesting and useful biomimetic transforms.' In *Proceedings of the 3rd International Biophysics Congress*, p. 297. 29 August–3 September 1969, Boston, Mass.

Scott P. (1999). 'A bug's lift'. *Scientific American*, April 1999.

Serruya M.D., Hatsopoulos N.G., Paninski L., Fellows M.R. and Donoghue J.P. (2002). 'Instant neural control of a movement signal.' *Nature*, **416**(6877): 141–2.

Southard L., Hoeg T.M., Palmer D.W., Antol J., Kolacinski R.M. *et al.* (2007). 'Exploring Mars using a group of Tumbleweed Rovers.' In *Proceedings of the 2007 IEEE International Conference on Robotics and Automation*, Roma, Italy, 10–14 April 2007.

Trotter J.A., Tipper J., Lyons-Levy G., Chino K., Heuer A.H. *et al.* (2000). 'Towards a fibrous composite with dynamically controlled stiffness: lessons from echinoderms.' *Biochemical Society Transactions*, **28**(4): 357–62.

Vincent J.F.V. (2001). 'Stealing ideas from nature.' In S. Pellegrino (ed.), *Deployable Structures*, pp. 51–8. Springer-Verlag, Vienna.

Vogel S. (2003). *Comparative Biomechanics: Life's Physical World*. Princeton University Press, Princeton, NJ.

Wilson J.L., Hartl A.E., Mazzoleni A. and DeJarnette F. (2006). 'Dynamics modeling of a Mars Tumbleweed Rover.' In *Proceedings of the 44th AIAA Aerospace Sciences Meeting and Exhibit*. Paper No. AIAA 2006-71. 9–12 January 2006, Reno, Nevada.

Yang X.F. (1995). 'A self-constraint strengthening mechanism and its application to seashells.' *Journal of Materials Research*, **10**: 1485–90.

3

Aircraft morphing technologies

W. W. HUEBSCH and S. D. HAMBURG, West Virginia
University, and R. W. GUILER, Physical Sciences Inc., USA

Abstract: This chapter gives an overview of both the history and current work in active aircraft morphing technologies. These adaptive structures have been utilized since the earliest days of manned flight to provide maneuver control. In the past they were neglected, as higher speed and power aircraft required strong, rigid structures, and no weight or volume could be spared for the more complex systems required to actuate morphing structures. However, recent advances in materials technology and actuator miniaturization, along with the increased use of small and micro unmanned aerial vehicles (UAVs), have brought these structures and technologies back into the spotlight for both maneuver control and expansion of the flight envelope.

Key words: wing morphing, shape memory actuator, adaptive structure, biomimetic flight, wing warping, active aeroelastic.

3.1 Introduction

When we admire the efficiency and elegance of bird flight in Nature, our aviation achievements seem to pale in comparison. Unlike natural fliers, typical aircraft are rigid and inflexible bodies that achieve their design goals through brute force. Inspired by gliding seeds, insects, bats and the flight of birds, work done by researchers in the area of aircraft morphing offers us a chance to achieve some of the efficiency of flight in Nature with aircraft that can operate from space to roads and water.

In the 1990s aircraft morphing focused mainly on improving the efficiency of wing and control surfaces. In the twenty-first century work continues on improving flight efficiency, but advances in lightweight materials, efficient innovative actuators and control systems have pushed the research and development focus to expanding the range of operational environments.

3.2 Early aircraft morphing developments

Advances in lightweight materials, actuator technology and modern aerodynamics have all come together in an attempt to imitate the agility and performance seen in bird flight. This technology is called wing morphing.

The ideas are as old as manned flight, but technological advancements are beginning to make them feasible on modern flight vehicles. One of the control mechanisms the Wright Brothers used in their original gliders and the Wright

37

3.1 The Wright Brothers' 1899 wing warping design. No photographs exist of the 1899 kite; only a sketch of it illustrating wing-warping, drawn by Wilbur Wright in 1912. (Source: Hallenberg, 2004.)

flyer was wing warping, which is the earliest documented wing morphing. In 1899 the Wright Brothers constructed a glider made of French sateen fabric, which covered a wooden frame. The fabric and the wood structure were sewn together to create a strong, lightweight, but flexible structure. They oriented the fabric weave at 45° to the aircraft flight direction, which allowed the structure to warp properly when pulled by actuation cables (Hallenberg, 2004). This 1899 glider was built solely to test the idea of wing warping. The design is shown in Fig. 3.1. This biplane configuration, using steel wire bracing with wing warping, was supposedly inspired by observing how a cardboard bicycle tube box could be warped yet still be rigid when returned to its normal configuration.

As aircraft began to approach and pass the speed of sound, wing sweep became an important factor in design. Wing sweep has always been a design compromise for high-speed aircraft. Straight wings are most efficient for the low-speed flight associated with landing or take-off, and swept wings are more efficient in the higher-speed flight regimes. Research conducted in Germany by Dr Adolf Busemann in the 1930s and 1940s on the theory of variable sweep wings eventually led to the 1950s flight of the Bell X-5, the first variable sweep aircraft. Research continued into the 1960s and culminated with aircraft designs such as Grumman's F-111 and F-14 aircraft (Busemann, 1971).

3.3 Keeping morphing alive – NASA research in morphing aircraft structures

The field of wing morphing was fairly quiet until the 1980s, when a joint US Air Force and NASA program called the Advanced Fighter Technology Integration (AFTI) began (Smith *et al.*, 1992). One aspect of this research focused on what was called a 'Mission Adaptive Wing' (MAW). The wing of a NASA F-111A was replaced with a MAW wing built by Boeing Aircraft, as shown in Fig. 3.2. This wing had an internal mechanism to flex the outer wing skin and produce a high

3.2 NASA-Boeing F-11 incorporating the mission adaptive wing (MAW). (Source: Curry, 2004.)

camber section for subsonic speeds, a supercritical section for transonic speeds, and a symmetrical section for supersonic speeds. The use of flexible wing skins to produce a smooth upper surface brought this wing a little closer in concept to that of a bird. A digital flight control system provided automatic changes to the wing geometry. The system had four automatic control modes:

- Maneuver camber control, adjusting camber shape for peak aerodynamic efficiency,
- Cruise camber control, for maximum speed at any altitude and power setting,
- Maneuver load control, providing the highest possible aircraft load factor,
- Maneuver enhancement alleviation, in part attempting to reduce the effects of gusts on the airplane ride.

The AFTI/F-111 MAW system was flown 59 times from 1985 to 1988. The flight test data showed a drag reduction of around 7% at the wing design cruise point to over 20% at an off-design condition. The four automatic modes were tested in flight with satisfactory results.

In 1996, the NASA Active Aeroelastic Wing (AAW) project began. This was a joint project between NASA's Dryden Flight Research Center, Edwards, CA, the US Air Force Research Laboratory (AFRL) and Boeing Phantom Works, which was investigating an adaptation of the Wright Brothers' wing-warping approach to aircraft flight controls using modern composites and actuators (Lokos *et al.*, 2002).

The main focus of AAW research was to develop and validate the concept of morphing (wing warping) for roll control in modern aircraft. The AAW research demonstrator was a modified F/A-18A (Fig. 3.3). Several of the existing wing skin

3.3 NASA-Boeing F/A-18A incorporating the active aeroelastic wing (AAW). (Source: Curry, 2005.)

panels along the rear section of the wing just ahead of the trailing-edge flaps and ailerons were replaced with thinner, more flexible skin panels and structure. The F/A-18's leading-edge flap was also divided into separate inboard and outboard segments, and additional actuators were added to operate the outboard leading-edge flaps separately from the inboard leading-edge surfaces. By using the outboard leading-edge flap and the aileron to twist the wing, the aerodynamic force on the twisted wing generated the desired roll forces. The F-18 AAW completed a two-phase flight test program with satisfactory results after a total of 50 flights.

3.4   Resurgence of morphing concepts

3.4.1   NASA biologically inspired aerodynamic geometries

A group of researchers at NASA Langley Research Center has investigated several directions in which future wing geometries could head. This research was based on biologically inspired hydrodynamic/aerodynamic structures and the control of wing morphing. Various structures in Nature have been considered as possible models, such as a seagull's wing, a hammerhead shark's head shape, a shark's fins and a hummingbird's wing in flight. We now have the technology to create simplified versions of these structures and the capability to begin to understand the complexity of structures and their control in Nature (Lazos and Visser, 2006). The Langley group studied these geometries, both dynamic and static, to learn how flight works in Nature and then attempted to apply these lessons through morphing aircraft. One new geometry, the hyper-elliptic wing, shown in Fig. 3.4, was inspired from Nature and showed that gains of 15% in lift over drag are possible. This research looked not only at biologically inspired geometries but

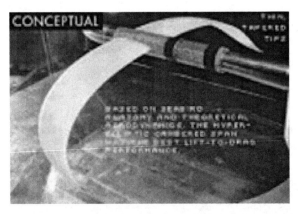

3.4 Hyper-elliptic cambered span biomimetic wing planform. (Source: Lazos and Visser, 2006.)

also at how we can duplicate these desired characteristics in actual air vehicles. This was a multidisciplinary effort, studying hydrodynamics, structures, controls, and actuation technologies.

Now that new technology has made complex wing morphing possible, there have been numerous research projects exploring the possibilities of future advances. A joint project between the Air Force Research Labs, Purdue University and Wright State University produced a report exploring many of the characteristics a customer such as the Air Force could ask for in a future morphing aircraft. A joint Air Force and Virginia Tech project looked at computer modeling of wing shape changes and what actuation forces would be needed to duplicate the control found in today's conventional wings.

3.4.2 DARPA morphing aircraft structures (MAS) program

Lockheed Martin's research under the Defense Advanced Research Projects Agency (DARPA) Morphing Aircraft Structures (MAS) program explored large planform changes for the wing of a tailless unmanned aerial vehicles (UAV) demonstrator, as seen in Fig. 3.5. This project used new piezoelectric actuators to fold and unfold the UAV wings in flight. Twist near the wing tips, thrust vectoring paddles and small areas of the leading and trailing edges near the center of the aircraft could also morph for elevon and pitch trim control.

The actuators for this program came from a joint DARPA and Penn State project called the Compact Hybrid Actuators Program (CHAP). These actuators are both highly efficient and quite small. They fit into small spaces and provide actuation power in specific locations as needed. Incorporating one large actuator requires a force-transmission system from a central location that adds both material weight and mechanical design complexity (Koopmann and Leiseutre, 2004).

3.5 Lockheed Martin morphing adaptive structures (MAS) demonstrator. (Source: Ivanco *et al.*, 2007.)

3.6 MFX-1 morphing demonstrator by Boeing Phantom Works. (Source: Warwick, 2007.)

Funded through the DARPA MAS program and Boeing's Phantom Works, the NextGen company also designed a morphing demonstrator UAV (Andersen and Cowan, 2007). NextGen's aircraft demonstrated a variable area, variable sweep wing on the MFX-1 vehicle, shown in Fig. 3.6. The wing was capable of changing area by 40%, span by 30% and sweep between 30–40°. These impressive planform changes were made possible by a novel silicone membrane/truss structure wing used on the aft inboard two-thirds of the wing. This variable geometry wing is unique in its ability to change chord independently of wing sweep. The MFX-1 was successfully flight tested in August 2008 at Camp Roberts Test Range in California.

The successful flight test program of the MFX-1 was followed by the design of a larger, more capable aircraft, the MFX-2. Wind-tunnel testing of the 90 kg aircraft configuration was conducted up to Mach 0.92, and a successful flight testing program occurred late in 2007.

3.4.3   University efforts in morphing research

University research efforts in morphing can be broken down into a series of six categories based on the extent of airfoil and planform deformation. These range from small deformations of the skin of an airfoil profile through complete planform morphing including independent control of wing area. The following categories have been arranged in order of increasing deformation of the fundamental wing structure from skin to ribs to spars, and the full structure.

Actuated surface roughness

Work by Huebsch *et al.* (2012) has explored the capability of sub-boundary layer actuated surface dynamic roughness elements, effectively localized skin morphing, to control transition and separation on a National Advisory Committee for Aeronautics (NACA) airfoil in collaboration with Air Force Office of Scientific Research (AFOSR) and NASA. This work has experimentally demonstrated that leading-edge flow separation can be eliminated at very high angles of attack, up to 14°, at chordwise Reynolds numbers ranging from approximately 50 000 to 150 000. Continuing work has focused on developing a physical understanding of this flow control mechanism as well as investigating dynamic roughness effects on transition development and the use of this morphing skin for aircraft maneuvering through asymmetric surface pressure alterations (Fig. 3.7).

Work by Folk and Ho (2001) made use of actuating micro bubbles on the leading-edge surface of a delta wing that was successful at maneuver control. This localized skin morphing was able to make global alterations to the leading-edge vortex on one wing to affect a maneuvering moment.

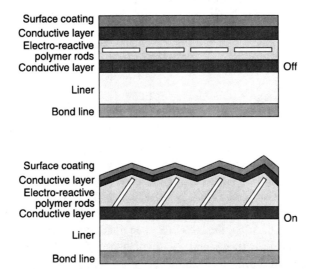

3.7 Dynamic roughness elements near the leading edge of a NACA 0012 airfoil showing suppression of the leading edge separation bubble. (Source: Huebsch *et al.*, 2012.)

Airfoil camber change

Several different approaches have been explored to achieve a smooth camber change at the leading and trailing edges of wings. Daynes *et al.* (2009) utilized a structure of bistable composite laminates to create a 'fan' type structure with two stable configurations. A sample segment of a helicopter main rotor blade was created with a trailing edge capable of deforming from neutral to a positive 10° deflection. The nature of the bistable structure provided both a smooth surface profile and a reduction in the actuator loads required to maintain the deformation due to structural stability in the two design positions.

A series of sealed honeycomb cells was used by Vos and Barrett (2011) to develop an airfoil section with a trailing edge capable of smooth and continuous deformation across a similar range of deflections. Designs were also proposed to apply this system to a leading-edge droop as well as a trailing-edge flap permitting a camber change across the length of the chord.

Bulk airfoil profile change

Generating shape changes in the mid-chord region of an airfoil has been more challenging than achieving camber changes at the leading and trailing edges. This has been primarily due to the difficulty of incorporating actuators and flexible structures into the regions usually used for primary load-bearing structures. Coutu *et al.* (2011) have used shape memory alloy actuator banks to deform the upper surface of an airfoil between the 3% and 74% chord positions with the goal of expanding the region of laminar flow in cruise flight. Linking these actuators to a closed-loop adaptive control scheme enabled an improvement of approximately 10% in lift-to-drag ratio.

Wing twist

In 2003 Garcia and colleagues at the University of Florida explored the use of active wing morphing for roll control of 61 cm, 30 cm and 25 cm (24″, 12″ and 10″) span UAVs (Abdulrahim *et al.*, 2004). Both aircraft were designed with the conventional wing followed by a tail with a rudder and elevator. The wing on the 61 cm aircraft had a torque rod running through the leading edge of the wing. Riblets of carbon roving trailed behind the torque rod and a latex membrane was stretched between the riblets. The 30 cm UAV used similar construction, but lacked the torque rod.

Control of the 61 cm UAV was achieved by rotating the torque rod with a fuselage-mounted servo. At the three-quarter span point the torque rod was bent from a spanwise to a chordwise orientation. Rotating the rod then twisted the outer quarter of the wing, as shown in Fig. 3.8. This wing twist created roll and yaw moments. The 30 cm aircraft had a Kevlar string that ran diagonally from the wing tip to the fuselage. A servo in the fuselage could provide tension or slack in the cable that both twisted and curled the wing tip to generate control forces (Fig. 3.9).

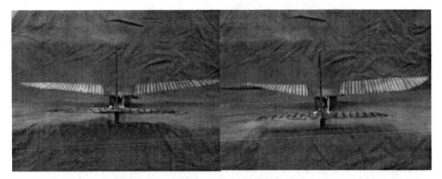

3.8 Active wing morphing on University of Florida 61 cm (24″) span MAV. (Source: Abdulrahim *et al.*, 2004.)

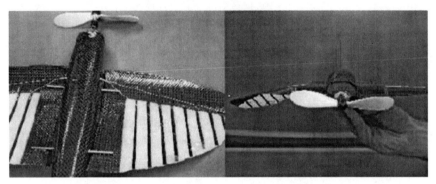

3.9 Active wing morphing on University of Florida 30 cm (12″) span MAV. (Source: Abdulrahim *et al.*, 2004.)

Both morphing strategies of twisting and curling proved well suited for roll control. Roll and yaw rates of close to 70°/s and 20°/s, respectively, were achieved. Both aircraft had good flight characteristics and proved that a significant roll rate could be generated with the twisting and curling morphing strategies.

Researchers at West Virginia University (WVU) conducted research to develop an adaptive washout morphing mechanism for the control of a swept-wing tailless aircraft (Guiler *et al.*, 2008). The adaptive washout morphing mechanism was able to provide effective roll, yaw and pitch control of a swept-wing tailless aircraft. This unique adaptive washout morphing mechanism utilized rigid airfoil segments that rotated around a leading-edge rod and were covered with a polypropylene skin. This new control technique was experimentally and numerically compared with an existing elevon-equipped tailless aircraft and has shown the potential for significant improvements over that system in terms of efficiency, providing much improved lift-to-drag performance. The feasibility of this mechanism was also validated by designing, fabricating and flight testing a flying prototype with the morphing system that produced similar maneuvering rates to a platform equipped with standard elevons, but with improved overall aerodynamic efficiency. The WVU flight demonstrator is shown in Fig. 3.10.

Global planform change, without direct control of wing area

The morphing RoboSwift aircraft, designed at the University of Delph, is characterized by the continuously variable shape of its wings, which are modeled on the wings of the Common Swift (Wageningen University and Research Centre, 2008). These wings make the aircraft, like its living model, very maneuverable and efficient (Fig. 3.11). As a result, the RoboSwift was the first aircraft in the world to have the wing properties of a living bird. Wind tunnel and flight tests

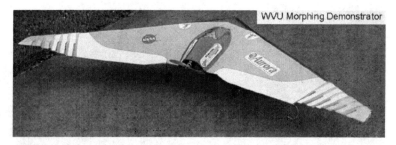

3.10 WVU swept-wing morphing demonstrator. (Source: Guiler *et al.*, 2008.)

3.11 RoboSwift by University of Delph. (Source: Wageningen University and Research Centre, 2008.)

have shown that it can come remarkably close to the exceptional flying ability of the swift. With a wingspan of approximately 50 cm and a weight of less than 100 g, the RoboSwift is significantly smaller than typical aircraft.

Work by Grant *et al.* (2010) explored radical dihedral/anihedral and sweep morphing on micro air vehicle platforms. Both configurations incorporated a conventional tractor electric motor and propeller, fuselage, and tail surfaces, but enabled changes of up to 30° in dihedral or sweep.

Global planform change, with direct control of wing area

Very little can expand the flight envelope of an aircraft as drastically or, temporarily neglecting structural issues, as efficiently as the ability to significantly increase or decrease wing area. Work by Vocke *et al.* (2011) in elastomer matrix composite (EMC) skins and deformable cellular structures has led to the creation of a skin, rib, and stringer structure supported by telescopic spars. Careful geometric design and material selection provided a skin and internal airfoil framework with a Poisson's ratio of 0 that allowed the wing to elongate up to 100% while preserving the original airfoil profile and chord length.

Jacob *et al.* (2007) have studied the use of inflatable wings for both compact storage of UAVs for on-site deployment as well as using the compliant structures for morphing. Both the utilization of twist for moment generation and wing area modification have been examined. The structures are composed of appropriately sized parallel tubes bonded together to form the basic structure of an airfoil. A smooth skin was attached to the outer surface of these tubes to improve aerodynamic performance. Span modification was achieved by combining the inflation system with constant force springs on the upper surface. These springs provided the retraction and stowage, while increasing the pressure in the spanwise tubes provided the deployment actuation. Functional deployable-wing UAV prototypes have been developed using this technique.

3.5   Current morphing component technologies

Achieving the fluid transformation from one wing shape to another as a bird does requires new actuation technologies. Most of the cables and hydraulics common in aircraft today would have to be replaced by local actuators, integral to the aircraft structure. Morphing aircraft structures emulate structures found in marine and airborne animals, where the muscles, bones, skin and feathers combine to form a continuously changeable biological control surface. Standard hydraulic or servomotor actuators have been used to flex or morph an appropriately designed composite aircraft structure. Pneumatic actuation can actually form a semi-rigid structure and actuate changes in shape. Other actuators act more like muscles in a living organism such as piezo or shape memory actuators. These new types of actuators can be embedded in an aircraft skin or structure and initiate a localized shape change when a voltage is applied. The lines between actuators, structures and skins have been blurred by a new family of hybrid skin/actuators that can morph a surface or provide subtle flow control of a local boundary layer, as do feathers, scales and flagella for many living organisms. This research is focusing on such things as embedded piezo-actuators, polymer liquid crystals and micro-electro-mechanical systems (MEMS). This section outlines some of the current areas of research for morphing component technologies.

3.5.1   Servo actuation

Electric servos have been in use for over 50 years, and their power and efficiency have dramatically improved in recent years with the current electronics revolution. Many modern servos are digitally controlled and can be programmed to modify their performance. Servos currently range in weight and power from 3 g (0.10 oz) and 500 g cm (6 oz in) torque to many kg in weight and thousands of Nm torque.

3.5.2   Pneumatic actuation

Recently the need has arisen for small UAV platforms to have deployable, controllable wings that may also change to fit a variety of flight regimes. In 2004 a group from ILC Dover conducted research efforts morphing inflatable wings to provide roll control by wing warping through actuation of the aft end of the wing to achieve changes in section camber (Cadogan *et al.*, 2004). Several approaches have been developed that lend themselves to camber control via locally altering the geometry of the wing. Apart from use as a stand-alone aerodynamic surface on a small UAV, the inflatable assemblies can also be used as an aspect ratio-increasing device on a larger aircraft to enable a more radical change in wing configuration. This approach serves to improve system efficiencies across changing flight regimes, allowing transitions from high-speed target approach to low-speed loitering. Shape changes of the wing can be created through the use of pneumatic chambers under a flexible skin.

3.5.3   Piezoelectric actuators

NASA Langley Research Center has also been conducting research into what materials can be used to create self-adaptive airframe systems and their utilization. This research has focused on two main classes of materials: high-performance continuous-fiber reinforced polymer matrix composites and piezoelectric films from high-temperature polyimides. Piezoelectric devices have been identified as a promising actuator technology for the implementation of active boundary layer control, high-bandwidth noise suppression and aeroservoelastic tailoring.

However, many potential aerospace applications require displacement performance larger than that which is achievable from conventional piezoelectrics.

NASA Langley has developed two high-displacement piezoelectric actuator technologies, Reduced And INternally-Biased Oxide Wafer (RAINBOW) and THin layer composite UNimorph ferroelectric DrivER and sensor (THUNDER), to meet these requirements (Simpson *et al.*, 1999). These devices are unimorph-type actuators, which consist of a piezoelectric ceramic layer bonded to one or more non-piezoelectric secondary layers. Because of the use of elevated temperatures during processing, internal stresses are created in the structures, which significantly enhance displacement through the thickness of the devices. Currently, the processing and characterization of these high-displacement actuators are under investigation. One recent characterization study involved the effects of electric field, load and frequency on the displacement properties of rectangular THUNDER devices. Results showed that individual actuators were capable of free displacements in excess of 3 mm when tested at ±9 kV/cm (see Fig. 3.12). Increasing device stiffness through metal selection and thickness resulted in improved load-bearing performance at the expense of displacement, allowing devices to be designed with a range of performance capabilities.

Currently, most piezo-actuators are not capable of the power or displacement required for practical use in aircraft, but a new actuator called the compact hybrid actuator (CHAP) has made great strides. In support of the Lockheed Martin MAS project, a Penn State–DARPA team has developed a new piezoelectric actuator that uses a novel motion accumulation mechanism for folding and unfolding the wings of the MAS demonstrator (Koopmann and Leiseutre, 2004). These actuators, shown in Fig. 3.13, consist of a piezoelectric stack, a lever arm, and a reversible ratchet-type drive. The novel mechanism of the CHAP piezo-actuator

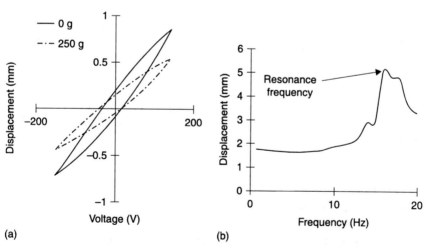

(a)   (b)

3.12 Response of THUNDER piezoelectric actuators. (Source: Simpson *et al.*, 1999.)

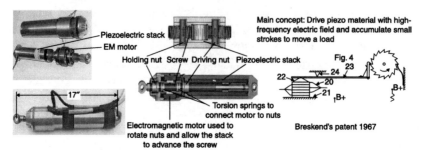

3.13 Images and schematics of compact hybrid actuator (CHAP) systems.

coupled with high-frequency piezoelectric membranes gives the power and displacement needed for practical aircraft applications.

Another technique for working around the low maximum strain of piezoelectric materials is to introduce bending, buckling or bistability into a structure to be used as an amplification device to generate useful deflections. Research currently being conducted at the University of Bath, UK, has demonstrated that unsymmetrical carbon fiber/epoxy composites bonded with piezoelectric actuators can produce useful actuation deflections (Bowen *et al.*, 2007).

3.5.4   Shape memory materials

Shape memory alloys (SMA) are metallic alloys that are able to undergo large reversible deformations under loading/thermal cycles and are able to generate high thermal–mechanical driving forces. The behavior of SMA is due to their native capability to undergo reversible changes of the crystallographic structure, depending on temperature and state of stress. These changes can be interpreted as reversible martensitic transformations between a crystallographic more-ordered parent phase, the austenite (A), to a crystallographic less-ordered product phase, the martensite (M). SMA can be produced as both wire and sheet (Auricchio and Sacco, 2000). Passing a voltage through the SMA wire or sheet generates the temperature change needed to produce a shape change. When the voltage is removed, the heat dissipates and the material returns to its original shape. SMA can be used alone as an actuator or incorporated into the matrix of a composite structure. Composites incorporating SMA are often referred to as smart composites. SMA use in smart composites is fairly new and is currently being studied by many research groups.

The Technical University of Berlin has been conducting research into the use of embedded SMA wires in aircraft structures. They have constructed SMA-controlled airfoil camber, winglets and motors (Mueller *et al.*, 2005).

Sacco from the University of Cassino in Italy is developing a micro-mechanical model for the evaluation of overall constitutive behavior of a composite material obtained by embedding SMA wires into an elastic matrix. A simplified thermo-mechanical model for the SMA inclusion, able to reproduce the super-elastic as well as the shape memory effect, is proposed (Auricchio and Sacco, 2001).

Research is also being conducted on a family of new polymers that exhibit behavior similar to the shape memory alloys. Cornerstone Research Group (CRG) is developing shape memory polymers, foams and dynamic composites for morphing aircraft with funding from NASA and DARPA (Cornerstone Research Group, 2011). CRG is also using these materials to develop morphing truss structures for aircraft. Both shape memory alloys and polymers show great promise, but these technologies must mature in order to alleviate excessive thermal and power consumption requirements.

3.5.5   Skins

The skins on morphing control mechanisms are very important because aerodynamic efficiency requires smooth shape changes. The Wright brothers used French sateen fabric, and NASA has used carbon-reinforced plastic skins on the F-111 MAW and the F-18 AEW. The desired properties of the skin are dependent on whether the skin is an integral component of the aircraft structure, as with the NASA aircraft, or just a smooth fairing material.

There are a number of elastic skin materials available today to cover a morphing mechanism. Research at Virginia Polytechnic Institute has begun to characterize the mechanical properties of a variety of possible materials. The skin of a morphing structure should be elastic and flexible, have high strain recovery rates, be resistant to weather conditions, abrasions and chemicals, and have a hardness high enough to resist aerodynamic loads. The materials studied at Virginia Polytechnic were thermoplastic polyurethanes, copolyester elastomers, shape memory polymers and a variety of woven materials (Kikuta, 2003).

Combining skins with structure has been another area of research interest. NextGen has demonstrated the application of a silicone skin combined with a morphing truss structure similar to that developed at Pennsylvania State University by Lesieutre (Ramrkahyani et al., 2004). Penn State's work was focused on segmented skins over a compliant cell truss structure, as shown in Fig. 3.14. Living organisms use skins, feathers and scales for control and sensing. Today these man-made mechanisms are beginning to simulate features found in Nature. Numerous studies are currently being done on the application of micro-electro-mechanical systems (MEMS) flaps, which can be much like the scales on a fish or the feathers of a bird. These devices can be used actively to modify a boundary layer or passively to sense changes in a boundary layer. Aerovironment, with funding from DARPA and NASA Dryden, flight-tested a delta-winged unmanned aircraft named the Gryphon, with leading-edge vortex control MEMS flaps that demonstrated the ability to control the leading-edge vortices (Huang et al., 2001).

3.14 Compliant cell morphing truss structure.

West Virginia University and Physical Sciences, Inc. have studied the use of dynamic roughness for boundary-layer control on aircraft and marine vehicles. This dynamic roughness could potentially be created by a novel membrane that utilized switchable polymer rods similar to liquid crystals in a LCD screen (Cairns *et al.*, 2010). When a voltage was applied these rods could create a controlled pattern of roughness on the surface of the membrane and manipulate the boundary-layer flow. This new type of membrane has the potential of also acting as a pressure-sensing device and a form of actuated or variable-rigidity skin for an aircraft. This boundary-layer control system is being designed as a smart skin to allow it to be added to existing aircraft (Huebsch *et al.*, 2012).

3.6   Conclusion: the future of aircraft morphing technologies

At the beginning of the twenty-first century the technologies of aircraft morphing have taken a dramatic change in direction. Most of the previous work had been to improve the efficiency of aircraft or allow them to operate in a wider range of flight regimes. With continued advances in lightweight structures, actuators and complex control systems, the current trend is to enable aircraft to operate in multiple environments.

The Biologically Inspired Robotics Laboratory of Case Western Reserve University teamed with the University of Florida to develop the morphing micro air and land vehicle (MMALV), shown in Fig. 3.15 (Bachmann *et al.*, 2009). The purpose of this vehicle was to serve as a device that could fly into hazardous/ hostile environments, then land and walk around to explore, gather and relay intelligence, so that decisions could be made about how to proceed in a dangerous situation without having to physically send people in to do the data collection. MMALV combines the Whegs™ terrestrial locomotion technology developed at Case with Florida's morphing wing technology. This is an aircraft that can fly and then crawl into a dangerous environment on the ground.

3.15 Morphing micro air and land vehicle (MMALV) (Bachmann *et al.,* 2009).

A similar effort to develop a vehicle that can operate both on the land and in the air has come from a group of Massachusetts Institute of Technology engineers who have come together to form a company called Terrafugia. The goal of this group is to develop a truly functional roadable aircraft. The Terrafugia Transition is a proof of concept roadable aircraft that received all necessary National Highway Traffic Safety Administration and Federal Aviation Administration special exemptions requested as of June 2011. At the time of writing customers may place orders and deposits for what is claimed to be the first functional flying car.

Aircraft now operate from insect-like environments to the edge of space; and the need to transition from one environment to another will grow. The future of morphing aircraft technologies is bright, as materials, actuators and controls will continue to advance exponentially.

3.7   References

Abdulrahim, M., Garcia, H. and Lind, R. (2004). 'Flight testing a micro air vehicle using morphing for aeroservoelastic control.' *45th AIAA/ASME/ASCE/AHS/ASC Structures, Structural Dynamics, and Materials Conference*, 19–22 April 2004, Palm Springs, CA.

Andersen, G.R. and Cowan, D.L. (2007). 'Aeroelastic modeling, analysis, and testing of a morphing wing structure.' *48th AIAA/ASME/ASCE/AHS/ASC Structures, Structural Dynamics, and Materials Conference*. 23–26 April 2007, Honolulu, HI.

Auricchio, F. and Sacco, E. (2001). 'Thermo-mechanical modeling of a superelastic shape-memory wire under cyclic stretching-bending loadings.' *International Journal of Solids and Structures*, **38**: 6123–45.

Bachmann, R.G., Boria, F.J., Vaidyanathan, R., Ifju, P.G. and Quinn, R.D. (2009). 'A biologically inspired micro-vehicle capable of aerial and terrestrial locomotion.' *Mechanism and Machine Theory*, **44**: 513–26.

Bowen, C.R., Salo, A.I.T., Butler, R. Chang, E. and Kim, H.A. (2007). 'Bi-stable composites with piezoelectric actuators for shape change.' *Key Engineering Materials* **334–335**: 1109–12.

Busemann, A. (1971). 'Compressible flow in the thirties.' *Annual Review of Fluid Mechanics*, **3**: 1–12.

Cadogan, D., Smith, T., Uhelsky, F. and MacKusick, M. (2004). 'Morphing inflatable wing development for compact package unmanned aerial vehicles.' *AIAA* 2004-1807.

Cairns, D.R., Shafran, M.S., Sierros, K.A., Huebsch, W.W. and Kessman, A.J. (2010). 'Stimulus-responsive fluidic dispersions of rod shaped liquid crystal polymer colloids.' *Materials Letters*, **64**(10): 1133–6.

Cornerstone Research Group (2011). 'Our contributions to SMP research.' Available at http://www.crgrp.com/technology/overviews/contributions.shtml. Accessed January 2012.

Coutu, D. Brailovski, V., Terriault, P., Mamou, M., Marki, Y. *et al.* (2011). 'Lift-to-drag ratio and laminar flow control of a morphing laminar wing in a wind tunnel.' *Smart Materials and Structures*, **20**: 3.

Curry, M. (2005). *NASA AAW fact sheet FS-061-DFRC*, NASA Dryden Flight Research Center. Available at http://www.nasa.gov/centers/dryden/news/FactSheets/FS-061-DFRC.html. Accessed January 2012.

Curry, M. (1998). *NASA AFTI/F-111 photo collection*, NASA Dryden Flight Research Center. Available at http://www.dfrc.nasa.gov/Gallery/Photo/F-111AFTI/index.html. Accessed January 2012.

Daynes, S., Weaver, P.M. and Potter, K.D. (2009). 'Aeroelastic study of bistable composite airfoils.' *Journal of Aircraft*, **46**(6): 2169–73.

Folk, C. and Ho, C.M. (2011). 'Micro-actuators for control of delta wing with sharp leading edge.' *39th Aerospace Sciences Meeting and Exhibit*, 8–11 January 2001, Reno, NV.

Grant, D.T., Abdulbrahim, M. and Lind, R. (2010). 'Design and analysis of biomimetic joints for morphing of micro air vehicles.' *Bioinspiration and Biomimicry*, **5**: 4.

Guiler, R.W. and Huebsch, W.W. (2008). 'Control of a swept-wing tailless aircraft through wing morphing,' ICAS 2008–2.7.1, ICAS 2008 Congress, September, 2008.

Hallenberg, J. (2004). 'The Wright Brothers, the invention of the aerial age.' National Air and Space Museum, Smithsonian. Available at http://www.nasm.si.edu/wrightbrothers/index_full.cfm. Accessed January 2012.

Huang, A., Folk, C., Ho, C.M., Liu, Z., Chu, W.W. *et al.* (2001). 'Gryphon M^3 system: Integration of MEMS for flight control.' *MEMS Components and Applications for Industry, Automobiles, Aerospace, and Communication, Proceedings of SPIE*, **4559**, 85–94.

Huebsch, W.W., Gall, P.D., Hamburg, S.D. and Rothmayer, A.P. (2012). 'Dynamic roughness as a means of leading edge separation flow control.' *AIAA Journal of Aircraft*, accepted for publication.

Ivanco, T.G., Scott, R.C., Love, M.H., Zink, S. and Weisshaar, T.A. (2007). 'Validation of the Lockheed Martin morphing concept with wind tunnel testing.' *48th AIAA/ASME/ASCE/AHS/ASC Structures, Structural Dynamics, and Materials Conference*, 23–26 April 2007, Honolulu, HI.

Jacob, J.D., Smith, S.W., Cadogan, D. and Scarborough, S. (2007). 'Expanding the small UAV design space with inflatable wings.' *SAE Aerotech Congress & Exhibition*, 2007-01-3911.

Kikuta, M. (2003). 'Mechanical properties of candidate materials for morphing wings.' Mechanical Engineering Master's Thesis, Virginia Polytechnic Institute and State Univesity, Blacksburg, VA.

Koopmann, G.H. and Leiseutre, G.A. (2004). *Center for Acoustics and Vibration Review '03*. Penn State Center for Acoustics and Vibration. Available at http://www.cav.psu.edu/newsletters.aspx. Accessed January 2012.

Lazos, B. and Visser, K.D. (2006). 'Aerodynamic comparison of hyper-elliptic cambered span (HECS) wings with conventional configurations.' *24th AIAA Applied Aerodynamics Conference*. 5–8 June 2006, San Francisco, CA.

Lokos, W.A., Olney, C.D., Crawford, N.D., Stauf, R. and Reichenbach, E.Y. (2002). 'Wing torsional stiffness tests of the active aeroelastic wing F/A-18 airplane.' *43rd AIAA/ASME/ASCE/AHS/ASC Structures, Structural Dynamics, and Materials Conference*. 22–25 April 2002, Denver, CO.

Mueller, I., Musolff, A. and Sahota, H. (2005). 'Adaptive winglet.' Available at http://www.smaterial.com/SMA/winglet/winglet.html. Accessed January 2012.

Ramrkahyani, D.S., Lesieutre, G.A., Frecker, M. and Bharti, S. (2004). 'Aircraft structural morphing using tendon actuated compliant cellular trusses.' *45th AIAA/ASME/ASCE/AHS/ASC Structures, Structural Dynamics & Materials Conference*. 19–22 April 2004, Palm Springs, CA.

Simpson, J.O., Wise, S.A., Bryant, R.G., Cano, R.J., Gates, T.S. *et al.* (1999). *Innovative materials for aircraft morphing*. NASA Langley Research Center Aircraft Morphing Program Report, Hampton, VA.

Smith, J.W., Lock, W.P. and Payne, G.A. (1992). *Variable-camber systems integration and operational performance of the AFTI/F-111 mission adaptive wing*. NASA Technical Memorandum 4370, April 1992.

Vocke, R.D. III, Kothera, C.S. and Woods, B.K.S. (2011). 'Development and testing of a span-extending morphing wing.' *Journal of Intelligent Material Systems and Structures*, **22**(9): 879–90.

Vos, R. and Barrett, R. (2011). 'Mechanics of pressure-adaptive honeycomb and its application to wing morphing.' *Smart Materials and Structures*, **20**(9): 9.

Wageningen University and Research Centre (2008). 'First flight of the RoboSwift micro-airplane is a success.' *ScienceDaily*, 4 March. Available at http://www.sciencedaily.com/releases/2008/03/080304192104.htm. Accessed 6 January 2012.

Warwick, G. (2007). 'NextGen's shape-changing UAV morphs in flight.' *Flightglobal*, 19 October 2007. Available at www.flightglobal.com.

4

Jet engine design drivers: past, present and future

R. SINGH, G. AMEYUGO and F. NOPPEL,
Cranfield University, UK

Abstract: The history of aero gas turbines is a history of innovation. Every challenge presented to the aero engine industry has been met with ingenuity, and today engineers continue to work on innovative concepts to meet the next set of challenges facing the aviation industry. This chapter will discuss past, present and future drivers for aero gas turbine development. It first looks at the motivation behind the development of the jet engine, and then outlines the technological and business drivers that have contributed to shape the first 50 years of aero gas turbine development.

Key words: jet engine, gas turbine, disruptive technology, technology drivers, emissions, contrail, innovation, alternative fuels.

4.1 Introduction

4.1.1 The jet engine as a disruptive technology

New technologies that result in substantial leaps are often referred to as disruptive technologies. These are innovations that change the rules of the game, introducing a new value proposition. They do not have to outperform the technologies they seek to replace in every sense, but they must offer some new valuable feature that will justify their implementation, at least at a small scale. If enough momentum can be gathered, the new disruptive technology will evolve to replace its predecessor.

The jet engine is a prime example of a disruptive technology. Early jet engines performed poorly compared with the well-established piston engines of the time. However, they opened the door to faster air travel at higher altitudes (see Fig. 4.1), allowing passengers to fly above the weather and resulting in shorter, more comfortable trips. These advantages allowed the gas turbine to replace the piston engine as the prime mover of choice in civil air travel. The initial fuel consumption, reliability and cost issues were overcome, making air travel accessible to ever-growing numbers of passengers.

In the late 1920s, Frank Whittle – an RAF officer – became interested in high-altitude, high-speed flight. He saw it as the key that would enable aviation to grow further, providing safer, faster and more comfortable travel above the weather. Whittle concluded that incremental improvements in existing piston engine–propeller configurations were unlikely to enable flight at higher altitudes and

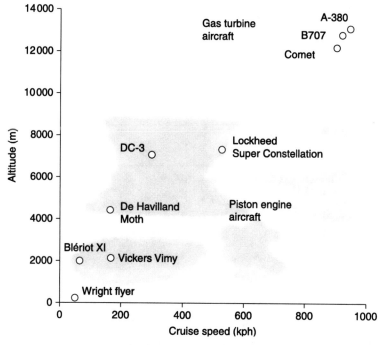

4.1 Gas turbines enabled aircraft to fly above the weather.

speeds, and this conclusion led him to begin the development of what would eventually become the first jet engine.

For civil aircraft, the jet engine transformed the 'earning capacity' of aircraft, forming the basis of mass civil aviation in the following decades. The world's three major gas turbine manufacturers, Rolls-Royce, Pratt and Whitney and General Electric, all started their gas turbine businesses based on Whittle's W2/700 gas turbine. Nevertheless, much work would still be required to bring gas turbines to the standard we know today. As Whittle himself said, 'inventing it was easy – making it work was the difficult bit!' (Whittle, 2007).

4.1.2   Nature of the industry: 1950 to present day

Launching a large civil gas turbine requires anywhere between US $500 m and $2.5 bn (Singh, 2007). It typically takes between 15 and 25 years to break even after the initial investment (see Fig. 4.2). In essence, every time a gas turbine manufacturer launches a new gas turbine, it is 'betting itself' in the process. The risks are so great that many famous companies – such as Bristol, Armstrong-Siddeley and De Havilland in the UK – have disappeared or been absorbed by other companies along the way.

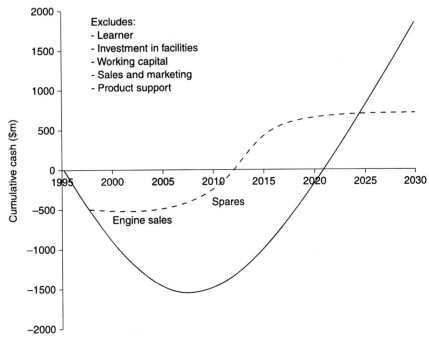

4.2 Entry into the large aero engine sector.

The timescales and capital investment required make it extremely difficult for new players to enter the aero gas turbine business. In fact, the long-term risks and exposure associated with the business are so great as to defy business logic, at least on the surface. What has driven large corporations to expose themselves repeatedly, putting their reputations and those of some of the world's best engineers and managers on the line?

The reason lies in the strategic importance of gas turbine technology within the wider industrial context. The development and use of gas turbines requires a substantial upfront investment, but the resulting technology and infrastructure will benefit a country's industrial base in the long term. The benefits are such that many governments are willing to heavily subsidise this industry. For this reason, some new companies have succeeded in breaking the Anglo-American dominance in aero gas turbines in certain sectors.

Once an aero gas turbine company has been in business for a number of years, it becomes increasingly possible for it to finance new developments by means of aftermarket revenues, which will be described later in this chapter. In addition, the technology developed for aero gas turbines can be transferred into other sectors with different exposure patterns to economic downturns, such as the marine propulsion industry or the power generation industry. This is the case, for example, of General Electric and Rolls-Royce, whose energy and marine propulsion revenues amounted to almost half the aero gas turbine turnover in 2008 (Rolls-Royce, 2009).

Another strategy adopted by gas turbine manufacturers that has helped the aerospace industry is the use of 'common cores'. As civil aviation grew, an increasing number of aircraft types were launched, and engine companies were expected to offer engines with thermodynamic cycles precisely suited to the aircraft duty and operation. Instead of developing new gas turbines from scratch, existing core designs were adapted in order to minimise the time, cost and risk of development. Airframers welcomed this, as it improved the earning capacity of the aircraft they were offering to airlines.

The efforts of aero gas turbine companies have been partly responsible for the increase in aircraft productivity over the last five decades. Large aircraft today are eight times more productive in terms of seat miles per year than airlines in the 1950s. How was this accomplished, and what was the role of gas turbine technology? The next section will examine some of the most important challenges and drivers for jet engine development over the last 50 years.

4.2   Technological drivers

The last 50 years have seen enormous changes within the field of aero gas turbines. The early 1950s saw the introduction of the first commercial gas turbine powered aircraft, the Viscount, powered by the Dart Turboprop gas turbine, followed by the early civil jet-powered aircraft, such as the Comet and the Boeing 707. The early 1970s saw the introduction of the supersonic civil airliner, the Concorde, powered by the Olympus 593 jet engine. The Anglo-French Concorde was an important example of international collaborative programmes, which were to become increasingly important for both military and civil gas turbines. The early 1970s also saw the launch of the wide-bodied aircraft: the Boeing 747, the DC10, and the Lockheed L1011. The Boeing 747, in particular, marked the start of the mass civil air transport market, with leisure travel becoming more important than business travel.

Throughout this period, aero gas turbine performance has never ceased to improve, and their commercial success has been mainly driven by technology. In fact, technology is so critical that a relatively small change in component efficiencies, cycle temperatures or pressure ratios can render an engine wholly uncompetitive.

4.2.1   Principles of operation

In order to understand the parameters that drive gas turbine development, it is useful to first define how aero gas turbines work. In its simplest form, an aero gas turbine is composed of an inlet, a compressor, a combustor, a turbine, and a nozzle. The air enters the engine through an inlet, and its total pressure is increased by the compressor before entering the combustion chamber. There, fuel is burned and mixed with the incoming air, increasing its total energy. The gases then pass through the turbine, which extracts the work necessary to drive the compressor. The exhaust gas is then expelled through the nozzle, 'pushing' against the atmospheric air to generate thrust.

Fuel consumption in aero gas turbines depends on two main parameters: thermal efficiency and propulsive efficiency. The former is a measure of how well the engine converts heat energy, and the latter describes how well the engine uses the kinetic energy of the jet to generate a thrust force. High thermal efficiency requires a high overall pressure ratio (OPR), high component efficiencies and a high turbine entry temperature (TET), whereas high propulsive efficiency results from low jet velocities.

The following section examines how the different design parameters have evolved over the years to meet different challenges in a quest for lower fuel consumption, weight, costs and environmental issues.

4.2.2   Cycle temperatures

A gas turbine's core specific power is the energy available to generate thrust (or drive the fan, in the case of turbofans). It is a crucial parameter to reduce the weight of the engine, and, if used in conjunction with a fan, it can lead to better propulsive efficiency. Over the past 50 years, core specific power has risen by a factor of five over the early Whittle engines. This improvement has resulted from increasingly high TET values. Investments in high-temperature technology have been directed at both the thermal capability of hot section materials (e.g. single crystal alloys and ceramic coatings) and turbine cooling (see Fig. 4.3). Whereas the TET of early engines only reached about 1000 K, today's turbofans operate with TET values of about 1800 K at takeoff and OPR values of around 45. The limit for hydrocarbon fuels is the stoichiometric temperature of about 2600 K, but NO_x emissions tend to rise with higher temperatures, and it might be necessary to limit TET to a value somewhere between 2000 and 2100 K for environmental reasons.

4.2.3   Component efficiencies

Considerable research has been undertaken to improve major components in the engine 'gas path'. At first, the efficiency and weight of components were improved by means of experimentation, but high-performance computing marked the advent of computational fluid dynamics (CFD), which allowed detailed simulation and optimisation of components. Some of the resulting improvements have been 3D blade shapes, 'wide chord' fan blades, and hollow and composite blades. The cumulative effect of these polytropic efficiency improvements has been very significant, increasing thermal efficiency and specific power. The thermal efficiencies of current engines in service are approximately 45%, based on

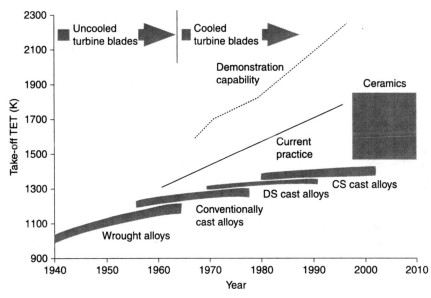

4.3 High-temperature capabilities.

polytropic efficiencies of 88% at cruise conditions. During the early part of the twenty-first century, it is likely that engines will enter service with thermal efficiencies approaching 50%, based on polytropic efficiencies of 92% and increases in cycle pressure ratio and turbine entry temperature.

4.2.4 Propulsive efficiency

Perhaps one of the most significant improvements to the gas turbine has been the turbofan. By the 1960s, aviation traffic growth made it imperative that jet noise be reduced. It was recognised that jet noise increases with the seventh power of velocity, and a solution was sought to decrease jet speed to an acceptable level. By that time, core specific power had increased enough to allow the excess power to be used to drive a low-speed fan. This fan would be driven by an additional turbine, and produce a similar amount of thrust by moving large masses of cooler air at lower velocities.

Early turbojets achieved overall efficiencies of about 20%. Early turbofans increased this value to about 25%, and the current generation of high bypass ratio turbofans achieve overall efficiencies of around 35%. The result is more efficient engines (see Fig. 4.4) that lead to a higher aircraft earning capacity. This in turn allowed air travel to become accessible to an ever-increasing proportion of the population.

However, there is a limit to how much air can be diverted to the fan: as bypass ratios are increased (hence, more air 'bypasses' the core), the overall diameter of

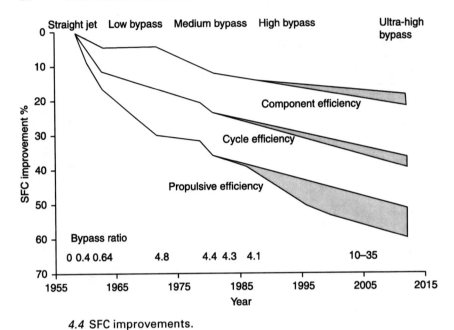

4.4 SFC improvements.

the engine increases, resulting in a drag and weight increase that eventually offsets the advantages offered by a higher propulsive efficiency.

4.2.5 The aftermarket

The aftermarket has undergone numerous changes in the last few decades, slowly evolving from a mere afterthought to a financial pillar for aero engine manufacturers.

In the early years, gas turbine maintenance was limited to run-to-failure strategies, visual inspections and oil monitoring. In the 1960s and 1970s, this strategy ceased to be acceptable, as growing numbers of passenger miles per year meant that failures were more likely to occur. The paradigm shifted accordingly to preventive maintenance: frequent overhauls were deemed the best solution, and engine manufacturers began generating substantial revenues from selling components to replace those deemed unfit during overhauls. In the 1980s and 1990s, advances in electronics and computing enabled the use of automatic engine health monitoring (EHM) techniques, revolutionising the way engine maintenance is carried out and increasing the safety and availability of aero gas turbines.

4.2.6 Developments in safety and availability

The early years of aero gas turbines were characterised by short mean times between failures (MTBF). Engine manufacturers began offering spares and overhaul services to aircraft operators, and planned maintenance was carried out at fixed intervals as recommended by the manufacturer. Despite the high safety margins, much of the maintenance was unplanned, and not much was done to improve the situation, since research emphasis was placed on achieving better performance.

The economic crisis brought about by the steep rise in oil prices in the early 1970s began to change the face of the aftermarket. Airlines, faced with lower profits and higher costs, demanded lower prices from engine manufacturers. As a result, the purchase prices of engines were driven down in a trend that has lasted to the present day. Faced with minute margins on engine sales, engine manufacturers found a means of survival in the aftermarket: spares were sold to airlines and third-party maintenance, repair and overhaul (MRO) outfits at increasingly high prices, recouping the losses made by offering low purchase prices.

The late 1980s and the 1990s were characterized by increased reliability and longer engine lives. Figure 4.5 illustrates the IFSD (in-flight shutdown) rate trend over the years for some of the main Rolls-Royce engines. As can be seen, the removal rate for new engines is less than a fifth of what it was in the 1970s.

The application of techniques such as gas path analysis and vibration monitoring to engine diagnostics allowed operators to carry out predictive maintenance, resulting in increased reliability and engine lives. MRO shop utilization dropped, and airlines began to outsource maintenance.

Engine manufacturers have realised the importance of the aftermarket, and they are expected to capture an increasing share. Partnerships between engine manufacturers, airlines, and third-party MROs are set to become the norm in the

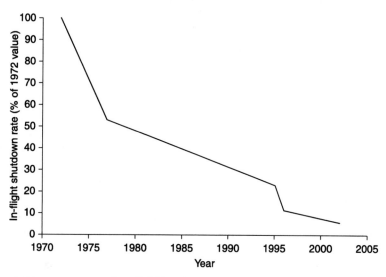

4.5 Improvements in reliability. (Source: data from Howse, 2004.)

industry. The concept of power by the hour, for example, allows airlines to pay a fixed fee for every hour the gas turbine is used, transferring the risks associated with reliability to the engine manufacturer, who also enjoys more predictable revenues. The trend seems to point towards asset management, where airlines could transfer ownership of the engine to the gas turbine manufacturer. In this case, however, manufacturers might have to finance their own engines upfront. This could lead to the emergence of an engine financing market similar to the one already existent for airframes.

There is a risk, as gas turbines become more and more reliable, that the aftermarket revenues will decrease sharply. If this happens, the business model will have to evolve again in order to ensure that the industry remains viable.

4.3   New challenges

Whilst speed, range and comfort were the primary drivers in civil aviation in the early twentieth century, the picture has changed with the new millennium. Two concerns have drastically increased the need for fuel-efficient technology over the past decade: depletion of natural fuel reserves and global warming. Scientists fear that, with ongoing levels of current greenhouse gas emissions, global temperatures could increase by several degrees by the end of the twenty-first century. Solar radiation absorbed by the planet's atmosphere and surface is converted into heat energy. The natural occurrence of greenhouse gases results in equilibrium conditions where terrestrial infrared radiation is offset by incoming solar radiation, and the energy fluxes of the ingoing and outgoing radiation are approximately equal on average, keeping the planet at an inhabitable temperature.

Pollutants of human origin, such as CO_2, impose a perturbation on the equilibrium state by absorbing additional heat energy. The amount of energy absorbed is generally denoted by radiative forcing (RF), which is the amount of radiation forced to remain within the 'Earth system', i.e. the atmosphere, oceans and soil. As more and more heat is absorbed, the Earth's temperature increases. A temperature rise will then occur, continuing until the outgoing and the ingoing radiation are in equilibrium again. The expected temperature change due to a particular pollutant depends on the strength of its radiative forcing. A positive radiative forcing will cause a temperature rise, whereas a negative radiative forcing will cause a temperature drop by shielding the Earth from incoming radiation. Furthermore, the temperature change is also determined by a pollutant-specific climate feedback parameter. It can be understood as a gain factor, accounting for three-dimensional atmosphere–ocean feedback mechanisms such as cloud or sea ice formation. The feedback parameter determines how effective the radiative forcing of the pollutant is and over what time period the temperature rise will take place.

Carbon dioxide is the dominant mode through which carbon is constantly transferred in the natural environment between a variety of carbon reservoirs (see

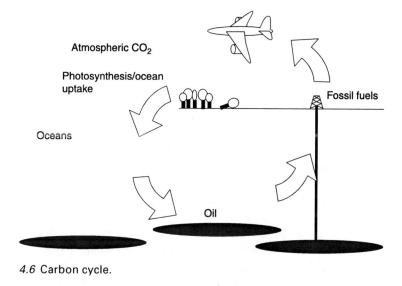

Atmospheric CO_2

Photosynthesis/ocean
uptake

Oceans

Fossil fuels

Oil

4.6 Carbon cycle.

Fig. 4.6). Burning fossil fuels, representing one reservoir, implies taking carbon from beneath the Earth's surface, converting it to CO_2 and emitting it into the atmosphere. Atmospheric carbon dioxide is absorbed by natural carbon dioxide sinks such as oceans, which take up carbon and facilitate the formation of carbonates and forests. The process of carbon transfer between carbon reservoirs is referred to as the 'carbon cycle'. The consequences of additional CO_2 in the atmosphere, causing a global temperature rise, are projected to be devastating: sea level rise, frequency and intensity of extreme weather events, changes in agricultural yields, and increases in the ranges of disease vectors, to name just a few. These consequences might have a catastrophic impact on our society, and current public consensus calls for an urgent reduction to counteract current trends. Although aviation only represents approximately 5% of man-made carbon emissions, projected passenger and cargo growth rates make it a potential major stakeholder in the drive towards lower emissions, and more environmentally friendly aircraft technology will be required.

However, CO_2 is not the only pollutant resulting from air traffic, and the fact that pollutants are released in areas of the atmosphere where their impact is potentially worse than at sea level results in environmental technology trade-offs between different pollutants. For example, higher turbine entry temperatures in aircraft engines result in higher thermal efficiencies and less CO_2 emissions, but also cause an increase in nitrogen oxide (NO_x) emissions. These contribute to the formation of ozone in the upper troposphere and decrease the concentration of methane, another greenhouse gas, implying a cooling of the atmosphere. Both effects, tropospheric ozone formation and methane depletion, approximately cancel each other out. Nevertheless, stratospheric NO_x emissions facilitate a reduction of ozone where it is needed to shield the earth from highly energetic cosmic radiation. When emitted at low altitudes, e.g. in the vicinity of airports, NO_x exhibits additional environmental and health hazards. NO_x can react to form nitric acid, which can penetrate into sensitive lung tissue and causes severe health damage and in extreme cases premature death. Inhalation of such particles may cause or worsen respiratory and heart diseases.

Water, which in its vapour phase is a relatively strong greenhouse gas, is also a combustion product. Water emitted by aircraft precipitates relatively shortly after emission, whereas stratospheric water can have longer residence times and hence cause radiative forcing, even though it is marginal compared with other aviation pollutants.

Elevated atmospheric aerosol concentrations caused by air traffic are relatively small compared with the contribution from surface sources. Soot emissions tend to warm the atmosphere, whereas sulphates have an opposite effect. Compared with other aircraft emissions, the direct radiative forcing of soot and aerosols is relatively small. Aerosols also play an important role in the formation of cirrus clouds that would not form in the absence of aviation, resulting in enhanced cloud formation and the modification of the radiative properties of natural cirrus clouds.

Contrails are artificial clouds composed of ice crystals. They form in the aircraft wake if water saturation occurs during the process of mixing the engine's exhaust gases with ambient air. An analogy can be found by comparing the formation of contrails to human breath in a cold winter day, which can become visible in the form of steam. Contrails can persist in the atmosphere if it is ice-supersaturated. Due to the shape and size of contrail ice crystals, contrails reflect incoming solar radiation to a lesser extent than terrestrial infrared radiation. As a result, heat energy is trapped in the atmosphere below the contrail. Although radiative properties of contrails vary depending on the ambient conditions, occasionally even leading to local cooling of the atmosphere below the contrail, the average effect of global contrail occurrences is a net contribution to global warming.

Moreover, dependent on ambient conditions, line-shaped persistent contrails can spread to form large-scale cirrus clouds, so-called contrail cirrus. Since they cover much larger areas than line-shaped contrails, and thus trap more heat in the atmosphere, it is estimated that their environmental impact is larger than that of line-shaped contrails. Furthermore, locally elevated atmospheric soot and aerosol concentrations due to air traffic can initiate water nucleation, leading to the formation of so-called secondary cirrus clouds. These clouds, with radiative properties similar to those of contrails, also contribute to global warming. It is important to note that they would not form in the absence of air traffic.

Recent climate assessments have emphasised the environmental impact from persistent contrails and contrail cirrus. Their radiative forcing might already exceed the radiative forcing from all other air traffic pollutants combined. This concern has initiated the search for strategies and technologies to avoid their formation.

Apart from environmental concerns, there is another driver away from the fossil fuel-based economy. The decrease of natural fuel reserves and its political implications demand alternatives. As hydrocarbon supplies diminish, fuel prices will increase and higher fuel prices are most likely to lead to increased alternative, renewable energy supplies. Renewable energy sources are currently more expensive than conventional fossil fuels sources, but may become economically viable in the near future. While there are enough coal reserves to make it a viable energy source well into the next century, associated environmental implications would create a need for drastic environmental protection measures such as CO_2 sequestration.

4.4   Meeting the challenges through innovation

The challenge of the environment has spurred the aero gas turbine industry into undertaking research programmes to achieve more environmentally friendly engines in the short to medium term. Figure 4.7 offers a view of the evolution in gas turbine fuel consumption from 1960 to the present day. It is possible to see a very large improvement in the early years, partly due to the adoption of the turbofan, and a constant reduction in fuel consumption ever since. Research and development programmes today are looking at concepts that could lead to step fuel efficiency improvements. In the short to medium term open rotor configurations might be seen to emerge, and recuperated cycles might become viable in some applications. However, long-term solutions could require more radical innovations.

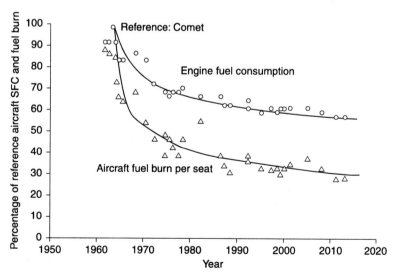

4.7 Aero gas turbine fuel consumption evolution. (Source: Green, 2005.)

The following section will consider the prospects in the near term, and the drivers and technologies that might affect aero gas turbines in the long term.

4.4.1   Near term

Research oriented towards near- to medium-term improvements is mainly directed at increasing thermal and propulsive efficiency of conventional aero engines. Aside from research into conventional supporting technologies, it is worth mentioning two promising lines of research: intercooling and recuperation to increase thermal efficiency, and open rotor technology to improve propulsive efficiency.

Core engine improvements

As discussed above, core power is principally a function of TET, which also affects thermal efficiency. Fig. 4.8 shows the specific fuel consumption of current engines as it relates to thermal and propulsive efficiency. Whereas current high-bypass ratio engines exhibit thermal efficiencies in the region of 0.45, it might be possible to reach values close to 0.6 by developing materials technologies that allow further increases in TET, along with higher component efficiencies and pressure ratios. Nevertheless, given that NO_x tends to rise with higher combustion temperatures, further developments in low-NO_x combustors would be necessary to reach this value in practice.

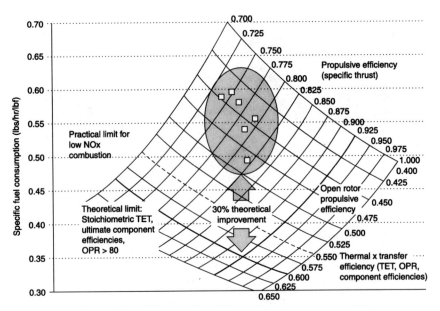

4.8 Propulsive efficiency for different types of aero engines.

Intercooling and recuperation

Recuperation techniques use heat exchangers to scavenge heat that would otherwise be lost to the atmosphere from the back of the gas turbine, using it to increase the temperature of the air before it enters the combustion chamber, and thus reducing the amount of fuel that must be burned to reach a given value of TET. Intercoolers reduce the temperature of the air between successive compressor stages, increasing the core specific power. A cycle making use of these techniques could in theory outperform current gas turbines. In practice, neither technology is used in aero gas turbines today because of the weight and volume associated with the heat exchangers they would both require. Nevertheless, much research is being carried out to manufacture lighter, more compact heat exchangers, and they should not be ruled out in the medium to long term, especially for power-producing applications such as helicopter engines.

Propulsive efficiency improvements

Propulsive efficiency depends on jet velocity (see Fig. 4.9), and can be improved by increasing the gas turbine's bypass ratio (BPR). However, higher BPR values mean that the linear velocity of the blades could be high enough at the tip to encounter supersonic flow, and generate shockwaves. This is not an insurmountable problem for current engines (with BPR values of about 10), since the inlet diffusers reduce the speed of the incoming air to values close to Mach 0.4. If attempts are made to increase BPR further, however, the weight and drag associated with the fan cowling become excessive. Without the help of the inlet diffuser, it is difficult to avoid shockwaves (in fact, this phenomenon has traditionally limited the maximum Mach number at which propeller-driven aircraft can fly).

Given the drive for lower fuel consumption, different ways of achieving higher BPR values are being explored by major manufacturers. Open rotors could potentially be the solution. Open rotor configurations eliminate the fan cowling

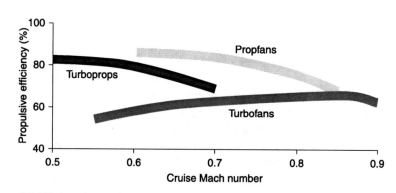

4.9 Wave rotor cycle.

and use advanced aerodynamic design techniques to reduce shock waves at the fan blade tips, allowing large fans to be used in free stream conditions and resulting in bypass ratios in excess of 30. However, open rotors have a significant drawback: noise. Without a cowling to absorb the noise generated by the fan blades, open rotor configurations can be unpleasant for both passengers and airport environments. Open rotors were already developed in the 1980s in the USA and Russia, but the resulting cabin noise prevented them from being adopted at the time. With a far stronger drive for low fuel consumption, open rotors might have a better chance this time. Noise reduction techniques are currently being studied, and alternative engine placement arrangements could allow better noise shielding.

As seen in Fig. 4.8, open rotors and high-temperature technologies could bring about a 30% improvement in fuel consumption. Considering intercooled–recuperated cycles, there is still plenty of scope for improving the conventional propulsion configuration. By combining current developments in aero gas turbines, airframes, and air traffic management, the environmental impact of aviation will be lowered significantly within the next two decades.

It would nevertheless be unwise to set our long-term goal for aero gas turbine fuel consumption reduction at 30%. Even if passenger growth rates only averaged 4% per annum, it would only take seven years for an aviation industry based on a 'greener' gas turbine to produce as much CO_2 as today.

Alternative architectures: distributed propulsion

Despite increasing environmental pressures, aero engine development cycles mean that alternative propulsion systems are unlikely to become available in the medium term. This means that gas turbine technology will continue to be the norm for decades to come. Once the predicted improvements in core and propulsive efficiency have materialised, there might still be room to increase efficiency further. This could be done by seeking synergies between the airframe and the propulsion system, and using alternative propulsion system architectures.

One such architecture could be distributed propulsion, which consists of spreading the propulsive force along the wings or airframe. A small gas turbine distributed propulsion (SGTDP) system could be composed of a large number of small gas turbines spread along the wings of the aircraft. Since the engines are now very small, they could be better shielded to reduce noise. Mass manufacturing might become a real possibility for large numbers of engines, and studies on land turbines indicate that manufacturing costs could be reduced by up to 90% through the implementation of appropriate large-scale manufacturing techniques. Safety could be enhanced by higher propulsion system redundancy, and this would also allow the redesign of aircraft around the propulsion system: less stringent airframe requirements could result in lighter, more efficient, and cheaper aircraft.

However, SGTDP is not feasible at the time of writing. A recent study at Cranfield University (Ameyugo and Singh, 2007) indicated that small gas turbine fuel consumption represents the single greatest hurdle to this concept. A solution could be found in heat exchangers, which improve small gas turbine efficiency to a larger extent than their large counterparts, and could thus become an enabling technology. Lighter, less bulky heat exchangers would allow the implementation of more efficient recuperated cycles.

In practice, the advantages of the resulting recuperated engine system might not be substantial enough to justify the large shift in industry structure and operations that would be required to bring about a change of configuration. Nevertheless, in a scenario where environmental concerns had driven the development of cheap near-zero emissions fuels for the land transport and power generation industries, aviation could benefit from the resulting shift in relative importance of fuel costs to operations by adopting a distributed propulsion configuration. In such a scenario, small gas turbine configuration distributed propulsion could provide a cheaper, more efficient, and safer way of travel.

4.4.2 Medium term

Improvements in engine efficiency are ultimately constrained by physical laws. Assuming stoichiometric turbine inlet temperature and open rotor propulsive efficiency, the theoretically achievable maximum overall engine efficiency is about 55%. This fact represents the dilemma of the aerospace industry today: every improvement in engine efficiency is a step toward the maximum achievable overall engine efficiency. As this value is approached, the law of diminishing returns sets in: improvements become more and more difficult to achieve, and take place at subsystem level with larger time spans between equivalent amounts of efficiency gains. Larger research and development spending is required for ever-smaller improvements, reducing profit margins in an effort to remain competitive in the eyes of the airlines. Medium-term solutions may exhibit a more disruptive character than short-term solutions, as more time for development is available.

Alternative fuels could constitute an interim solution to reduce the environmental impact of aviation in the medium term. Two main avenues of research are currently open: Fischer-Tropsch fuel synthesis and biofuels. The Fischer-Tropsch (FT) process is a chemical process that transforms coal into hydrocarbon fuels. However, FT processes require high temperatures and pressures, and release CO_2 into the atmosphere during the transformation process. Although CO_2 sequestration could be an option, biofuels might offer an interesting alternative.

Biofuels

Biofuels rely on using biological matter – such as plants – to produce hydrocarbon-based fuels, creating an additional loop in the carbon cycle. The biomass used to generate fuel would absorb CO_2 from the atmosphere while it is growing, before it is harvested. For this reason, the use of biofuels as a substitute for kerosene has been discussed many times over the past few years.

Two main methods could be used to synthesise biofuels: vegetable oil hydrotreating and synthetic biology processes.

Hydrotreated renewable jet fuel (HRJ)

Hydrotreated renewable jet fuel (HRJ) is an option that could be implemented within the next few years. This process removes oxygen and other contaminants from vegetable oil, and generates a fuel with performance properties exceeding current jet fuel requirements.

Synthetic biology

Another option to generate biofuels is to convert lignocellulosic material, such as forest and agricultural waste products, into a biofuel cocktail. This process would be very attractive from an environmental point of view, and it is already being used to generate ethanol for ground transportation. Research and development efforts are progressing rapidly to make this process economical.

Although biofuels remain a long-term possibility, there are several hurdles to their application in aerospace propulsion systems. The use of sugarcane-derived ethanol for road transportation has been successfully proven over the last two decades in Brazil. However, it has been necessary to devote a significant part of the country's arable land to provide enough sugarcane for the fuel used for road transportation. Ethanol itself is not well suited for aviation use, as it contains a significant amount of oxygen, decreasing its energy density and resulting in bulkier, heavier fuel tanks. Other alternatives are possible, but the amount of land required to provide enough biofuels for even 15% of the US commercial fleet would be equivalent to the size of Florida. Microalgae are being studied as a promising source for vegetable oils suitable for biofuel applications. However, current production systems are small, and production costs would still have to be reduced anywhere from 10–100-fold (Dagget, 2009) for algae to be cost-competitive with predicted oil prices for the next decades.

A more practical limitation lies in the freezing point of biofuels: whereas conventional fuels can be used in aircraft without complications, biofuels will freeze at the operational temperatures experienced by aircraft in cruise. Possible solutions could include the use of insulating materials, anti-freeze agents in the fuel, or even heating the fuel with the excess heat from the engines.

Constant volume combustion

Current jet engines are based on the Brayton cycle, where the working fluid successively experiences isentropic compression, isobaric increase in temperature, and isentropic expansion. Deviations from the ideal cycle in real engines occur through component inefficiencies. The maximum achievable thermodynamic efficiency of the Brayton cycle increases hand-in-hand with its peak cycle temperature. Since the peak cycle temperature is limited by material properties of the turbine, the maximum cycle efficiency of current jet engines is limited by the laws of thermodynamics. Hence, efficiency improvements of jet engines beyond what is possible with conventional designs are only feasible through novel thermodynamic cycles that incorporate features such as constant volume combustion.

Technologies enabling constant volume combustion for aircraft propulsion include wave rotors, pulse detonators and piston engines. Wave rotors are non-steady flow pressure exchangers based on wave processes; they act as a compressor and turbine without blades (see Fig. 4.10). Two scenarios for wave rotor applications are conceivable. The wave rotor could take the place of a set of high-pressure compressor stages, for example to make the engine more compact. Alternatively, a wave rotor can enable higher turbine delivery pressure without the need to increase the turbine entry temperature.

In its most common form, the wave rotor is an array of cylindrically arranged tubes in the form of a drum. In the four-port configuration, the wave rotor is located between the compressor exit, the inlet as well as the outlet of a combustion unit, and the turbine inlet. The rotation of the drum causes a periodical opening and closing of both sides of the tubes in such a way that the working fluid can either enter or exit the tubes.

The wave rotor works in four steps. In the first step, low-pressure air from the engine compressor enters the tubes of the wave rotor. Through the rotation of the drum, the ports of the tubes become connected with the inlet and outlet of the

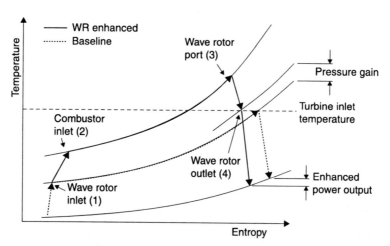

4.10 Practical limits to engine efficiency. (Source: Green, 2005.)

combustor containing high-pressure gas. This causes compression of the low-pressure air via shock waves. In effect, the high-pressure air enters the combustor at its inlet. That leaves the tube with compressed gas at an elevated temperature. Through further rotation, the tube is now connected with the turbine and the gas leaves the tube to drive the turbine.

The general advantage of wave rotor topping is an increase of the overall pressure ratio achieved in the engine, increasing the output work and therefore thermal efficiency. The weight of a particular engine is believed to increase unless the wave rotor is used as a substitute for the final compressor stages. Therefore, weight penalties associated with the wave rotor would have to be taken into account to assess the installed performance of a wave rotor-topped engine. The performance enhancement via wave rotors is particularly effective if turbomachinery component losses are high. Generally, component efficiencies decrease with decreasing engine mass flow, which can be found in engines with low power output. Therefore, the implementation of wave rotor-supported performance enhancements is probably more effective in low-power engines.

Pulse detonation is a non-steady flow quasi-constant volume combustion process. As opposed to deflagrations, detonations are supersonic combustion waves. Across a detonation front, the pressure increases while the specific volume decreases. A detonation taking place in a tube that is closed at one end leaves hot, pressurized gas behind. In its simplest form, a pulse detonation engine is a tube where one end is repeatedly opened and closed while a combustible gas is ignited while the tube is closed. Hot, pressurized gas exits the tube at its closed end shortly after ignition takes place.

A wide variety of applications have been proposed for pulse detonation engines. Due to the high specific impulse of the exhaust gas, thrust-generating pulse detonation engines exhibit much higher fuel consumption than turbojets in the low Mach number regime. However, pulse detonation engines would outperform turbojets above Mach numbers of about 3.5. An application of pulse detonation engines for subsonic, commercial airliners is hence less attractive in the light of the current environmental debate. Pulse detonation engines can be used in hybrid mode to increase the performance of conventional aero engines. In the ultimate embodiment of this concept, the pulse detonation engine takes the place of the combustor and, depending on the overall design, a set of high-pressure compressor stages. The air leaving the high-pressure compressor stage enters a pulse detonation tube, which delivers hot gases at increased pressure to a turbine. Combining pulse detonation with conventional aero engine technology would hence enable higher turbine delivery pressure without the need to increase the turbine entry temperature.

Lastly, piston engines, generally based on the Otto cycle featuring constant volume combustion, could replace the combustor of a jet engine. This would enable higher combustor delivery pressure, leading to improved thermal efficiency. The idea is not new: the Napier Nomad engine, which was in commercial production in the early 1950s, had a piston engine instead of a combustor in its core.

Hydrogen

When in 1937 one of the first jet engine prototypes was tested in a German laboratory it ran with pressurised hydrogen. However, hydrocarbon-based fuels being more abundant, more accessible and easier to store, kerosene eventually succeeded as aviation fuel and the first jet aircraft were fuelled exclusively with kerosene. Subsequently, kerosene became the common aviation fuel in both civil and military aviation. Some research programmes were undertaken to investigate hydrogen as fuel for military aircraft. For example, in the late 1950s a B57 was modified to fly with hydrogen and tests were successful; but as of today, with the exception of some research aircraft, the global fleet flies with kerosene.

Hydrogen is the first element in the periodic table. With just one electron in its shell it is also the lightest element. Hydrogen is very reactive and very rare in its natural, diatomic form (H_2) on Earth. It is a constituent of many known materials, gases and fluids e.g. water (H_2O). Hydrogen is highly flammable, reacting exothermically with oxygen to form water. It burns in air with concentrations between 4 and 75% by volume.

Hydrogen is not a natural resource and can be seen as an intermediate energy carrier. Hydrogen would have to be produced by consuming energy before it could be utilised to provide energy, e.g. to power aircraft. The probably most viable approach for large-scale hydrogen production is electrolysis, where water is broken down into its constituents – oxygen and hydrogen – by means of electrical energy. An alternative to electrolysis of water is provided through gasification of biomass. When hydrogen is burned with pure oxygen the only combustion product is water. Hydrogen combustion with air yields secondary combustion products, such as NO_x, forming in the presence of high temperatures.

With hydrogen being used as an energy carrier, other, more abundant energy sources than crude oil could be used to power aircraft. Biofuels, nuclear power or renewables such as solar or wind power could provide electrical energy to produce hydrogen. If the production of hydrogen itself were not a source of greenhouse gases, hydrogen would certainly be a more environmental aviation fuel than kerosene.

Hydrogen being a very light gas, it contains 2.8 times more energy than kerosene in terms of weight. In terms of energy per volume, kerosene is more advantageous: liquefied hydrogen stored at −253 °C still requires four times more volume than kerosene for the same amount of energy. Although hydrogen exhibits large weight advantages, its volume requirements place technical challenges around storage and fuel systems, especially because liquid hydrogen would have to be stored at very low temperatures. Hence, if hydrogen were to be used as an aviation fuel, large fuel tanks would add to the drag and weight of the airframe, and the complexity of the fuel systems would increase. The gross aircraft weight may be reduced due to the low energy specific weight of hydrogen. Hence, the trade-off is an increase in wetted area adding to drag against aircraft weight. This trade-off depends on size and configuration of the aircraft. Revolutionary aircraft configurations, such as the blended wing body, or engine technologies such as the fuel cell, may even be more suitable for hydrogen as aviation fuel.

While hydrogen could imply less energy-efficient aircraft, this would not necessarily involve an increase in aviation's environmental impact or economical viability if hydrogen could be produced and distributed without greenhouse gas emissions and in an economical way. The most important consideration in this case would probably be economical viability.

The absence of infrastructure for hydrogen distribution would require enormous investments to achieve a network coverage that would enable refuelling of aircraft at any airport. Hydrogen could therefore be distributed in gaseous form to airports and then liquefied via cooling, a process that would require extra energy. Alternatively, hydrogen could be directly produced at the airport via electrolysis or gasification. Establishing the hydrogen network would most likely require more time than designing the first hydrogen-powered, commercial airliner.

Hydrogen could be used to power conventional turbofan engines. Major modifications to the power plant would be required in the fuel supply and combustion section. Liquid hydrogen would have to be heated before injection into the combustion chamber, requiring additional heat exchangers. The pumps delivering the hydrogen to the engine would have to be modified to operate at very low temperatures. Low NO_x emissions combustors compatible with hydrogen would be required to keep NO_x emissions at a minimum.

Alternative solutions may be found in fuel cell technology. Hydrogen would be a suitable candidate to power low-temperature fuel cells. With higher efficiencies than gas turbines and the potential of being more economical, fuel cells may be used to produce electricity to power electrically driven fans. Whilst the practical feasibility is currently being investigated, fuel cells may become substitutes for gas turbine-based auxiliary power units in the shorter term.

In burning hydrogen with air, the only combustion products would be water and NO_x. Since NO_x occurs also during the combustion of kerosene, there would not be an additional impact through NO_x emissions. With water having a far lower atmospheric residence time compared with CO_2, this would have a positive impact in terms of environmental compatibility. Although emissions from hydrogen-powered aircraft would not include CO_2, they could contribute to additional contrails. The additional amount of water in the engine exhaust would facilitate contrail formation over a wider range of atmospheric conditions and altitudes. Studies suggest that the global annual mean contrail coverage from hydrogen-fuelled aircraft could increase by as much as 1.56 times. Due to the lack of soot particles in the exhaust of hydrogen-powered aircraft, contrails would consist of fewer but larger particles, having an impact on the optical properties of contrails, reducing their radiative forcing and hence impact on the environment. Even though more contrails would appear in the sky, the overall effect may be a reduction in the contribution of contrails to global warming.

4.4.3 Long term

The sustainability of aviation in the long term is restricted by two factors: environmental compatibility and limited natural fuel resources. Both factors can impede the growth in passenger kilometres, but continuous growth is vital for the development of sustainable technology in the aviation industry sector. The importance of the above-mentioned factors has changed over time, with environmental concerns dominating the current debate. Fuel availability, pricing and price stability will also become important in the coming decades. Considering the forecast growth rates in passenger kilometres, sustainable aviation is believed to be achievable only if certain criteria regarding fuel economy are met. The changes that may prove necessary to meet the challenge of the environment imply the introduction of disruptive technologies and levels of sustained investment and risk large even in comparison with the norm for this industry. Though it is not straightforward to foresee the future of aviation beyond the next three decades, it is possible to begin to form an image of what future airframes and propulsion systems might look like by predicting the drivers of future technological development.

Long-term drivers

The main drivers for change in the aero gas turbine industry in the long term are likely to be the environment, profitability, and the emergence of disruptive technologies. The importance of achieving environmental sustainability for continued growth of the industry has been discussed above. Airline operating costs are likely to be increasingly affected by environmental regulations in the future, and the emergence of competition from developing countries will increase the pressure to reduce costs. Increased competitive pressure could also drive the industry towards more risky, innovative concepts, and might eventually lead to the introduction of new disruptive technologies in an effort to achieve step improvements in environmental and cost performance. In the coming decades, the latter could include nano-technology, micro-electromechanical systems, and superconductivity.

Nanotechnology

Breakthroughs in nanotechnology are leading to lighter materials with ever-increasing strengths (potentially over 30 times stronger than steel with much lower material densities). The application of carbon nanotubes in aero gas turbines is still some years away, but promises to drive their weight down considerably. Carbon nanotubes have also been proposed as a means of safely storing liquid hydrogen with minimum losses, and might therefore play a part in alternative fuel development.

Micro-electromechanical systems

A second result of future developments in nanotechnology might be low-cost micro-electro-mechanical systems (MEMS). By using MEMS pressure sensors with smart materials or even MEMS actuators to control boundary layer flow, it might be possible to reduce drag and therefore achieve considerably higher aerodynamic efficiency for propulsion system turbomachinery. MEMS technology could also be used to control shockwave formation on blade tips, helping, for example, to lower open rotor noise during taxiing, take-off and landing.

Superconductivity

Another disruptive technology is that of superconductivity: by minimising electrical losses, this technology could result in more efficient electrical systems. Electric motors could one day provide a more efficient way of transferring fuel energy to the propulsive elements than conventional drives. By making use of this technology, more environmentally friendly aircraft without the need for alternative fuels could be envisaged.

Future airframes

Long-term propulsion systems might be electrically driven. A first step towards this more efficient aircraft could be a change towards alternative airframes. Of the airframes proposed so far, the blended wing body (BWB) seems the most promising for future applications. The BWB concept is discussed elsewhere in this book, but, apart from its inherent benefits to aerodynamic efficiency, BWB configurations could enable a higher degree of integration between the airframe and the propulsion system. This could lead to lower drag, and lighter, more efficient propulsion configurations.

Electrical propulsion and superconductivity

Electrical propulsion systems would be too heavy to power a commercial airliner today. A possible configuration could use a gas turbine core to generate enough power to drive an electrical generator (see Fig. 4.11). Power could be routed through a set of transmission lines, and transferred to a number of low-velocity electrical fans. If the propulsion system is integrated within, say, a BWB airframe, it might be possible to increase the thermal efficiency of the core to a similar level

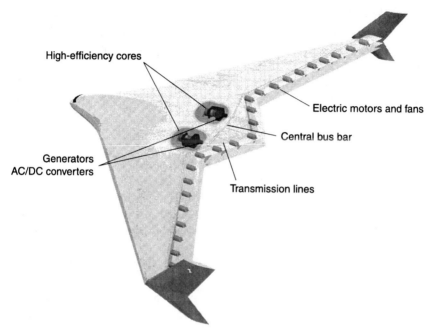

High-efficiency cores

Electric motors and fans

Central bus bar

Generators
AC/DC converters

Transmission lines

4.11 Distributed electrical fans concept. (Source: Ameyugo and Noppel, 2007.)

to industrial gas turbines (e.g. by using recuperated cycles). Such systems would offer unparalleled flexibility to achieve unprecedented levels of overall efficiency, and could eventually afford reductions in fuel consumption over conventional gas turbine technology of up to 40% (Ameyugo, 2007), allowing emissions to be minimised. However, in order to achieve this goal, a fundamental disruptive technology would be needed: superconductivity.

Superconductive motors have recently come under the spotlight as the aerospace industry turns towards the search for an emissionless aircraft. Although some high-temperature superconducting materials have been known for some time (Edmonds *et al.*, 1992), most operate at very low temperatures. A recent superconductive motor concept for application to light aircraft, for example, had an operating temperature of 30 K (Masson and Luongo, 2005).

The implication of the low temperature required to operate superconductive components is that, in general, a cryogenic source would be required onboard. Hydrogen fuel could provide the solution, but the date at which distributed driven fans might become feasible could be set back several decades. Nevertheless, it might not be necessary to run the main propulsion system on hydrogen fuel: if auxiliary power units (APUs) are replaced with fuel cells in the medium term, it might make sense to operate them with hydrogen. This would eliminate the need for a reformer, making them lighter, and the only emission from the APU would be water. Although hydrogen should enter the fuel cell in gaseous form, the most efficient way of storing it is in liquid form. It is not difficult to envisage a liquid-H_2 tank for the APU, with fuel lines that cool the generators and main power transmission lines, at the very least, enabling superconductive elements to be used in those components and reducing the weight and losses of the propulsion system.

4.5   Conclusion

In this chapter have been seen the factors that influenced the development of gas turbines. The main motivation behind aero gas turbines was flying faster and above the weather. Subsequent development efforts helped reduce the cost of flying, allowing ever-increasing numbers of people to enjoy the benefits of air travel, and turning aviation into one of the main driving forces of economic progress and globalisation.

The challenge currently facing the aero engine industry is that of supporting a new generation of environmentally friendly aircraft. Making aviation sustainable is a critical task, and the industry's research and development efforts are opening up a number of avenues to achieve this.

While evolutionary improvements to current gas turbines, biofuels, and alternative engine and aircraft configurations are set to lead the way into a new, cleaner paradigm for air travel, evolutionary improvements cannot sustain aviation indefinitely, and we must not lose sight of our long-term aim. Fig. 4.12 shows alternative fuel consumption scenarios for different performance improvements over the next century. At current passenger growth rates, a reduction

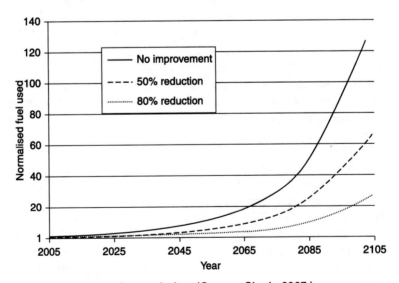

4.12 Fuel consumption evolution. (Source: Singh, 2007.)

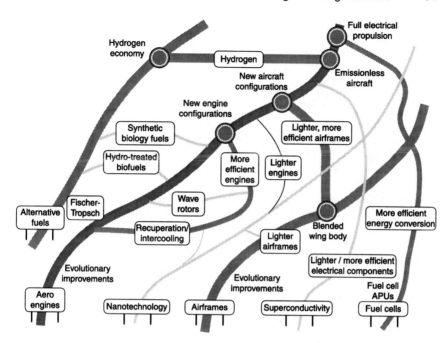

4.13 Technology roadmap.

of even 80% in block fuel consumption would still lead to a 20-fold increase in aviation fuel consumption by the end of the century. It is therefore clear that alternative fuels such as hydrogen could one day play a central role in achieving sustainability. A land-based hydrogen economy might very well be a prerequisite to develop the necessary infrastructures for generating, distributing and storing hydrogen.

The road towards sustainable aviation was, is, and will continue to be supported by innovations in aero engines. Fig. 4.13 shows a (literal) road map that highlights some of the main avenues that might affect aero engine development over the next few decades. There is plenty of scope for synergies and creativity: the coming years will certainly be an exciting period for aero gas turbine engineers.

4.6   References

Ameyugo G and Noppel F (2007). *From Three to Zero and Beyond: Clean Propulsion for Civil Aviation*, Whittle Reactionaries 2007. Institution of Mechanical Engineers, London.

Ameyugo G and Singh R (2007). 'Advanced cycles for distributed propulsion'. In *Proceedings of the 18th Conference on Airbreathing Engines*, 2–7 September 2007, Beijing, China, ISABE 2007-1154.

Dagget D (2009). *Innovations in Aircraft Alternate Fuels*, Cranfield University, Cranfield.

Edmonds, JS, Sharma, DK, Jordan, HE, Edick JD and Schiferl RF (1992). 'Application of high temperature superconductivity to electric motor design'. *IEEE Transactions on Energy Conversion*, 7(2): 322–9.

Green J (2005). 'Greener by design'. In *Innovative Configurations and Advanced Concepts for Future Civil Aircraft*. Von Karman Institute for Fluid Dynamics, Rhode-Saint-Genèse.

Howse M (2004). *Aero Gas Turbines – An Ever-changing Engineering Challenge*, 3 February 2004, Whittle lecture, Royal Aeronautical Society, London.

Masson P and Luongo C (2005). 'High power density superconducting motor for all-electric aircraft propulsion'. *IEEE Transactions on Applied Superconductivity*, **15**(2): 2226–9.

Rolls-Royce (2009). *Annual Report 2008*. Rolls-Royce Group plc, London.

Singh R (2007). 'Civil aerospace propulsion: the challenge of the environment'. Presented at the *6th International Defence and Aerospace Exhibition, Aero India 2007*, 7–11 February, Bangalore.

Whittle I (2007). 'The Dawn of the Jet Age'. In *Whittle 2007*. Cranfield University, Cranfield.

5

Innovation in avionic systems: developments underpinned by digital technologies

M. HIRST, Independent consultant, UK

Abstract: This chapter reviews the cardinal aspects of avionics development of recent decades, and considers the orientations that will play the greatest part in determining the degrees to which there will be change in the future. In considerations of cost, technical improvements affecting the overall capability, the influence of demand, and aspects affecting timing, this forward-looking chapter delimits what is still undefinable. The potential for change is charted in terms of the way that these parameters are expected to interact. A case is made for investigating wider-ranging innovation paradigms, based on external, not just internal, influences. An example to cite is that any operational personnel with an influence on safety will, in the future, find that they are far from alone; science having brought to them a capability that, in current terms, is unthinkable, but already recognised as desirable.

Key words: integration, synthesis, change, cost, capability, demand, timing.

5.1   Introduction

There have been two major transitions in avionics. The first was technological, and involved the change from analogue to digital. It started in the late 1960s, and the transition was virtually completed by year 2000 (although analogue systems still remain in niche applications). The second transition was architectural, involving the shift from segregated to integrated systems. The changes marking this transition were first evident in the 1970s, then gathered pace through the end of the twentieth century, and this remains an ongoing, but increasingly less innovative, area of endeavour.

Viewed as onion skins, the bulb has grown from its central core (technology); it has been wrapped in the second skin (integration); and the challenge for innovative avionics producers is to deliver next the elements that will be the building blocks of a third skin. In that it will be neither architectural nor technological, it is heralded as an extension that takes on totally new perspectives, albeit depending upon digital technology, and it will rely on the integration that had preceded it. For the rest of this chapter it will be called the synthesis transition. Synthesis will take place in stages. First, it will involve the diffusion of boundaries between elements of existing systems; next a dilution of the boundaries between the synthesised avionics and the aircraft itself; and eventually there will be synthesis between the aircraft and its many associated system elements in the operational system in which its daily work is immersed (see Fig. 5.1).

83

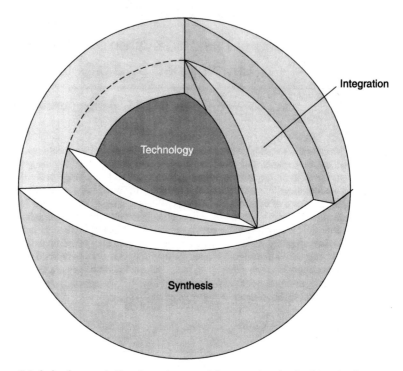

5.1 Avionics evolution has stemmed from technological beginning, through a period of integration, and is entering the phase of seeing integrated systems synthesised to address more wide-ranging and wide-spread operational capabilities.

This is true with respect to military and commercial aircraft implementations, in the near term and in the long term. Aircraft-wide synthesis has been implemented already, albeit rather piecemeal compared to what will be desirable in the long term, in the military air-vehicle sector. The biggest test of success will be in civil aviation, and it will be in this arena, as there are fewer security-shrouded topics, where most debate will be concentrated.

An exercise that has broached the boundaries of synthesis has been fly-by-wire (FBW) flight control systems (FCS), but it has been largely an extension of pre-existing FCS technology, with particular synthesis, into aircraft handling qualities and one safety topic; flight-envelope protection. Whilst successful in military and civil applications for almost 40 years, FBW has not diffused boundaries between systems. It has demonstrated the need to proceed cautiously within a well-structured set of requirements to retain safety-criticality, and it has placed benchmarks in the process arena that will have an influence on hardware and software evolution in the future.

Within modern aircraft, the boundaries around each system are, to a great extent, artificially imposed, and their presence is largely justified by a mix of operational and commercial considerations. The definition of form, fit and function (F^3) requirements in a system specification assures a manufacturer of their design responsibilities, and it promises to the user interoperability of equipment supplied from different sources. The boundary becomes a point at which attributes are artificially limited, however. Whilst the boundary is most clearly defined in terms of data ports (both inwards and outwards), it is also power; applying equally again to energy in and out. It means that cabling, connectors and cooling are all design responsibilities for the designer and builder. The emergence of integrated cabinets in the 1980s was a move that enabled some synthesis in this regard, but its application has been far from global.

The reticence to support the implementation of synthesis is understandable from an industrial perspective. Essentially, a unit (from a relatively simple VHF communications radio through to a complex flight management system (FMS)), is an avionics package that is so thoroughly standardised that, if there are a number of suppliers, the playing field is 'level'. Suppliers can compete on life-cycle cost (LCC), being the combination of initial cost and running costs throughout the system's lifetime, which in turn is a function of such attributes as price, reliability, repairability, and customer support. The user can opt for the cheap option, but can be hampered by the ongoing cost imposed by poor reliability. Seeing the user's experience may, in turn, influence the suppliers to improve the qualities of their products.

On the whole, most suppliers have viewed synthesis as something they will address, at some point, on the road leading to future generation devices, but they have been stymied too by clear evidence that the investment would be considerable, and the field of expertise is beyond that which exists in most avionic design and supply organisations. Industrial reorganisation is already underway that will potentially remedy this, with avionics and aircraft firms already merging to form modern 'systems' companies. Synthesis is being addressed, albeit cautiously.

Life-cycle cost will remain the most vital attributes on the system designer's priority list, whilst technology will press more capability into their hands as time goes by. Where and when new capability will be introduced, and why and how, are amongst the most difficult questions to address. It is widely expected that change will come through the fusion of cost, capability, demand and timing. None of these are easy to determine. That would have been true in previous times, and it is simply exacerbated by the new 'ownership' model in combined aircraft/ system design organisations, which makes design responsibility and operational implementation more associated than they were in the past. As well as having new, internal, working procedures, such organisations will benefit from having access to new forums that are receptive to new ideas, and that have much broader mandates than hitherto. These are beginning to emerge, and to serve as ideologists, motivators and stakeholders, but they are still largely anchored by stakeholder constraints. The most active fora are research-orientated, multi-national, and multi-disciplinary. They are having to strive to identify new ways of thinking about what the systems do, on the aircraft, and in respect of their role in the bigger picture of total operations, and in this review the perspective is largely aligned to the mandates that are being adopted by the most proactive of the 'systems thinking' community. The four change parameters that they need to address, and that are likely to be the greatest influence of the future, have been mentioned already – they are cost, capability, demand and timing.

5.2   Cost

Cost is the most important commercial attribute, and LCC is the most influential of the financial benchmarks for suppliers to control. An aspect, in commercial markets, that has changed over time is that there is less and less to import from the military sector. Often the military sector is no longer the market leader in technology.

Conducting research has become increasingly complex, and one can reflect that it always does as an industry matures. One consequence is that the proportion of price that is charged nowadays for avionic products, and that is attributable to research and development, is considerably higher than it was a few decades ago. Tracking what can be detected of the trend reveals several stages, and places an interesting perspective on the future.

In the analogue avionics era (up to 1970) most research was inherited from the military. The radio navigation, communications and surveillance technologies had been analysed and tested, and the best had been proven, in military operations. Civil avionics producers picked up the technology, innovation costs conveniently written off in government-led research programmes, and had production lines that shared military and civil technology. Where a unique attribute was required, and the commercial case for automatic landing is a good case to cite, it was usually investigated through a government-sponsored programme, with work shared between the research teams and industry. This worked well in the US and in Europe, but sadly, where government-led research was linked to nationalised production, the inertia in the development programme was often far too much, and time left some ambitious plans moribund. The Soviet system of industrial development, whilst it had some successes, proved the inadequacies of the fully nationalised approach in this regard, never challenging, and having to be content to follow, the West (both militarily and in the civil sector). There were some brilliant concepts, and some fine products, but their integration into fully fledged systems was a rarity in the Soviet Union. Even so, by the late 1980s, in most Western nations, the idea of the government sponsoring work that would benefit the operational performance of national industry was almost dead, and initiatives that were technically sound were often grounded by political expediency. The new paradigm was market forces, and in the East and the West the time was ripe for innovative change; but it many respects it has yet to come.

Over the last 20 years all research has been driven by the need to devise ways of applying the burgeoning power of the microprocessor and making a profit from the endeavours. The limits to which synthesis have been addressed reside in the integration of it (the hardware) with the programs within it (the software). Whilst this makes it all sound very simple, the development has been far from easy, and the volume of industrial effort poured into the digitalisation of all things possible, electronically, has been a vast industrial undertaking. Whilst the software industry has burgeoned, some safety-critical aviation and industrial applications have regarded total hard-wired functionality (effectively inaccessible and therefore secure programs) as essential. Nevertheless, the majority of computers in use today in avionics, and indeed on desks throughout offices and homes worldwide, have a minimal amount of hard-wired programming (it is there; for example, the built-in operating system). They maintain flexibility by using software that tailors the hardware to application and purpose, and integrate using protocols that allow data to pass freely across the globe, and through space. The vast majority of this development, which leans heavily towards low-LCC solutions, has been conducted in commercial enterprises. Government-led research in these areas, secure military project applications apart, has been largely overtaken by commercial enterprise.

Thus, commercial avionic system developers for airliners today have to invest heavily in research. As the applications are so limited, the production has to be spread widely in order to keep individual product example prices down, and the consequence of this has been the conglomeration of the plethora of small businesses (albeit some were sizeable, multi-thousand-employee enterprises) to a mere handful of significant, and comparatively mega-scale, avionics firms worldwide today. The process has been almost uncharted but was evident in the late 1970s and early 1980s as sensor suppliers were swallowed by their own customers. This meant that the integrators, whose expertise now extended from raw data to the processed data for instrumentation (or display), became the dominant players in the avionics market. They changed the shape of the industry across the decade that straddled the millennium, and they were the eventual winners, largely because they were the companies that had the largest influence on avionics systems architecture.

Whilst commercially attractive, and indeed essential to survival, the widely applied solution has led to the concentration of research, and a reduction in time devoted to the study of genuinely innovative research opportunities. Even what purports to be innovative in avionic systems terms today tends to be based on the incremental extension of existing infrastructure and capability, because businesses control the risk to themselves. Some of the new generation avionic businesses are offshoots from aircraft companies, either working as an entity within an industrial conglomerate or bearing a name that was an aviation legend in the past, and is now a supplier, and often with strong links into other areas where there is good synergy. They see synthesis as a way ahead, but as industrial entities that have much less governmental support than at any time in their past, and they also have shareholders to satisfy. Without taking unprecedented risks they are unable to nurture, sow and germinate seeds that require a lot more external than internal influence. As the implications of synthesis are examined later, the multi-disciplinary needs, and the challenges they set, will be examined in terms of how the avionics world perhaps needs to change its innovation paradigm.

5.3   Capability

Capability has had little to do with the commercial realignment of businesses. It has been improving throughout the last 60 years, and the avionic systems in commercial airliners provide a stark example of how radically things are changing in nature. In some museums, individuals can sit at the flight deck of a 1960s or 1970s airliner, and gaze at what may be referred to as a 'clockwork' environment. It is rarely clockwork, but it is genuinely largely mechanical, and the ingenuity of the instrumentation (if it is capable of being demonstrated; and sadly in most cases it is not) is often quite breathtaking. The precision engineering of the components in a 1960 technology cross-wire or V-bar multi-mode flight director was indeed akin to the proverbial 'Swiss watch'. That attention to detail, and the wide use of circular instruments with needle and scale presentation of data, led to the euphemism of the 'clockwork cockpit'. Even in this era there had been significant change. Systems had been improved over time, in the sense of how they handled the information that they collected and conveyed. This had led to a reduction in the flight-deck complement from a crew of five – captain, first officer, flight engineer, navigator and radio operator – to the first three crew members alone. It is not the case that radios were no longer used, or that navigation was no longer practised, but that through ingenious design the functions that these specialist crew members had performed had been integrated into the procedures, and the workspace, of the three remaining crew members.

Around 1980 the flight deck changed dramatically and, to see examples of how post-1980s flight decks appeared, one has only to look in a modern airliner, as, whilst there have been some substantial changes, the principles have been retained. The 'clockwork' clutter has been replaced with the smooth surfaces of electronic displays, albeit interspersed in places with some chunky old-age control knobs and levers. Above all, the flight engineer has gone, and a two-place crew compartment has been adopted. The engineer's role has not disappeared, but, like those of the navigator and the radio operator, it has been integrated into the roles of captain and first officer. This time the redundant flight-deck crew member was usurped by the digital computer.

A flight engineer's role was to monitor and adjust systems, making decisions when critical situations were observed, and co-ordinating the availability of functions with the two pilots. It was a critical job, in respect of impact on safe and economical operations, but it was largely a well-understood routine that could be programmed by competent software developers. The modern electronic display, associated with some ergonomically driven control design knowledge, led to the 'automation' of the flight engineer's role, with the large and expensive panel that faced the engineer (usually in the rear right-hand portion of the flight deck) replaced with a more compact and very skilfully designed overhead panel. The ingenuity of the newer generation overhead panel, combining human–machine interface knowledge with operational effectiveness considerations, is the all-too-often overlooked major contributor to the cost-effectiveness of the modern flight deck. It, and many similar design conundrums, brought psychology to centre-stage in the design process regarding the long-standing electronic and instrumentation engineer disciplines, and this was perhaps the point in avionics development history when the role of the psychologist, in addition to the ergonomist, was initially set. The design process depended on understanding not just what would 'look and feel' good for a crew, but what display and control attributes would have a positive impact on the crew's operational efficiency.

Flight deck changes over the last 20 or so years have been largely in terms of technological application. The cathode-ray tube (CRT) display has been usurped by the liquid-crystal display (LCD), the background colour of panel facias were changed in some cases to 'soften' the stark appearance of the flight deck, and legends were illuminated more discreetly, albeit often applying quite complex lighting concepts. Changes sought to minimise the amount of information striving for attention, and presentation assessments extended from being centred solely on facia design to using integrated lighting policies, such that the illumination level is often now under computer control. It minimises glare and vision adjustment time as the ambient conditions affecting the flight deck change, happening slowly in the day and night transition periods and more quickly, but not suddenly, as an aircraft flits between cloud and sun-filled skies. No one sees these advances unless they work in the environment and, even when it works satisfactorily, it is barely noticed. It is another positive psychological impact, achieved through technological capability.

Effectively, for the time being, the flight-deck control and display technology domain has reached a quiet spell in its development. Tight regulatory regimes, whilst essential in the way they contribute to maintaining attributes essential for safety, have done much to stifle the attitude towards change. They do not necessarily stand in the way of change, but their increasingly vehement demand for justifiable processes can lead to design teams being involved in enormous paper chases, and the greatest impact they have is that they diminish the intellectual routes through which innovative designers can seek simplification. Capability has, in many respects, hit the buffers, as so many changes of a physical nature now proposed are all too often turned down because they affect principles protected in the name of safety. This is a time for some radical thinking, to address the inherent limitations of the designs that have evolved, discovering where it is clear that incorporating change was always a dichotomous challenge across both commercial and safety risk boundaries, and seeing what regimes remain ripe for improvement. There is evidence of this from the way that multi-disciplinary attitudes essential to achieve such change are addressed amongst the desirable wide audience, increasingly, in aviation technical and operational symposia and conferences.

5.4   Demand

Demand drives commercial incentives more than anything else. In maritime history, the large passenger-carrying liner was consigned to the history books as soon as aircraft showed their worth as the faster intercontinental alternative, but the liner has been reinvented, albeit now as a vehicle that meets leisure rather than essential travel needs. The reinvention has been largely as a result of diversifying the technologies that meanwhile had been invested into large ocean-going commercial goods ships, such as very large container carrier (VLCC) vessels. In the maritime business, where every vessel is almost a 'one-off', the process is very different from that in aviation, where a slick production line of aircraft that are homogeneously cloned is still the necessity for profit. It might not always be the case, however, and the shared interest in companies that have joint aircraft and systems investments makes the possibility of such a radical re-think more possible than ever.

Consider the possibility that the way airliners are developed, produced and serviced undergoes changes, driven by purchaser demand, that makes bespoke production vehicles more of a likelihood. Just as some large business jet operators have made the transition to fractional-ownership organisations, and in doing so became more customer- than fleet-orientated, it is possible that the airline of the future will also change its allegiance, and need fewer, and more diversified, fleets. Commercial airliner production is about 1000 hulls annually, and it seems unlikely to rise above this level by much more than 20% (on average) for the next few decades. Manufacturers confidently forecast that annual production figures will continue to rise, but there is always the possibility that green awareness will cause the next decade or two to be the peak, and that demand, and annual production, will fall thereafter. This could trigger the airline revolution. Whilst it is common wisdom today that the manufacturing basis will be a small number of companies with busy production lines, the prognosis for the future might be for lower numbers of aircraft, with each more capable of meeting a wider range of tough demands. Potentially, avionics will be an arena where much of this operational flexibility will reside. It will have to be introduced, and managed, according to policies that address a wider range of influences than anything considered so far.

Finally, fundamental capability is almost certainly already out of reach to avionics company research teams. It is very likely that the breakthrough into the next generation of avionic systems will come from the redistribution of technology that originates from elsewhere. The green shoots of such change are evident to some extent. The data-transmission systems used to convey information between systems on airliners, and increasingly within military aircraft too, have gradually matured from being industry-based specifications to some very 'open' specifications. These are being drawn in from the information technology (IT) sector, whose turnover is the most significant of all technological businesses, and where research has spawned specifications that are not solely based on implementation, but on the quantity (data rate) and quality (integrity) of applications. In describing the specifications as 'open', the distinction drawn with regard to what has been used in the past is that the technologies of the data-transfer device, in terms of medium, data rate or functionality, are no longer fixed. The system protocol is fixed, however, and a data-transfer node can be based on one of any number of technologies, or even a set of them. The ubiquitous wire has been usurped by the fibre-optic cable, and the cable itself is now challenged by an electromagnetic medium; either in the radio, visible or even higher frequency bands (somewhere within and around the visible-light portion of the spectrum). The data-transfer application, in terms of the integrity, capacity and timeliness needs of the receiver, is what will define the medium used. If wire is best suited to transferring the medium in any specific application, and especially as this is a proven technology where the commercial outcome is almost certainly attractive, it will be used. But it will be evident, increasingly, that this is the way that systems were linked in previous days and, as time goes by, wires will begin to disappear, because the demand for what were hitherto vast data-rates (to which they are often ill-suited) is forever on the increase. The clear relationship that has always linked a physical architecture with the capability concept will disappear.

5.5   Timing

Timing will be crucial too. When the first generation of electronic flight-desks were designed, in the late 1970s, and introduced in 1982 on the Boeing 757/767 and Airbus A310, the data-transmission system they used was based on a mid-1960s specification (Arinc 429), which had seen the light of day, and been proven in service, in the specialised role of conveying data from the Boeing 747 inertial navigation system (INS). It was tempting to go further (higher data-rates and autonomous control protocol), and the research was done, but the technical risk was judged to be too high. Almost two decades later Airbus side-stepped change, and retained the same fundamental architecture on the A320 and A330/A340, but Boeing reached into the unknown on the Boeing 777 with a more technically challenging system (Arinc 629). It served well, but with no other takers the supply of systems to the programme became bespoke (although not so naively, as the interfaces of units that are common to other airliners are modifiable to suit the 777). The architecture affects functionality too, and the higher data-rate, bus controller-less Arinc 629 system should have allowed a lot more coalescence of devices than actually happened. Boeing did use integrated cabinets in the 777, and where it was applied (largely in utility systems) the mass and system complexity advantages were considerable. Notably, this was a demonstration of the way that applying an aircraft/systems synergy can be beneficial. In other areas – communication, navigation, surveillance (CNS), data processing and display and control – technology infusion brought reliability and LCC advantages, whilst the unit itself reduced in mass, but not size. An aircraft carrying black boxes filled largely with fresh air is not using technology as efficiently as one that has the latest electronic technology more appropriately packaged; but, ironically, it is easier to cool, and more reliable. It is hard to say that either of the two philosophies explained – essentially Arinc 429 and Arinc 629 based – is in any way better than the other. Boeing will argue that the step they took is one pace down a road along which avionics development must proceed. They may well be right, and their courage in stepping out first should bring benefits that accrue from experience. What the examples do illustrate is that timing is as critical as technology, and timing is not always kind to the innovative designer.

Boeing's architecture has adapted, through its inherent flexibility, in ways that have paid dividends when they have sought to pioneer data-hungry crew information initiatives, such as terrain awareness and warning system (TAWS) and enhanced ground-proximity warning systems (EGPWS). The devil is in the detail, as Airbus, and other aircraft manufacturers, are not barred from using such devices, but for them the installation is more tedious and the commercial advantage is somewhat annulled. Technical risk is not an issue, however.

This chapter has looked at the leading requirements over a period of avionics history that has spanned an era of 'horses for courses'. This is not the kind of route map that will extend into the future; the time for a fundamental change has been reached. If, in the future, the role of the avionics system is different in every aircraft, or even different in terms of what it contributes in every flight, the existing door has to open wider, and new doors have to be investigated too – even invented. In prospect is potentially the most challenging of all fundamental aerospace development, and the most exciting era in avionic systems development.

5.6   Future requirements

Civil aviation's safety arguments do not change over time, but the desire for better safety performance has always been prevalent, and that will not change. When safety is expressed in terms of frequency of operations, the trend has been gradually improving over time. The rate of improvement is becoming more difficult to resolve, with mechanical and system failures apparently under control, but with human factor-induced accidents relatively static. It is recognised that there is nothing better than an experienced and well-motivated crew to get the best safety performance. There are publicly acknowledged circumstances of airlines, or even aircraft types, having perceived safety relationships that allow them to be judged as good, indifferent or bad. In the perfect world, there would be no such distinctions, and thus the challenge to the next breed of avionic system designers is to contribute in ways that will lead to a lower floor to accident statistics, and in the predictable world of operations to the idyllic, and so far elusive, goal of a 100% (accident free) safety record – year after year. It will take decades, or even a century, to achieve, and it can easily be argued that the goal is impossible, but philosophically, in the quest for near ultimate perfection, it is still a sane objective to adopt.

5.7   Current safety processes

The safety criteria already in use to test the veracity of safety management systems (SMS) processes will retain their principal configuration unaffected. These depend on the attribution of failure events, and the effect of failure, to acceptable probability levels. This is where the philosophical debate can rapidly be brought to a halt, as any regulation that says there is an acceptable probability of a catastrophic failure implies that accident-free operations are accepted to be impossible. This may be good pragmatic practice; but there is nothing to stop the regulators, when they see innovation offering the prospect, of realigning their requirements to reflect that progressive desire for change. It cannot be done without a lot of debate, and that argument is perhaps a long way in the future; but it is an argument that it is reasonable to assume will take place, and will be one that, while the safety regulators preside over the outcome, will be shared with all the incumbents in the safety arena.

The current safety process assumes that the technical prowess of systems designers will rise to the challenge of existing regulatory requirements in terms of 'designing out' significant safety failure cases. Equivalently, however, they impose operational safety demands that in the more 'human factor' riddled world of the flight deck depend largely on the carriage of responsibility, in such a manner that equivalent capability is embedded in the people-orientated functions to minimise their contribution to accident statistics.

The way this affects current operations is through process and procedures. Considering flight crew only (the arguments can apply equivalently to aerodrome operational safety officers, air traffic controllers, and aircraft maintenance crews), a crew follows a well-defined set of procedures for every flight. For example, they brief on arrival at the airport for a flight, or a set of flights, if they are operating successive sectors. They may be required to do an external check, but may abrogate, or share, this role with the ramp engineers, and likewise they will be aware of passenger manifest issues, insofar only that it is regarded as important that they do know of circumstances in the cabin – from the presence of heads of state to medical emergency cases, etc. Keeping to the avionics systems theme of this report, the crew are the sole conveyors of aircraft systems knowledge, in that they have had the training to understand them, and are provided with the display and control features that allow them to manage the systems.

As soon they sit in an aircraft at the commencement of a flight, the flight crew have 100% responsibility. They take on board the state of aircraft systems, the loading of payload and fuel, and the statutory requirements of the journey ahead. If there is inadequate fuel reserve, given the weather forecast provided, if there is a system that is unserviceable but that is not rated as mandatory on the aircraft's minimum equipment list (MEL), and so on, they will make the appropriate adjustments to plans. There are procedures and communication systems that are there to assist in this regard. Getting all this to happen, to be certifiable, and virtually impossible to abuse, has been the most challenging objective facing safety regulators for many decades.

If it is right to believe that this is the time to consider moving the safety paradigm even further forward, in terms of being able to influence pilot efficiency and effectiveness, training captains offer potentially the best knowledge base to address. They are selected for examination, as it is they who assess and approve standards, and they have to view all crews – young and old, novice and veteran, to the aircraft type and the operator's procedures, and so on. Training captains are themselves trained to employ a baseline of essential attributes, each examined or observed and rated, using analytical methods that will reflect not just expertise, but also motivation. They have to enumerate, too, how this information will affect the crew as well as the individual, because on the modern flight deck it is essential to harmonise the qualities of the individuals in order to cope with all the demands they face in daily operations.

Their judgement processes employ both cognitive and prescriptive components. The 'better' the training captain, the more successfully they will fulfil their role, not just in terms of passing on appropriate knowledge from crew to crew, but also in leaving an assessed crew feeling that their session, even one that ends in a negative assessment, has been a positive experience. The SMS approval process goes some way to achieving this – but the final frontier is the personal contact itself.

Some airlines 'score' their crews in terms of their in-service performance and, as much as possible without prejudice, by monitoring their flying performance through daily flight. This may be through analysis of downloads from data recorders that have been specified, in the main, for post-accident analysis. Some of this information is kept for training sessions, perhaps informing a training captain that the pilot they are assessing lacks competences, or may harbour a bad habit (or two!). These may have developed for some innate reason, or might have come simply through lack of opportunity. Airline crews know that they are assessed from the records they leave on aircraft recorders, and they are at liberty to express their views with a training captain who asks to investigate these issues. The overall desire in such examinations is to be positive.

5.8   The system of the future

In the future these human aspects will not be divorced from the avionic systems designer's role. There is a wistful belief that one day in the not too distant future the crew will arrive, insert a card (please – not yet another one!) and the system will know who they are, and it may even greet them by name; like the imperious and ingratiating 'computer of the future' in many a science-fiction novel. Yes, it could do that, but consider if the system, in knowing who was there (and through whatever medium), was as adept as a third, and very experienced, crew member. To be that good it perhaps should have some very human sensory perceptions too. Ideally it should be able to determine such personal attributes as demeanour, mood and condition. How this is done without physical probes is something to ponder. What is detected will depend on what is determined to be necessary.

Electronics are being imagined that will integrate information from the many sources on the aircraft and assimilate them from the content of a temporal matrix of data that means something to someone or something: but who or what that might be is to be considered in a while. The temporal matrix is the ultimate synthesis. It is difficult to visualise, but Fig 5.2 offers a glimpse of a potential configuration.

The idea implies a set of data of almost amorphous form, with data that evolve on several levels. The most basic level will be that which is measured and stored. This is akin to the systems that replaced the flight engineer, and that monitored aircraft systems, not people. It has the ability to assimilate data in terms of the significance of value: perhaps ascribing a state (typically described as green, amber or red) to a value. The processing power associated with the database will be able to interpret causes based on sets of readings, and might even be able to announce a solution to a problem. The best possible outcome is a solution where the cause has been deduced to be somewhere other than where the initial warnings appeared. An example might be when an engine gearbox failure causes accessories driven by the gears to fail to deliver energy, and the annunciation to the crew is to disable the gearbox, before it fails catastrophically, rather than have them first of all reconfiguring hydraulics and electrical power supplies. An automatic system might do the latter, and faster than the crew, who make the most critical decision, to disable the mechanical gearbox.

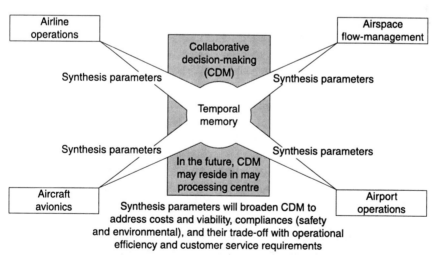

5.2 The temporal memory envisaged for future systems may not be wholly flight-deck orientated. This diagram addresses airline, aircraft, airport and airspace issues as needing to be addressed simultaneously, and does not commit to saying where, if there is a single place at all, that temporal memory will be located.

The second, and subsequent, levels of a temporal matrix are much more difficult to examine and define. To appreciate what is anticipated, imagine the training captain's circumstances, as well as their role, on a check ride. With the two best-qualified crew members in the company, a gin-clear day, no significant weather, a flight between two relatively quiet airports, and through uncrowded airspace en route, the ability of the observer to deduce overall performance is actually impaired compared with an occasion when there is any reduction in the quality of any of the attributes quoted. The check ride, in these ideal flying circumstances, will be far from uneventful, however. Every time there is a check list, there are many ways in which the ability and efficiency of a crew can be assessed, and not least through listening to their exchanges. This is viewing the crew's psychological attitude, as defined by the day, the circumstances, and the external influences that are relevant and affecting their lives. It is a consideration of flight-dependent circumstances.

5.9   The ultimate avionics computer

It is evident that the processing heart of a system that will fulfil such demanding requirements has to be akin to the human brain, and the most suited of modern technical processing and analytical techniques to employ is likely to be a neural network-based solution.

Natural neural networks are human brain functions that receive and supply data to the human nervous system, and a modern computer-based neural network uses engineered items to mimic the properties of the biological system. The widest range of applications to date, whilst acknowledged not to harness the full potential of this technology, have been used to achieve adaptive status, changing the logical path from data input to output, as if the device had acquired the ability to apply reasoning. They are used in image and speech recognition situations, where speed is of an essence and the quality of output desired leaves little margin for error. The better the system can adapt, the more likely it will not depend on what is given to it, but initiate its own search for information. This is akin to data mining using conventional electronic processors today.

The most suitable solution is one that will exhibit an instinct to know where to look for information, and will learn how to integrate it. A concept that is vital to achieving this capability is a mimicry of what neuroscience calls synaptic plasticity. The process is expressed as being akin to having a vast (billions-plus) number of links between neural nodes and ascribing a value to the usefulness of a link. The processor thus navigates along a path that will be changed, over time, as the properties that determine the usefulness or strength of a link change.

Decision-making is so complex that it needs to be handled by more complex rules than those of simple logic, and the most eminent candidate for the role is fuzzy logic. This is a branch of mathematical endeavour that has not yet been fully accepted; its value is not questioned, but its methodology is regarded by many as being a replication of what can be achieved by well-understood statistical analysis. The idea is to identify 'truth', and in this regard there is a suggestion of probability theory in the way it makes the link between inputs and outcome. Fuzzy logic might be quicker and, if it is, and especially if it is more likely to provide a correct answer, it will be certain to find a niche within the systems that evolve.

The physical architecture of the solution is almost a guess at this point in time. It would be natural to assume it is going to be at a point, in space and time, but there is no reason why it should be. There is no reason why it should be contained to the aircraft. The temporal matrix could be partly within an aircraft, and partly within other aircraft, or the processing entities of associated systems; at airports, ATC centres, airline operation centres, etc. The failure of any set could be inconsequential to the whole system, and this could be what drives the aspirations towards that ultimate goal of 100% safety. In the worst of circumstances a crew, or any other hands-on operational personnel within the complete system, might find that they are far from alone, and science has brought to them capability that, in current terms, is unthinkable; to the point of miraculous in modern reportage.

5.10   System–crew interaction

In a traditional processing application, the end result is a definite solution, but in a neural network the outcome can be a set of exit points. The route map from where this chapter began, with the view that synthesis is the latest skin around the evolved core of technology and integration, begins to lack the spirit to express how information, having been sensed and processed, will be passed on; indeed, to whom it should be passed on, when and how.

The concept is no different from a light-emitting diode (LED) matrix, which can be lit selectively so that it spells a word, and even arranged to scroll the word, so that successive words making a phrase appear to the observer; but the translation from fact to reality of a multi-output situation, where the outputs make up an almost infinite set of possible presentations, is where conventional engineering falls apart.

The most general way to describe a flight deck is that it is a crew workstation that comprises display and control elements. View a flight deck, and the displays tend to dominate the viewer's impression. The controls, on the whole, are very visible; such as the primary control (the stick, central or side-mounted, and rudder pedals), or switches and levers buried between display elements, in the centre console, on the glareshield area, or on the overhead panel. It is control that the crew most craves. Provide a crew with a fully active set of displays and no control authority and their impotence is all too evident; whereas remove the displays and leave them with control, and they will know that they have a chance of survival.

This tells what is required very vividly; not another display, but a cognitive over-riding device that will enhance the use of controls. The best, and intrinsically simple, example of this, to date, has been the ground-proximity warning system

(GPWS), which has no instruments: just a voice output that tells the crew what to do, and a cryptic word emphasising the reason for the demand, e.g.: 'terrain, terrain, pull up!'

A voice-based command system is not new, but there is doubt that it should even be oral. There are other senses to address, but they are often difficult to stimulate: touch, taste, smell are very different, in human interaction, from sight and sound. The only clear solution is to recruit the expertise of psychologists, who can associate the human senses, and reactions, to a wider range of stimulus than engineers have ever considered it is necessary to provide. A range of stimuli, including changing ambient conditions, be they temperature, lighting, air movement, and so on, might be within their orbit of interest. The perfect flight deck processing support system will communicate almost humanly, and thus will have a voice, language, intonations and in these terms, if not physically, a 'presence'. The designer's desire has to be to assist the crew to make and implement the correct decisions, or to remedy faulty decisions. The avionics computer does not take over: it assists.

It is a tall order, but the training ground could be in education, where not dissimilar requirements are being addressed in systems that have levels of interaction, and tailor their responses to the user's needs, using information from a wide range, and as non-invasive as possible, of sources.

Inevitably, the flight deck, because it has been used as an example, has become a boundary to the debate. But this process has to be taken out into the wider field, and thus embrace all other operational specialists. On the basis of the way that radio operators, navigators and flight engineers have disappeared, and equivalently within the air traffic control community there has been rationalisation and re-orientation of responsibility boundaries, and the 'licensed' engineer requirements have shifted dramatically over the last few decades, these are the specialists who could even themselves be usurped. There is no reason to associate a task, or set of tasks, with a particular job, and to assume that the job will be so coveted that it will remain sacrosanct in the era of change. Not only pilots, but even personnel staff could be as much a memory in the future as the candlewick maker is in society today.

5.11 Conclusions

The avionics system of the future will have an extra skin. It will be outside the existing skins, but possibly it will not be recognisable as a skin as such. Its beginning will be amongst the regions occupied by the current outer 'integration' skin, and its domain will stretch out into the human interface. It will have attributes that are so personalised to its application that it is perhaps blue for boys and pink for girls, and simply expressing its almost indescribable consistency in this way illustrates immediately how difficult it will be to describe it at all. The chameleon-like nature of the system will bring into future systems what has become imaginable in science fiction games, where beasts and beings change their physical, intellectual and spiritual persona so seamlessly that their many guises seem unrelated. The temperament of the synthesised avionics system has to be well above reproach, however.

How this will be achieved is where research will begin. There are many concepts for 'friendly' or 'user-associate' style human–machine interfaces that will present some useful fundamental baselines. These are essentially 'machine-intelligence' systems. The synthesised avionic system will be 'applicant-intelligent'. The best way to imagine it would be as an additional, integrated and not benign, crew member. It should not 'appear' or have a persona, or impose a will. It should influence the way that the thought processes of an incumbent crew (and perhaps 'associates' – in air traffic centres, maintenance hangars, and so on) start to view situations, whether they are subliminal, and perhaps slow in terms of their developing influence; or up-front and fast-moving, as would be the case in an emergency, for example.

The avionics system of the future will require a challenging development programme, more multi-faceted in terms of the technologies and disciplines it integrates than any previous system development programme in aviation. The elements of the project will have overlaps with other ground-breaking initiatives, but not necessarily into the laboratories from where such systems have usually emerged. The flight deck components of the new avionics, indeed, might be only a part of an even greater scope system, which supports air crews, and other aviation operations personnel, whose common goals allow a lot more common ground to be explored than is possible in current implementations.

5.12   Further reading

Coombs, L.F.E. (2005). *Control in the Sky.* Pen & Sword, Barnsley.

Lloyd, E., and Tye, W. (1982). *Systematic Systems.* Civil Aviation Authority, London.

Moir, I., and Seabridge, A. (2001). *Aircraft Systems.* Professional Engineering Publishing Ltd, Bury St Edmunds.

Moir, I., and Seabridge, A. (2003). *Civil Avionics Systems.* Professional Engineering Publishing Ltd, Bury St Edmunds.

Pallett, E.J.H. (1985). *Microelectronics in Aircraft Systems.* Pitman Publishing, London.

6

The environment as the key design driver in aeronautics

D. I. A. POLL, Cranfield University, UK

Abstract: The aviation industry has the potential to make a significant contribution to the level of carbon dioxide in the atmosphere and to introduce short-term perturbations to other contributors to the Earth's radiation balance. There is a direct link between the economics of aviation and its environmental impact. From the Bréguet range equation, nine theorems relating to aspects of an aircraft's energy liberated to revenue work ratio (*ETRW*) are derived, the ratio becoming large when distances flown are small. The *ETRW* for aircraft in service is estimated and the best aircraft from an environmental impact point of view identified. The effect of alternative fuels is considered.

Key words: emission index, energy liberated to revenue work ratio, contrail, Bréguet range equation, alternative fuels.

6.1 Introduction

During a 25-year working life, a typical commercial aircraft will fly for 100 000 hours and cover some 40 million miles. It will support wealth creation by providing rapid access to traditional markets and open up new markets that are completely inaccessible by other forms of transport. Aviation brings significant benefits to some of the world's poorest countries through tourism (the world's largest industry) and by giving local producers access to markets in richer parts of the world. Travel fosters better international understanding and facilitates cultural exchange, both of which help to reduce the risk of dispute or conflict. It also allows important social links to be maintained and, last but by no means least, it contributes to improved quality of life by providing a wider, affordable choice of holiday destinations.

Not surprisingly, aviation's benefits have resulted in an ever-increasing demand for air transport. In 1995, the global airline industry flew approximately 3.5 trillion available seat kilometres for passengers and approximately 200 billion available tonne kilometres for freight, with 50% of the freight capacity being provided by dedicated freighter aircraft. By 2025, the forecasts[3,4] suggest that these figures will have risen to 12 trillion available seat kilometres and 1 trillion available tonne kilometres – a very substantial increase by any standards. This growth is also expected to vary from region to region; e.g. China is expected to grow at almost 9% per annum and Southeast Asia by 8% per annum, whilst North America is expected to grow at 3% per annum and Europe by 3.5% per annum.

Aircraft operators and manufacturers expect that, in the process of delivering this increase in capacity, 50% of the current fleet of 26 000 aircraft will be replaced

100

and 17 000 additional aircraft will be needed. According to Airbus[1] and Boeing[2] these new aircraft are expected to be distributed along the following lines: regional jets (10%), single aisle (65%), twin aisle (20%) and very large aircraft (5%).

However, over the past 20 years, there has been a growing awareness of, and a growing concern about, the impact of anthropological activity on the environment. In 1988, the World Meteorological Organization and the United Nations Environment Programme established the Intergovernmental Panel on Climate Change (IPCC). This body provides independent information through a series of reports based upon scientific evidence and it reflects the consensus (or as near consensus as possible) views of the broad scientific community. The IPCC links scientists, and other experts, from all over the world, and their work is subject to strenuous, and wide-ranging, peer review.

Since 1988, the IPCC has issued four assessment reports, with the most recent being published in 2007. These have charted a progressive convergence of evidence-based, scientific opinion that anthropological emissions of greenhouse and other gases are having a major impact on the characteristics of the Earth's atmosphere. It has been concluded that the greenhouse effect is responsible for the observed steady increase of the global average surface temperature; the parameter that is a measurable physical characteristic of climate change. The inescapable conclusion is that the continued release of greenhouse gases through human activity is producing a situation in which major and irreversible changes are likely to occur, e.g. the melting of the Arctic ice and the Greenland ice sheet. These events will be accompanied by a rising sea level and will probably produce major climate change.

According to the IPCC,[3] carbon dioxide is the most important anthropogenic greenhouse gas, with a current annual emission rate of about 38 gigatonnes (3.8×10^{10} tonnes). Approximately 73% of this comes from the burning of fossil fuels, and about 20% of all fossil fuel is used to power transportation systems.

Carbon dioxide emissions are important because a significant percentage of the gas from each emission event (of the order of 50%) stays in the atmosphere for a very long time, possibly for centuries. Hence, for all practical purposes, carbon dioxide accumulates in the atmosphere and its climate effects persist long after the emission itself took place.

The total mass of air in the global atmosphere is about 5200 teratonnes (5.2×10^{15} tonnes) and chemical analysis of present-day air indicates that it contains about 3 teratonnes of carbon dioxide. The IPCC research suggests that, if global mean surface temperature rise is to be stabilised at about 3 °C above pre-industrial levels, with a corresponding rise of about 1 m in the mean sea level, then the amount of carbon dioxide in the atmosphere should not be allowed to exceed approximately 3.8 teratonnes. This is an absolute level for carbon dioxide from all sources. The clear implication is that a simple reduction in the rate of emission of anthropological carbon dioxide may not be sufficient to contain the problem and that emissions may need to go to zero at some point in the not too distant future.

At present, civil aviation uses the kerosene-burning gas turbine as its prime mover. The kerosene that is used is derived, in the main, from oil, but a little is synthesised from coal. Combustion of kerosene produces carbon dioxide, water vapour, a mixture of nitric oxide and nitrogen dioxide (NO_x) and some aerosols in the form of carbon and sulphur particulates. The relationship between the mass of fuel consumed and the mass of the individual species produced is determined by the chemistry of the combustion process. The ratio of masses is usually expressed as an emission index (EI), and in the case of carbon dioxide:

$$EI_{CO_2} = \frac{massCO_2}{MassFuel} = \frac{44.01}{\left(12.01 + \left(Y/X\right)\right)}$$ [6.1]

where Y is the average number of hydrogen atoms and X the average number of carbon atoms in a 'molecule' of hydrocarbon fuel. In the case of kerosene, (Y/X) ≈ 1.91 and

$$EI_{CO_2} \approx 3.16$$ [6.2]

Therefore, for every kilogram of kerosene burned, 3.16 kilograms of carbon dioxide are produced and emitted into the atmosphere.

This year the civil aviation industry will consume about 220 million tonnes of kerosene and this will produce 700 million tonnes (0.0007 teratonnes) of carbon dioxide. This is equivalent to a 0.023% increase in the level of carbon dioxide in the atmosphere.

If we now consider the anticipated demand for air transport, the industry forecasts suggest a global average traffic growth rate of 5% per annum. However, as new, higher technology aircraft are introduced and older aircraft are retired, the global fleet becomes progressively more efficient and so the growth rate for fuel burn is expected to be somewhat lower. Historical data suggest the growth rate for fuel burn is about 1.5% less than that for the traffic, i.e. a fuel burn growth rate of 3.5% is expected. Therefore, by 2030, we could see the annual aviation fuel burn increase to 470 million tonnes and, if this growth rate continued until 2050, the fuel burn rate could be almost 1 billion tonnes per annum. In terms of carbon dioxide emissions, the annual rate would be 1.5 billion tonnes in 2030, potentially rising to over 3 billion tonnes by 2050. Furthermore, since the carbon dioxide would be accumulating, by 2050 aviation could have increased the total amount of carbon dioxide in the atmosphere by some 70 billion tonnes (0.07 teratonnes), or a 2.3% increase over the current level. Therefore, if there was an international agreement to limit the total amount of carbon dioxide in the atmosphere to 3.8 teratonnes, then between the time of writing and 2050, if kerosene continues as the fuel of choice, global aviation alone could account for up to 8% of the remaining 0.8 teratonnes quota.

The other emissions from the kerosene-burning gas turbine do not remain in the atmosphere for long and, consequently, only have an environmental impact for a short time. Nevertheless, while they last, the magnitude of the impact can be large. NO_X has an indirect effect on global warming because it affects the concentrations of two other greenhouse gases, namely ozone and methane. Methane is destroyed and this produces a cooling effect, whilst ozone is created and this has a warming effect. At a global level, these opposing effects almost cancel out. However, the methane imbalance appears in both the northern and southern hemispheres, whilst the ozone imbalance appears mainly in the northern hemisphere. The impact of an emission of NO_X on global warming largely dissipates after a few months, although the small residual perturbation to methane persists for a long time.

Perhaps a more significant fact is that NO_X is harmful to humans and there is already EU legislation in place that specifies the maximum safe exposure limits. Therefore, NO_X emissions near the ground are important, since they affect airport 'local air quality'. Consequently, NO_X generation and its local dispersion during ground operations and during the landing and take-off cycle need to be understood and quantified. There are also health concerns associated with particulate material (PM). So-called primary PM comes directly from engine soot and tyre smoke, while secondary PM is formed in the atmosphere by the action of the emissions of NO_X and the oxides of sulphur (SO_X). Therefore, the sources and dispersion characteristics of PM also need to be understood and quantified.

In terms of the emission index, the amount of NO_X produced depends upon both the fuel and the design of the engine combustion chamber. Therefore, the precise relationship between the amount of fuel burned and the amount of NO_X produced depends upon the type of engine. However, typically,

$$EI_{NO_X} \approx 0.015 \qquad\qquad\qquad [6.3]$$

Although water vapour is itself a greenhouse gas, the amount emitted by the gas turbine is relatively small and the resulting direct impact upon global warming is low. Similarly, soot particles have a small warming effect, whilst sulphur particles have a small cooling effect. However, when the atmospheric conditions are right, i.e. when the air through which the aircraft is flying is supersaturated with respect to ice, the water efflux from the engine may produce a contrail.[4] Relative to the size of the aircraft, contrails are massive features and, although the water from the combustion of the fuel initiates the contrail, the vast majority (99%) of the visible ice comes from water in the atmosphere. The contrail influence on radiative forcing depends upon whether it is day or night, but on balance the net impact is a warming effect. If the contrail is persistent, it may evolve over time into 'contrail cirrus' cloud, or the particulates emitted from the engine may trigger the formation of a cloud without the need for a contrail. Either way, the formation of clouds produces a warming effect that may be very strong.

It is useful to note that contrail formation could be virtually eliminated by avoiding air that is supersaturated with respect to ice. At present, the optimum cruise altitudes for civil transport aircraft coincide almost exactly with the altitudes most likely to carry supersaturated air. Over the UK, this critical altitude is about 10 km and the likelihood of creating a contrail is considerably reduced by flying significantly higher or lower.

6.2   Economic efficiency

The operation of airlines is dominated by economic considerations, and the economics of airline operations is an extremely complex subject. However, there are some parameters that are guaranteed to be important. A clear case in point is the ratio of the revenue generated to the cost of energy expended during the flight, i.e.

$$\frac{revenue\ generated}{cost\ of\ energy\ used} = \frac{A}{B}\left(\frac{MP.g.R}{MMF.LCV}\right) \tag{6.4}$$

where A = revenue/unit payload weight/unit distance travelled, MP = payload mass, g = acceleration due to gravity, R = the great circle distance between departure point and destination, B = the fuel cost/unit of energy released, MMF = the mass of fuel consumed, LCV = the fuel lower calorific value.

The elements A and B are dependent upon market forces and will vary during the working life of the aircraft. However, the term in brackets is something that is fixed by the aircraft's designers and is constant throughout the life of the aircraft.

In general, the best economic operating condition is determined through a balance between the cost of fuel and the cost of time, as currently exemplified through the Boeing cost index (CI) approach.[5] However, in the future and for a whole variety of reasons, the likelihood is that the price of fuel will rise and so fuel will become an increasingly large element in the aircraft direct operating cost ($CI \rightarrow 0$). Therefore, achieving a high value of revenue generated over energy cost will become progressively more important and the market will demand aircraft with progressively smaller values of the energy liberated to revenue work ratio ($ETRW$), where

$$ETRW = \frac{MMF.LCV}{MP.g.R} \tag{6.5}$$

This parameter has already been identified by some authors, e.g. Hileman *et al.*,[6] as a key metric, not only for the investigation of airline economics, but also for comparisons between different modes of transport.

6.3   Environmental impact

In the context of an aircraft's impact on the environment, we may define a coefficient of environmental performance, CEP, as the amount of emission released per unit of revenue work performed, i.e.

$$CEP = \frac{emissions\ mass.LCV}{useful\ work\ done} \tag{6.6}$$

In terms of the chemical species emitted, this becomes

$$CEP = \left(ETRW\right)\left(a.EI_{CO_2} + b.EI_{NO_X} + c.EI_{H_2O} + d.EI_{SO_X} + ...\right) \qquad [6.7]$$

where the constants a, b, c, d, etc. and the emission indices depend upon both the fuel being used and the level of engine technology. For example, if the fuel is oil-derived kerosene, a, b, c and d will be unity, but if biomass-derived kerosene is being used, a could be less than unity, since the net carbon dioxide addition to the atmosphere could be lower than that for oil-derived kerosene. Similarly, if the fuel was hydrogen then the emission indices for carbon dioxide and SO_X would be zero.

Notwithstanding the values of the constants and the emission indices, the *CEP* is lowest when the value of *ETRW* is as low as possible. However, in the context of aviation, it is sometimes necessary to consider both the amount of emission and the altitude at which it takes place. The altitude is an important parameter for NO_X chemistry. It is also important for water vapour-induced contrails, since the appropriate atmospheric conditions tend to occur at specific altitudes.

6.4   The characteristics of the aeroplane

From the very beginning of aviation, the design process of any aircraft has begun with a customer specification. For a given level of technology, one group of parameters that is sufficient to specify a complete civil aircraft configuration is

- A passenger payload,
- A range at the specified payload,
- The engine characteristics and number of engines, and
- The length of runway to be used for take-off and landing.

In addition, there will be a number of constraints that have to be satisfied. The most important, and the most comprehensive, of these are the regulatory requirements for safety, i.e. compliance with the provisions of an appropriate design code. However, in certain instances, there may be others, e.g. a specific cruise Mach number, airport limits for wingspan, pavement weight or noise restrictions or en-route height limitations for particular customers.

Once these have been taken into account, the design process is an optimisation involving one or more target functions. Simple examples of possible target functions are maximum payload fraction, i.e. the smallest aircraft mass to carry a given load, or minimum fuel burn.

In the past, the target functions and the constraints have varied depending upon the economic and regulatory conditions in force at the time. However, in the future, economic and environmental forces are likely to become strongly aligned in favour of minimising the *ETRW* and flying in a way that avoids, or at least minimises, the formation of contrails.

Published by Woodhead Publishing Limited, 2012

6.5   What determines the value of the energy liberated to revenue work ratio (*ETRW*)?

The *ETRW* depends upon the amount of fuel burned to carry a given payload a given distance. As the flight progresses, the total mass of the aircraft decreases as fuel is burned, i.e.

$$\frac{dM}{dt} = -\frac{dm_f}{dt} = -\dot{m}_f \qquad\qquad [6.8]$$

where M and m_f are the instantaneous masses of the total aircraft and the fuel respectively. In addition, the overall thermodynamic efficiency, η_o, of the propulsion system is given by

$$\eta_0 = \frac{(T.V_\infty)}{(\dot{m}_f.LCV)} \qquad\qquad [6.9]$$

where T is the thrust developed and V_∞ (= dS/dt) is the flight speed.

When the aircraft is in the cruise condition, the aerodynamic lift is equal to the total aircraft weight and the engine thrust is equal to the drag, D, i.e.

$$-\frac{dM}{dt} = -\frac{dM}{dS}\frac{dS}{dt} = -\frac{dM}{dS}V_\infty = \frac{(Mg.V_\infty)}{LCV.(\eta_o.L/D)} \qquad\qquad [6.10]$$

The value of $\eta_o(L/D)$ depends upon the Mach number and the altitude at which the aircraft is flown. For a given aircraft, there is a Mach number–altitude trajectory that gives a constant value of $\eta_o(L/D)$ in cruise. For a turbo-fan powered aircraft, this is a constant Mach number, constant lift coefficient (cruise climb) trajectory. With $\eta_o(L/D)$ constant, the equation may be integrated to obtain the total fuel consumed as the aircraft cruises between two points separated by a total great circle distance, R

$$\frac{(MF)_{cruise}}{M_1} = 1 - EXP\left(-\frac{(g.R/LCV)}{(\eta_o.L/D)}\right) \qquad\qquad [6.11]$$

where M_1 is the mass of the aircraft at the beginning of cruise.

However, any flight between two points on the surface of the Earth involves a take-off, a climb, a cruise, a descent and a landing. This means that, in reality, more fuel will be consumed in flying from the departure point to the destination than would be required to cruise the same distance. If this additional fuel is designated by Δmf, where

$$\frac{\Delta mf}{MTO} = \varepsilon = 1 - k \qquad\qquad [6.12]$$

and *MTO* is the mass of the aircraft at the beginning of the take-off run, the total (mission) fuel, *MMF*, consumed on a trip between two locations separated by a great circle distance, *R*, is

$$\frac{MMF}{MTO} = 1 - kEXP\left(-\frac{g.R}{LCV\left(\eta_o\,L/D\right)}\right)$$ [6.13]

In practice, ε has a value of about 0.025. However, the performance of the aircraft is very sensitive to this parameter.

If we introduce a non-dimensional range, X, where

$$X = \frac{g.R}{LCV.\left(\eta_0.L/D\right)}$$ [6.14]

then,

$$\frac{MMF}{MTO} = 1 - kEXP(-X) = \alpha$$ [6.15]

This is a form of the classic range equation originally derived during the First World War and frequently attributed to Louis Bréguet, the French aviation pioneer, although other authors have been associated with it.[7] It is important to note that this equation is exact, for the conditions specified, and it applies to all aircraft irrespective of their configuration.

Turning now to the mass of the aircraft, the total mass at the beginning of the take-off run, MTO, is made up of the following components,

$$MTO = MOE + MP + MMF + MF_{NC}$$ [6.16]

where MOE is the operational empty mass (mass of everything except payload and fuel), MP is the payload mass (passengers + 'belly' freight), MMF is the mass of mission fuel (fuel consumed during flight) and MF_{NC} is the mass of fuel carried, but not consumed, i.e. reserve fuel plus tankered fuel, where

$$MF_{NC} = \beta \cdot MTO.$$ [6.17]

The total mass at the end of the landing run, ML, is given by

$$ML = MOE + MP + MF_{NC} = MTO(1 - \alpha)$$ [6.18]

Similarly, the zero fuel mass, MZF, is

$$MZF = MOE + MP = ML - MF_{NC} = MTO(1 - \alpha - \beta)$$ [6.19]

Hence, the fuel burn per unit payload per unit non-dimensional distance flown is given by

$$\frac{MMF}{MP.X} = \left(\frac{MZF}{MP}\right)\left(\frac{1}{X}\frac{\left(1 - kEXP(-X)\right)}{\left(kEXP(-X) - \beta\right)}\right)$$ [6.20]

and this is related directly to the $ETRW$ since

$$ETRW = \left(\frac{MMF}{MP.X}\right)\bigg/\left(\eta_o.\frac{L}{D}\right)$$ [6.21]

These equations are also exact for the conditions specified. They do not depend upon the aircraft configuration, i.e. they apply to all flying machines, and it is possible to use them to derive a number of theorems that are relevant to the optimisation problem.

6.5.1   Theorem 1

Consider a given aircraft travelling a fixed distance and cruising at constant $\eta_o(L/D)$, i.e. a is constant. It follows that

$$\frac{MMF}{MZF + MF_{NC}} = \frac{MMF}{ML} = \frac{\mathrm{d}MMF}{\mathrm{d}ML} = \frac{\alpha}{(1-\alpha)} \qquad [6.22]$$

and, hence,

$$\frac{\mathrm{d}MMF}{MMF} = \frac{\mathrm{d}\left(MZF + MF_{NC}\right)}{\left(MZF + MF_{NC}\right)} = \frac{\mathrm{d}ML}{ML} \qquad [6.23]$$

Therefore, if the mass of the aircraft is changed in any way, the percentage change in the fuel actually burned to carry out the flight is equal to the percentage change in the sum of the aircraft zero fuel mass and the fuel carried, but not burned.

6.5.2   Theorem 2

If the load factor, LF, is defined as the ratio of the actual payload mass to the maximum permitted payload mass, MMP, i.e.

$$LF = \frac{MP}{MMP} \qquad [6.24]$$

then

$$\frac{MZF}{MP} = \frac{MOE + MP}{MP} = \left(\frac{MOE}{MMP}\right)\left(\frac{1}{LF}\right) + 1 \qquad [6.25]$$

Hence, for a given aircraft travelling a fixed distance, cruising at constant $\eta_o(L/D)$, the minimum value of $ETRW$ occurs when the payload has its maximum possible value (LF is unity) and the fuel carried, but not consumed, has its minimum possible value, i.e. the minimum reserve required by law.

6.5.3   Theorem 3

Consider what happens to the $ETRW$ of an aircraft cruising at constant $\eta_o(L/D)$ with a fixed ratio of zero-fuel mass to payload mass, MZF/MP, i.e. constant MOE/MP, as more fuel is added and the aircraft flies further and further.

A typical result is shown in Fig. 6.1.

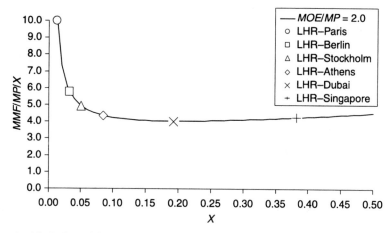

6.1 Variation of fuel burn per unit payload per unit non-dimensional distance with non-dimensional range. $MOE/MP = 2$, $\varepsilon = 0.025$ and $\beta = 0.05$.

Under these conditions, the best value of *ETRW* occurs at a value of X that depends only upon ε (the fuel used over and above the cruise value for the same range) and β (the fuel carried, but not used).

Since ε is small (≈ 0.025) and X is typically less than 0.4, using the expression derived in the Appendix

$$\frac{MMF}{MP.X} \approx \frac{\left(MZF/MP\right)}{\left(1-\beta\right)}$$

$$\left(\frac{\varepsilon}{X} + \left(1 + \varepsilon\frac{\left(1+\beta\right)}{\left(1-\beta\right)}\right) + \frac{X}{2}\left(\left(\frac{1+\beta}{1-\beta}\right) + \varepsilon\left(\frac{1+4\beta+\beta^2}{\left(1-\beta\right)^2}\right)\right) + \dots\right) \quad [6.26]$$

Therefore, the range for best *ETRW* is

$$R \approx \left(\frac{LCV\left(\eta_o\, L/D\right)}{g}\right)\left(1-\beta\right)\left(\frac{2\varepsilon}{1-\beta^2}\right)^{1/2}\left(1-\left(\frac{1+4\beta+\beta^2}{4}\right)\left(\frac{2\varepsilon}{1-\beta^2}\right)\right) \quad [6.27]$$

and, if β is small (< 0.1),

$$R \approx \left(\frac{LCV\left(\eta_o\, L/D\right)}{g}\right)\left(2\varepsilon\right)^{1/2}\left(1-\frac{1}{2}\left(2\beta+\varepsilon\right)\right) \quad [6.28]$$

6.5.4   Theorem 4

For an aircraft cruising at constant $\eta_o(L/D)$ and carrying the maximum permissible payload, *MMP*, the best value of *ETRW* is given by

$$ETRW \approx \left(\frac{MMZF/MMP}{\eta_o \, L/D(1-\beta)} \right)$$

$$\left(1 + (1+\beta) \left(\left(\frac{2\varepsilon}{1-\beta^2} \right)^{1/2} + \frac{(1+\beta)}{2} \left(\frac{2\varepsilon}{1-\beta^2} \right) + \left(\frac{1+4\beta+\beta^2}{4} \right) \left(\frac{2\varepsilon}{1-\beta^2} \right)^{3/2} \right) \right) \qquad [6.29]$$

or, if β is small,

$$ETRW \approx \left(\frac{MMZF}{MMP} \right) \left(\frac{1+\varepsilon+\beta}{\eta_o \, L/D} \right) \left(1 + (2\varepsilon)^{1/2} \left(1 + \frac{1}{2}(2\beta-\varepsilon) \right) \right) \qquad [6.30]$$

For a typical optimum, long-range flight

$$ETRW \approx 1.3 \left(\frac{MMZF/MMP}{\eta_o \, (L/D)} \right) \qquad [6.31]$$

Under these conditions, the influence of the aircraft characteristics and the operational characteristics are separated and the absolute minimum value of *ETRW* is obtained when the product of the structural efficiency (*MMP/MMZF*), the propulsion efficiency (η_o) and the aerodynamic efficiency (*L/D*) is maximised.

In general, these three efficiencies are interrelated in a complex way.

6.5.5 Theorem 5

Consider what happens to the *ETRW* of a given aircraft operating at a fixed take-off mass, i.e. constant *MOE/MTO*, and cruising at constant $\eta_o(L/D)$ as more fuel is added and the aircraft flies further and further. However, since the take-off mass is fixed, as more fuel is added, the payload mass must be reduced. Hence,

$$LF = \frac{MTO}{MMP} \left(1 - \left(\beta + \alpha + \frac{MOE}{MTO} \right) \right) \qquad [6.32]$$

Therefore,

$$\frac{MZF}{MP} = \frac{1 - (\beta+\alpha)}{1 - \left(\beta + \alpha + \frac{MOE}{MTO} \right)} \qquad [6.33]$$

and

$$\frac{MMF}{MP.X} = \left(\frac{1}{X} \right) \left(\frac{1 - kEXP(-X)}{\left(kEXP(-X) - \left(\beta + \frac{MOE}{MTO} \right) \right)} \right) \qquad [6.34]$$

In this case, the best *ETRW* occurs when *MTO* has its largest possible value, i.e. the maximum take-off mass, *MMTO*, and β has its minimum possible value.

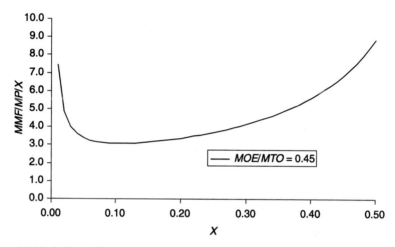

6.2 Variation of fuel burn per unit payload per unit non-dimensional distance with non-dimensional range. *MOE/MTO* = 0.45, ε = 0.025 and β = 0.05.

6.5.6 Theorem 6

The variation of ETRW with range for an aircraft with a fixed value of *MOE/MTO* is given in Fig. 6.2.

For a given aircraft with a fixed take-off mass, the best *ETRW* occurs at a value of *X* that depends upon ε (the fuel used over and above the cruise value for the same range), β (the fuel carried, but not used) and the ratio *MOE/MTO*.

The range for best *ETRW* is found to be

$$R \approx \left(\frac{LCV\left(\eta_o\, L/D\right)}{g} \right)\left(1 - \left(\beta + \frac{MOE}{MTO}\right)\right)$$

$$\left(\frac{2\varepsilon}{1 - \left(\beta + \dfrac{MOE}{MTO}\right)^2 + \varepsilon\left(1 + 4\left(\beta + \dfrac{MOE}{MTO}\right) + \left(\beta + \dfrac{MOE}{MTO}\right)^2\right)} \right)^{1/2} \quad [6.35]$$

6.5.7 Theorem 7

For an aircraft operating at its maximum take-off mass and cruising at constant $\eta_o(L/D)$, the best value of *ETRW* is given by

$$ETRW \approx \left(\frac{1}{\eta_o L/D}\right)\left(\frac{\left(1-\left(\beta+\dfrac{MOE}{MMTO}\right)+\varepsilon\left(1+\dfrac{MOE}{MMTO}\right)\right)}{\left(1-\left(\beta+\dfrac{MOE}{MMTO}\right)\right)^2}\right.$$

[6.36]

$$\left.+\frac{\left(2\varepsilon\left(1-\left(\beta+\dfrac{MOE}{MMTO}\right)\right)^2+\varepsilon\left(1+4\left(\beta+\dfrac{MOE}{MMTO}\right)+\left(\beta+\dfrac{MOE}{MMTO}\right)^2\right)\right)^{1/2}}{\left(1-\left(\beta+\dfrac{MOE}{MMTO}\right)\right)^4}\right)$$

Under these conditions, the influence of the aircraft characteristics and the operational characteristics are not separated and the absolute minimum value of *ETRW* is obtained when the structural efficiency, *MMTO/MOE*, and the product of the propulsive and aerodynamic efficiencies, $\eta_o(L/D)$, have their largest values.

6.5.8   Theorem 8

As shown in Fig. 6.3, the complete *ETRW* variation with range for any real aircraft is determined by a combination of curves of the types shown in Fig. 6.1 and 6.2. When the payload and the take-off masses have their maximum values, the solid line defines the operating boundary for the aircraft.

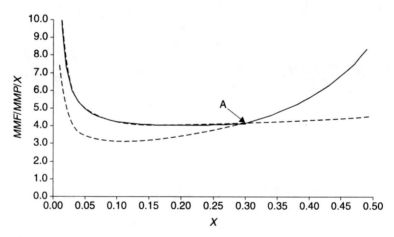

6.3 Variation of fuel burn per unit payload per unit non-dimensional distance with non-dimensional range. *MOE/MMP* = 2, *MOE/MMTO* = 0.45, ε = 0.025 and β = 0.05.

Published by Woodhead Publishing Limited, 2012

The point of intersection of the two curves is labelled point A. Here,

$$\alpha_A = (1-\beta) - \frac{MMZF}{MMTO} \qquad [6.37]$$

and

$$X_A = \ln\left(\frac{1-\varepsilon}{\beta + \dfrac{MMZF}{MMTO}}\right) \qquad [6.38]$$

By expanding $EXP(-X)$ as a power series in ascending powers of X and neglecting terms of order X^3 and above, an approximation to X_A is given by

$$X_A \approx 1 - \left(\frac{2}{(1-\varepsilon)}\left(\beta + \frac{MMZF}{MMTO}\right) - 1\right)^{1/2} \qquad [6.39]$$

Therefore, for an aircraft operating at fixed values of $\eta_o (L/D)$, ε and β the curves of $ETRW$ versus X at maximum payload and maximum take-off mass cross only once. The point of intersection is determined by the ratio of the maximum zero fuel mass to the maximum take-off mass, $MMZF/MMTO$.

6.5.9 Theorem 9

When the distance flown is less than X_A, the minimum $ETRW$ is determined by the curve for maximum permissible payload, $LF = 1$, whilst for distances greater than X_A the minimum $ETRW$ is determined by the curve for maximum take-off mass, $LF < 1$. In the space above the solid line, both the payload mass and the take-off mass are always less than the maximum permitted values.

At point A,

$$\left(\frac{MMF}{MMP.X}\right)_A = \frac{MMTO}{MMP}\frac{1}{X_A}\left(1-\beta-\frac{MMZF}{MMTO}\right) \qquad [6.40]$$

i.e.

$$(ETRW)_A \approx \frac{(MMTO/MMP)}{(\eta_o L/D)}\left(\frac{1-\beta-(MMZF/MMTO)}{1-\left(\dfrac{2}{(1-\varepsilon)}\left(\beta + \dfrac{MMZF}{MMTO}\right)-1\right)^{1/2}}\right) \qquad [6.41]$$

For an aircraft operating at fixed values of $\eta_o(L/D)$, ε and β, the value of *ETRW* at point A is determined by the ratio of maximum zero fuel mass to maximum take-off mass, *MMZF/MMTO*, and the ratio of maximum zero fuel mass to maximum payload mass, *MMZF/MMP*.

6.5.10 Theorem 10

For a given aircraft, the minimum achievable *ETRW* occurs at point A when the value of X for minimum *ETRW* at the maximum permissible payload mass (Theorem 3) is greater than the value of X at point A (Theorem 8). Therefore, the best possible value of *ETRW* occurs at point A when

$$(1-\beta)\left(\frac{2\varepsilon}{1-\beta^2+\varepsilon(1+4\beta+\beta^2)}\right)^{1/2} \geq 1-\left(\frac{2}{(1-\varepsilon)}\left(\beta+\frac{MMZF}{MMTO}\right)-1\right)^{1/2} \qquad [6.42]$$

or, if ε is small compared with 1,

$$\frac{MMZF}{MMTO} \geq (1-\beta)\left(1-\left(\frac{2\varepsilon}{1-\beta^2}\right)^{1/2}-\beta\left(\frac{2\varepsilon}{1-\beta^2}\right)\right.$$
$$\left.+\left(\frac{3+4\beta-\beta^2}{4}\right)\left(\frac{2\varepsilon}{1-\beta^2}\right)^{3/2}\right) \qquad [6.43]$$

For a given mission, the location of the point of minimum *ETRW* on the payload range diagram is determined by the ratio of maximum zero fuel mass to maximum take-off mass, ε (the fuel used over and above the cruise value for the same range) and β (the fuel carried, but not used).

6.6   Observations on the *ETRW*

Whilst the *MOE* is a mass that can be determined by weighing the aircraft, the *MMZF* and *MMTO* are 'certification masses' whose values are agreed with the Regulator. These masses, $\eta_o L/D$ and the minimum value of ε are determined by the design process. The minimum permissible value of β is specified by the Regulator, whilst the actual value of β and the value of *LF* are determined by the aircraft operator and the actual value of ε is determined by the air traffic system that is managing the flight. Once the values of these parameters have been fixed the aircraft *ETRW* for a given flight is determined.

For a given aircraft, lines of constant *ETRW* (constant *MMF/MP/X*) can be plotted on a normalised payload range diagram, as shown in Fig. 6.4. Here the payload is normalised with the maximum payload and the range is normalised with the maximum range at maximum payload, R_{MMP}.

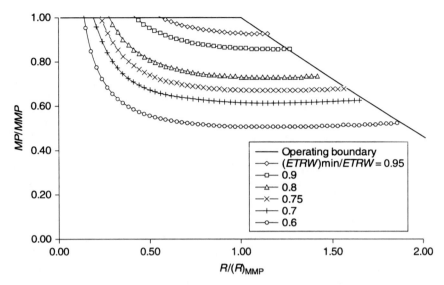

6.4 Variation of normalised *ETRW* with normalised range. *MOE/MMTO* = 0.55, ε = 0.025 and β = 0.05.

The smallest value of *ETRW* occurs along the line *MP/MMP* = *1* (*LF* = *1*) at a value of $R/(R)_{MMP}$ less than, or equal to, unity. When the aircraft is operated with payloads below the maximum permitted and over ranges above and below the range for best *ETRW*, the economic efficiency of the flight is reduced. The figure shows that load factor is a very important parameter for flight efficiency and that operating an aircraft over short ranges can result in very large efficiency penalties.

The data given in Fig. 6.1 show that, for current conventional kerosene-powered aircraft, all flights conducted within Europe are 'short', i.e. less than the range for best *ETRW*.

It is also possible to determine the sensitivity of *ETRW* to variations in the governing parameters. In order of increasing importance, these are

$$\frac{\partial ETRW}{\partial \left(\eta_o\, L/D\right)} = -\frac{ETRW}{\left(\eta_o\, L/D\right)} \qquad [6.44]$$

$$\frac{\partial ETRW}{\partial LF} = -\left(\frac{1}{1 + LF\, \dfrac{MMP}{MOE}}\right)\frac{ETRW}{LF} \qquad [6.45]$$

$$\frac{\partial ETRW}{\partial \varepsilon} = \left(\frac{\varepsilon\left(1-\beta\right)EXP(-X)}{\left(kEXP(-X)-\beta\right)\left(1-kEXP(-X)\right)}\right)\frac{ETRW}{\varepsilon} \qquad [6.46]$$

and

$$\frac{\partial ETRW}{\partial \beta} = \left(\frac{\beta}{(kEXP(-X) - \beta)}\right)\frac{ETRW}{\beta} \qquad [6.47]$$

All aircraft have an operating condition at which the $ETRW$ is a minimum. This corresponds to a particular value of $\eta_0 L/D$ that is achieved at a particular combination of flight Mach number and altitude. If the aircraft cannot fly at this optimum condition, there is a fuel burn penalty. In percentage terms, this penalty is equal to the percentage increase in $\eta_0 L/D$ that results from the changes to the flight Mach number and altitude.

The load factor has a maximum value of unity, but in normal operations it is likely to be closer to 0.7. If the load factor is increased above 0.7, e.g. by the addition of 'belly freight', the $ETRW$ is decreased. Typically, a 10% increase in LF will reduce the $ETRW$ by about 7.5%, making the flight more fuel efficient. This change is independent of all parameters except MOE/MMP.

When extra fuel is consumed due to non-optimum climb and descent profiles, or due to additional miles being flown for any reason, e.g. air traffic directives or adverse weather, ε is greater than its minimum value and a fuel burn penalty is incurred. For a flight distance equal, or near, to the optimum given in Theorem 3, a 10% increase in ε relative to its optimum value ($\varepsilon_{opt} \approx 0.02$) causes $ETRW$ to increase by about 1.5%.

Finally, it is possible to identify the impact of reserve fuel. The minimum reserve fuel is equal to approximately 4.5% of the take-off mass, i.e. $\beta \approx 0.045$. Therefore, for flights at, or near, the optimum range given by Theorem 3, the addition of minimum reserve fuel increases the $ETRW$ by about 6%. Therefore, the environmental cost of safety, in the form of reserve fuel, is a 6% increase in fuel burn per unit payload per unit distance. If the aircraft carries more fuel in the form of additional reserves, or if the aircraft is tankering fuel for the onward or return trip, a 10% increase in β relative to its minimum legal value will increase the $ETRW$ by about 0.7%.

It is important to note that all the theorems and the observations apply to any flying machine using a propulsion system that consumes fuel as the flight proceeds.

6.7   Aircraft performance

The economic and business considerations that determine the requirements of airlines are both complex and subject to variation. Therefore, the aircraft in service today have not, in general, been specifically designed for minimum $ETRW$ since, although fuel consumption has always been an important consideration, it has not been the dominant consideration. Therefore, the obvious questions are 'what are the values of $ETRW$ for current aircraft?' and 'how much better could the values be in future?' In order to provide answers it is necessary to resort to empiricism. Therefore, it should be noted that the following analysis involves approximations.

As noted above, the relationship between the three key masses, *MOE, MMP* and *MMTO*, is central to the determination of the *ETRW*. In reality, the relationship is complex and determined by many factors. However, it is possible to make some progress by recognising that the masses of some components of an aircraft are related to the maximum take-off mass, whilst the masses of others are related to the maximum payload mass. A rough breakdown is:

- Lifting surfaces $\approx 0.14\ MMTO$,
- Propulsion system $\approx 0.08\ MMTO$,
- Undercarriage $\approx 0.05\ MMTO$,
- Avionics, instruments and electronics $\approx 0.02\ MMTO$, and
- Flight controls and hydraulics $\approx 0.02\ MMTO$,

with

- Fuselage $\approx 0.50\ MMP$,
- Furnishings $\approx 0.27\ MMP$,
- Air conditioning $\approx 0.03\ MMP$, and
- Auxiliary power unit $\approx 0.01\ MMP$,

plus

- Operational items $\approx 0.13\ MMP$, and
- Crew $\approx 0.02\ MMP$.

Therefore, summing the components gives

$$\frac{MOE}{MMTO} \approx 0.31 + 0.96\,\frac{MMP}{MMTO} \qquad [6.48]$$

In Fig. 6.5, this relation is compared with data from a wide range of Boeing and Airbus aircraft currently in service and, although there is some variation, it is clear that there is a correlation between *MOE/MMTO* and *MMP/MMTO*.

It follows that

$$\frac{MMP}{MMTO} \approx \frac{1}{1.96}\left(\frac{MMZF}{MMTO} - 0.31\right) \qquad [6.49]$$

and, from Theorems 8 and 9,

$$\left(\frac{MMF}{MMP.X}\right)_A \approx \frac{1.96}{X_A}\left(\frac{\alpha_A}{\left(1-\beta-0.31\right)-\alpha_A}\right) \qquad [6.50]$$

or

$$\left(\frac{MMF}{MMP.X}\right)_A \approx \frac{1.96}{X_A}\left(\frac{1-kEXP\left(-X_A\right)}{kEXP\left(-X_A\right)-\left(\beta+0.31\right)}\right) \qquad [6.51]$$

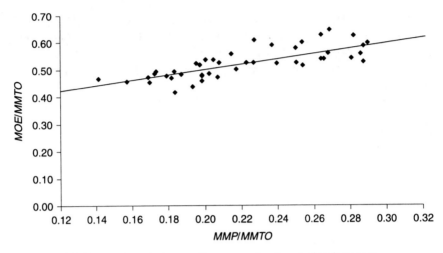

6.5 Comparison between the approximate relation between
MOE/MMTO and *MMP/MMTO* and data for Boeing and Airbus aircraft.

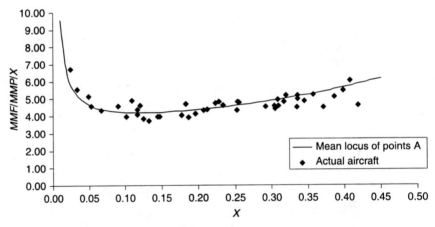

6.6 Comparison between the mean locus of points A (*MTO = MMTO*
and *MP = MMP*) and the data for Boeing and Airbus aircraft.

This last equation describes the approximate locus of points A for current aircraft, i.e. the 'state of the art' for aircraft performance. This locus is shown in Fig. 6.6, together with data for current Airbus and Boeing aircraft. These data have been derived from information contained in published payload-range diagrams. It can be seen that the aircraft follow the general trend of the mean line, but that the value for any individual aircraft may differ from the estimate by as much as 20%.

The payload-range diagrammes can also be used to estimate the *ETRW* at point A, and a set of values for aircraft currently in service is given in Fig. 6.7. It is immediately apparent that, in this form, the data exhibit considerable scatter and

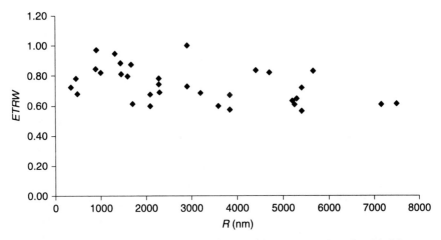

6.7 The variation of *ETRW* with range for current aircraft with *LF* = 1 and *MTO*/*MMTO* = 1.

there is no clear minimum value for *ETRW*. The reason for this loss of coherence is that, whilst the masses *MOE, MMP* and *MMTO* are reasonably well correlated over a wide range of different aircraft, $\eta_o L/D$ can exhibit large variations from aircraft to aircraft for a number of reasons. Firstly, as already noted, aircraft are not usually optimised for minimum *ETRW*. Therefore, different specifications result in different values of $\eta_o L/D$. Secondly, $\eta_o L/D$ depends upon the design cruise Mach number and this varies from aircraft to aircraft, and, thirdly, $\eta_o L/D$ has been the subject of continuous improvement over time due to advances in propulsion efficiency and high-speed aerofoil and wing design. Therefore, different versions of the same aircraft may have very different values of $\eta_o L/D$.

Nevertheless, it is safe to conclude from Fig. 6.7 that the *ETRW* for current aircraft is roughly constant, with the average value being about 0.75 (±0.15), whilst the best aircraft achieve values of around 0.6. Therefore, from Fig. 6.6,

$$ETRW \approx 0.75 \approx \frac{5}{\eta_o\, L/D} \tag{6.52}$$

giving $\eta_o L/D$ a current fleet average value of approximately 6.5. Also from Fig. 6.6, the *ETRW* for the best aircraft is

$$ETRW \approx \frac{4}{\eta_o\, L/D} \tag{6.53}$$

provided that

$$0.075 \le \left(\frac{g.R}{LCV.\eta_o\, L/D} \right) \le 0.25 \tag{6.54}$$

This aircraft has the following characteristics:

- Operational empty weight: $MOE/MMTO \approx 0.50$,
- Disposable load (including reserve fuel): $M_{disp}/MMTO \approx 0.50$,
- Payload: $MMP/MMTO \approx 0.25$,
- Mission fuel: $MMF/MMTO \approx 0.20$

and it will deliver its best $ETRW$ for a trip, with maximum permitted payload, of between 1500 and 5000 nautical miles (nm). The longer range is equivalent to a flight from London to Los Angeles, Rio de Janeiro, Cape Town, Bangkok or Beijing, whilst the shorter range is equivalent to a flight from London to Reykjavik, Tromso, Moscow, Bucharest, Ankara, Crete, Casablanca or the Azores.

On these flights, the fuel consumption per unit payload per unit distance flown would be about 0.00028 kg/kg/nm or, taking the average mass of a passenger plus baggage to be 95 kg, 0.026 kg/passenger/nm. By way of comparison, this is equivalent to a family car, with four passengers and luggage, i.e. a load factor of unity, achieving 45 statute miles per Imperial gallon.

Having obtained an estimate for the optimum performance of the aircraft currently in service, it is useful to compare this with the global fleet average fuel burn. According to Airbus,[1] on average 0.45 L of kerosene are burned to take 1000 kg of payload 1 km. This translates to 0.00067 kg/kg/nm, which is equivalent to an $ETRW$ of 1.6.

Therefore, the global fleet average $ETRW$ is over 2.5 times the optimum value for the best aircraft and over twice the average value for all aircraft.

The obvious questions that follow this observation are 'where does it all go?' and 'can we reduce this ratio by better use of the fleet?'

6.8   Where does it all go? Explaining the discrepancy between energy liberated and revenue work

It has already been shown that $ETRW$ is very sensitive to the load factor. The publicly available data[6,8] indicate that the overall load factor, LF, is only about 60%. From Theorem 2, we can see that, with an LF of 60%, the $ETRW$ is 50% greater than would be the case for a load factor of 100%. Therefore, half of the discrepancy is attributable directly to low average load factors.

The majority of the flights taking place inside Europe and inside the US are less that 1500 nm and therefore are defined as short. Short flights are intrinsically less efficient than long flights and the $ETRW$ is more sensitive to extra fuel burned due to air traffic requirements, though slightly less sensitive to variations in tankered fuel carried. Referring to Fig. 6.1 and 6.6, the average $ETRW$ for short flights could be 15–20% higher than the average $ETRW$ for all flights. Therefore, operations probably account for 70–80% of the discrepancy.

The air traffic management (ATM) system is in place to make sure that operations are conducted safely. However, if an aircraft is required to fly at an altitude that is not optimum from the point of view of fuel burn for any part of the journey, or if the distance actually flown is larger than the great circle distance between the departure point and the destination, the fuel consumed will be greater than the minimum, i.e. ε will be increased and, consequently, *ETRW* will be increased.

Whilst the amount of fuel wasted in non-optimum climb profiles has not yet been quantified, the extra distance flown due to non-optimum routing is known. According to Reynolds,[9] on average an extra 9 nm is flown in the departure local area, an extra 27 nm is flown in the destination local area and an extra 20 nm plus 2.5% of the great circle distance is flown en route. The maximum deviations can be up to 2.5 times these values. For short flights (<1500 nm), the result is an increase in *ETRW* of approximately 10%, whilst for long flights the increase is about 5%. When the effects of non-optimum climb and descent are included, ATM is likely to be responsible for a 10–15% increase in *ETRW*.

The combination of the effects of low *LF* (factor 1.5), large numbers of short flights (factor 1.2) and the wastages in the ATM (factor 1.125) fully account for the discrepancy between the average aircraft and the global fleet average *ETRW*, since $1.5 \times 1.175 \times 1.125 \approx 2$.

6.9   Improving the discrepancy between energy liberated and revenue work

On the assumption that, in the future, there will be a requirement to reduce *ETRW* to the lowest practical value, the objectives will be to:

- Operate with the largest possible passenger and cargo load,
- Reduce the mass of the aircraft structure for a given payload,
- Increase the propulsive system efficiency, η_o,
- Increase the airframe lift-to-drag (*L/D*) ratio,
- Increase the energy per unit volume of the fuel, and
- Optimise the combinations of the above to give minimum *ETRW*.

In the context of payload, each passenger requires a certain amount of physical space. For economy class, the floor space per passenger is about 0.85 m^2. An analysis of current civil aircraft reveals that the maximum permitted payload mass per unit area is about 140 kg/m^2, i.e. about 118 kg/passenger. For long-haul flights, the standard passenger plus bags has a mass of 95 kg, whilst for short haul the value is closer to 82 kg. Therefore, it is immediately apparent that the maximum number of passengers that can be accommodated in an aircraft is determined, not by the maximum permitted payload mass, but by the available cabin floor area.

If the passenger load factor, *LFP*, is the actual number of passengers divided by the total number of seats and the cargo load factor, *LFC*, is the mass of cargo divided by the maximum available cargo mass, then, for long haul,

$$LF = LFP(0.8) + LFC(1 - 0.8LFP),$$

[6.55]

and, for short haul,

$$LF = LFP(0.7) + LFC(1 - 0.7LFP).$$ [6.56]

Therefore, if no cargo is carried, but every seat on the aircraft is full, the maximum *LF* for long haul is 0.8 and the maximum *LF* for short haul is 0.7. Unfortunately, it is very unlikely, and probably undesirable, that all flights will carry the maximum number of passengers. A practical upper limit for *LFP* is likely to be closer to 0.8. Therefore, the load factors without cargo are probably closer to 0.65 and 0.55. Clearly, the only way to extract more revenue work from the fuel burned during a flight is to add as much cargo to the aircraft as possible. This is particularly important for short flights.

Aircraft structures have to be designed to meet strict codes defined by the Regulator and, given that aircraft design has always focused on weight minimisation, there would appear to be little scope for reducing structure mass through novel structural concepts. Therefore, the most effective approach is to adopt new, lightweight materials.

The recent move towards carbon fibre composite for primary structure, i.e. wing boxes and fuselages, offers the opportunity for considerable weight saving. However, there are a number of challenges to be overcome before this benefit can be realised to its fullest extent. Amongst these are:

- The availability of the raw material at an economically viable cost,
- The achievement of a safe structural design,
- The provision of reliable and affordable through-life maintenance, and
- An environmentally acceptable disposal process for the structure at the end of its working life.

Unfortunately, none of these are trivial issues.

When all the structural design processes have matured, and this may take another decade, the maximum likely impact of carbon fibre composite is a reduction of about 25% in the mass of the wing and the fuselage of a conventional civil aircraft. However, this does not mean that the *ETRW* will decrease by the same percentage.

The impact of reduced component mass can be estimated by modifying the (very) approximate relationship between MOE/MMTO and MMP/MMTO. When a lighter fuselage and wing are introduced the relationship becomes

$$\frac{MOE}{MMTO} \approx 0.27 + 0.84\frac{MMP}{MMTO}$$ [6.57]

The locus of possible 'all composite' aircraft is shown in Fig. 6.8, together with the corresponding result for current 'metal' aircraft. For *X* greater than 0.1, the all composite aircraft have values of *ETRW* that are between 8% and 15% lower than current metal aircraft, with the largest improvement being achieved with the

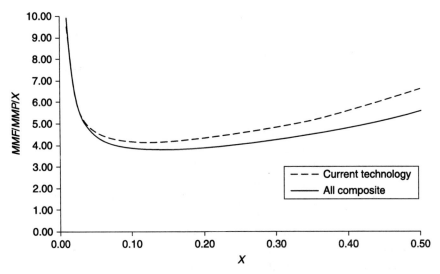

6.8 Comparison of the mean locus of points A (*MTO* = *MMTO* and *MP* = *MMP*) for current aircraft and future 'composite' aircraft.

aircraft that are designed to fly the longest ranges. However, the observation that such a large change in the properties of two major components delivers a much reduced improvement in the *ETRW* is disappointing and it highlights the need always to assess the implications for the total system when a change is made at the component level.

The engine efficiency, η_o, can be subdivided into two parts

$$\eta_o = \eta_C.\eta_P = \frac{\Delta KE}{\left(\dot{m}f.LCV\right)} \cdot \frac{\left(T.V_\infty\right)}{\Delta KE} \qquad [6.58]$$

Here, η_C is the thermodynamic cycle efficiency, which is the ratio of the rate of mechanical energy generated to the rate of thermal energy liberated, and η_P is the propulsive efficiency, which is the ratio of the rate at which work is done on the aircraft to the rate of generation of mechanical energy.

The thermodynamic cycle efficiency can be improved by increasing the overall pressure ratio across the compressor stages or by increasing the gas temperature at the turbine entry station. Unfortunately, both these measures tend to increase the production of NO_X and an increase in the thermal efficiency of the engine also increases the likelihood of forming a contrail in a given set of meteorological conditions.[6] Therefore, an environmentally optimum balance needs to be struck between the emissions of CO_2 and NO_X and the formation of contrails. However, at present, the understanding of the environmental impacts of these three pollutants is not yet sufficiently clear to be able to design aircraft for minimum environmental impact.

However, it is also possible to improve the propulsive efficiency of the engine by replacing the turbo-fan design with either an advanced turbo-prop, open-rotor or unducted fan design. The propulsive efficiency of a high bypass ratio turbo-fan engine cruising at a Mach number of 0.85 is about 75%. However, advanced turbo-props and unducted fans offer propulsive efficiencies of about 85% at the same Mach number. Unfortunately, before this improvement can be realised, there are several severe technical challenges to be overcome. For example, without the bypass duct, there is nothing to contain the fragments in the event of a rotor burst. This is a major safety issue. Also, without the bypass duct, there is nothing to absorb the noise radiating from the rotor and nowhere to put acoustic shielding material. Therefore, noise is also an issue. Solutions to these problems will probably involve increased mass and this will reduce the overall performance benefit. Nevertheless, simple theoretical considerations suggest that the open rotor system has the potential to deliver a one-off improvement in engine efficiency of the order of 10–15%. Since the *ETRW* is directly proportional to engine efficiency, this would also improve by the same amount and there would be no secondary environmental penalties.

For a conventional aircraft with a classical drag polar, the aerodynamic efficiency, *L/D*, is largely governed by the wing aspect ratio, *AR*, and the zero-lift drag coefficient, Cd_0, since

$$\left(\frac{L}{D} \right)_{max} \propto \left(\frac{AR}{Cd_0} \right)^{1/2} \qquad\qquad [6.59]$$

Hence, the *ETRW* can be improved by increasing aspect ratio and decreasing Cd_0. However, the cruise lift coefficient, C_L, also depends upon these parameters, since

$$\left(C_L \right)_{cruise} \propto \left(AR.Cd_0 \right)^{1/2} \qquad\qquad [6.60]$$

Therefore, an increase in the aspect ratio will improve *L/D*, but the required cruise C_L will also increase. This can lead to problems since, for safety reasons, there is an airworthiness requirement for the aircraft to be able to execute a buffet-free, 1.3g manoeuvre in the cruise condition and, for a given wing, this sets an upper limit on C_L. In addition, with no other changes, an increase in cruise C_L will increase the wave drag and this will reduce *L/D*. Therefore, an increase in aspect ratio has to be accompanied either by an increase in the wing sweep angle so that the drag penalty is avoided or by the introduction of a new wing section that has the same, or lower, wave drag at the original sweep angle. Unfortunately, changing the sweep angle has structural implications and usually leads to a heavier wing. Moreover, there is a safety issue related to the pitch-up characteristics when the wing stalls at low speed. This condition is such that, for a given sweep angle, there is a maximum aspect ratio below which the stalling characteristics are deemed to be acceptable. This aspect ratio versus sweep boundary limits the use of high-aspect ratio wings (*AR* > 10) to situations in which the wing sweep is less than 20° and this poses challenges for aircraft with cruise Mach numbers in excess of 0.8. Therefore, the research challenge is to design high-aspect ratio wings that have acceptable low-speed stalling and high-speed manoeuvre characteristics at sweep angles that are sufficiently large to eliminate cruise wave drag penalties.

A decrease in the zero-lift drag coefficient, Cd_0, is clearly beneficial, since L/D is increased and the cruise C_L is reduced. One obvious target is to reduce, or even eliminate, the interference drag between the various components, e.g. the engine and the wing. However, thanks to the increasing sophistication of computer-based design tools, modern aircraft have very low interference drag and there is probably not much more performance improvement to be extracted from the current configuration. At an operational level, more precise rigging of the aircraft to avoid misalignment of surfaces and, in an extreme case, polishing the outer surface could lead to measurable reductions in Cd_0. Beyond these simple measures, the technology that has the potential to deliver large performance improvements is laminar flow control. This has been studied continuously since the end of the Second World War. It is well understood and it has been demonstrated repeatedly in full-scale flight. However, this is a high risk technology with significant development costs and it will be difficult, and expensive, to integrate into routine airline operations. Finally, we should note that one way of changing the value of Cd_0 dramatically is to move from the conventional configuration to the flying wing, or blended wing body, layout. This reduces Cd_0 by a factor of two. However, the apparent advantage is offset by a great many complex design considerations, many of which lead to a much reduced aspect ratio, and this lowers the potential benefit dramatically. Nevertheless, significant net performance improvements are expected and so a change of configuration must remain a possibility in the medium to long term, i.e. beyond 2020.

Recently, there has been considerable interest in the possibility of using fuels other than kerosene for aviation.[12] From an environmental point of view, there is much to be said for bio-fuels. The advantage is that, whilst it was growing, the material from which the fuel is made has absorbed carbon from the present-day atmosphere. Therefore, when the fuel is burned the emitted CO_2 effectively returns that carbon to the atmosphere, thus forming a closed cycle in which the total amount of carbon is fixed. This contrasts with the use of fossil fuel which, when burned, releases CO_2 that has been trapped under ground for millions of years, thereby increasing the total amount of CO_2 in the present-day atmosphere. Unfortunately, in the case of aviation, all of the easily obtainable bio-fuels, e.g. the alcohols and the bio-esters, are deemed to be either unsatisfactory or unsafe, for a variety of reasons, e.g. the basic energy density is too low, the freezing point is too high or they may be contaminated with traces of metallic elements that can seriously damage very hot engine components. However, it may be possible to use blends of oil-derived kerosene and small percentages of bio-ester fuel.

The most promising approach is to make synthetic kerosene from non-food bio-mass using the Fischer-Tropsch process. This would be a so-called 'drop-in' fuel that could be used in current aircraft engines without the need for any modifications.

Synthetic kerosene also has the advantage that it produces less primary and secondary PM. Unfortunately, the production of synthetic kerosene requires a substantial chemical engineering plant that consumes a considerable amount of energy, and this naturally casts doubt upon the economic feasibility of the process. Nevertheless, it has been suggested that, with an aggressive research and development programme, 30% of jet fuel requirements could be derived from biological sources by 2030.

Hydrogen has frequently been proposed as an alternative to kerosene. It has some operational advantages over kerosene, the most notable being the complete absence of carbon. However, a hydrogen-powered engine will still produce NO_X and, because the EI_{H2O}/LCV for hydrogen is 2.6 times the value for kerosene, contrails are more likely over a wider range of meteorological conditions.[6] Moreover, there are two enormous disadvantages. The first is that the density of liquid hydrogen is very low, approximately 11 times smaller than that of kerosene. This means that new aircraft, with a very substantial increase in the available fuel capacity, would be required. Secondly, hydrogen does not occur freely in nature and so it has to be made. Strictly speaking, hydrogen is not a fuel; rather, it is a medium for storing energy that can be generated in one place, e.g. in a ground-based power station, for use in another place, e.g. in an aircraft. The making of hydrogen in the quantities required for aviation would be a very substantial engineering project in itself and it also has to be stored and delivered under cryogenic conditions, bringing enormous infrastructure costs, handling costs, transportation costs and safety issues. It is for these reasons that hydrogen is unlikely to be used for aircraft propulsion in either the short or the medium term.

In addition to all the possible technical advances, it is important to examine the potential for optimisation as a means of reducing the value of *ETRW*. This is relevant because current designs were not optimised for best *ETRW*. In order to gain an indication of the possible improvements, a simple optimisation exercise has been undertaken.

The requirements for the aircraft were:

- 295 passengers in a three-class configuration (440 in single-class configuration) with a maximum permitted payload (*MMP*) of 52 000 kg,
- A maximum range of 3000 nm at maximum zero fuel mass,
- A specified cruise Mach number,
- A twin-engine configuration using a typical, current technology, power unit with a by-pass ratio of about 6; η_0 is a function of M_∞ only and the engine mass is proportional to sea-level static thrust,
- A minimum of 1750 m of runway available, and
- A diversion distance of 200 nm.

The airframe technology level was defined by specifying the following:

- Component mass models based upon the characteristics of current Boeing and Airbus aircraft with the option to vary the masses of the components individually,

- Wing aerodynamics:

    ○ Low-speed characteristics with, and without, high-lift systems deployed,
    ○ Transonic drag rise characteristics modelled,
    ○ Low-speed, pitch-up boundary defined in terms of aspect ratio versus wing sweep $(AR.Tan(\Lambda) = 5.8)$,
    ○ Transonic maximum manoeuvre C_L without buffet,
    ○ Option to vary boundary layer transition location on the wings to simulate laminar flow control technology,

- Brake characteristics,
- Glide slope angle, and
- Fuel – LCV and density.

The model was run for a range of cruise Mach numbers and the results for fuel burn per unit payload for the 3000 nm trip are given in Fig. 6.9. These indicate that there is a cruise Mach number at which fuel burn per unit payload and, hence, $ETRW$ are minimised. Therefore, we may conclude that there is an $ETRW$ penalty paid for cruising at higher Mach numbers and that, in this case, if the aircraft is designed to cruise at Mach 0.85 the $ETRW$ is about 10% greater than if the aircraft is designed to cruise at Mach 0.65.

If minimum $ETRW$ were the absolute requirement, the best aircraft with the current technology level would have the following characteristics:

- Cruise Mach number 0.65,
- Initial cruise altitude 30 000 ft,
- $MMTO \approx 220\,000$ kg,

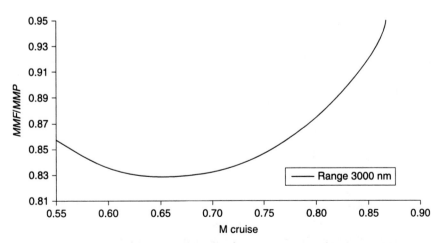

6.9 The variation of *MMF/MMP* with cruise Mach number for an aircraft optimised for minimum *ETRW* with 295 passengers (three-class) and a design range of 3000 nm.

- Wing span $\approx$ 67 m,
- Aspect ratio $\approx$ 14,
- Sweep angle $\approx$ 15°,
- *MMP/MMTO* $\approx$ 0.23,
- *MMF/MMTO* $\approx$ 0.2,
- *MOE/MMTO* $\approx$ 0.53, and
- *ETRW* $\approx$ 0.65.

In this case, the aspect ratio is limited by the buffet-free 1.3g cruise manoeuvre requirement.

Not only does this aircraft have a significantly better *ETRW*, but the initial cruise altitude is only 30 000 ft with maximum payload. Consequently, this aircraft would be less likely to encounter the atmospheric conditions necessary for the formation of contrails.

As an additional point, it was found that by moving boundary layer transition to 50% chord on both top and bottom surfaces of the wing, but with no allowance for increased system mass, the *ETRW* was reduced by a further 20%. This is an indication of the upper limit for the reduction in *ETRW* that could be obtained by the application of laminar flow control technology.

Therefore, by re-optimising for minimum *ETRW*, engineering the full weight reduction benefit from composite materials and introducing the open–rotor propulsion system, the *ETRW* could be reduced by as much as 30% relative to current levels. This could be achieved in a relatively short time scale and would not require the use of any high-risk technologies such as laminar flow control.

6.10   Addressing the climate issue

In the short term, i.e. over the next ten years, the priorities should be to tackle the operational and ATM inefficiencies and to maintain the flow of new technology, particularly weight reduction though improved use of composite material on the Boeing 787 and Airbus A350 programmes and the development of the open-rotor concept.

In the medium term, 2020–2030, the highest priority is to minimise the *ETRW* of the Boeing 737 and Airbus A320 replacement aircraft. There will be a large number of these aircraft in service and they will be operating for over 20 years. They will be the principal source of aviation emissions during the period 2020–2050 and it is vital that they are as environmentally friendly as possible. This will be achieved if the new technologies being developed by The Advisory Council for Aeronautics Research and Innovation in Europe (ACARE) programmes are brought through. Towards 2030, serious consideration must be given to the introduction of new configurations, such as the blended wing body (BWB). Initial studies have suggested that the BWB could offer another 25% improvement in *ETRW* without the introduction of any high-risk technologies.

Finally, looking to the long term, it is clear that the only way to deliver an air transport system that can grow without limits to meet the demands of global wealth creation is to eliminate the dependence on oil-derived kerosene completely. This is possible, but it is not easy. Bio-fuel containing carbon from the present-day atmosphere is one possibility, but we should not forget nuclear power. Today, it is possible to travel across France in a nuclear-powered train. This is because the train is powered by electricity that is generated by a nuclear power station. In the context of aviation, nuclear power could be used to generate hydrogen from sea water, with very little environmental impact. The cryogenic liquid hydrogen would need to be supplied to airports and loaded onto aircraft. This would require new infrastructure. Nevertheless, if it could be made available at about 200 major airports, this would probably be sufficient to address a large part of the problem. The gas turbines on the aircraft would then burn the hydrogen. However, as already noted, special aircraft would need to be designed to provide the high storage volume that hydrogen needs and the enhanced contrail formation issue would need to be eliminated by better weather prediction and an air traffic supersaturated air avoidance system. Alternatively, a small airborne nuclear heat source could be developed for the gas turbine. In this case, the aircraft could take off using hydrogen fuel and then switch to the nuclear-powered gas turbine for the cruise phase. Clearly, the mass of the radiation shielding would be a major issue but, of course, this would be offset by the reduction in the mass of the fuel. This idea was studied extensively in the 1950s when it was taken very seriously. However, the concept raises many challenges and not all of them are technical.

6.11   Conclusions

The aviation industry has the potential to make a significant contribution to the level of carbon dioxide in the atmosphere and to introduce short-term perturbations to other contributors to the Earth's radiation balance. These effects contribute directly to global warming. The consensus scientific view (IPCC) is that increasing global warming is leading to climate change and this is set to become an increasingly important political and economic issue over the next few decades.

Since the cost of kerosene is a large, and growing, element in aircraft direct operating cost, there is a direct link between the economics of aviation and its environmental impact. It has been argued that this link is characterised by *ETRW*, the thermal energy liberated per unit revenue work delivered for an individual flight.

By using the Bréguet range equation, nine theorems relating to aspects of an aircraft's *ETRW* have been derived. These theorems are valid for all flying machines. They allow the aircraft parameters that govern *ETRW* to be identified, the *ETRW* values to be calculated and the sensitivity of *ETRW* to changes in these parameters to be determined. In particular, it is demonstrated that *ETRW* becomes very large when the distances flown become small.

The *ETRW* for aircraft currently in service have been estimated and the best aircraft from an environmental impact point of view has been identified. This has half its maximum take-off mass available for payload and fuel and a maximum permissible payload mass equal to 25% of *MMTO*. It operates at close to minimum *ETRW* over ranges of 1500–5000 nautical miles.

It has been shown that the global fleet average fuel burn is currently more than twice that of an average aircraft operating at maximum payload and maximum take-off mass and more than 2.5 times that of the best aircraft. This difference is attributed to the large number of flights over short routes in Europe and North America (factor ≈ 1.2), low load factors (factor ≈ 1.5) and inefficient ATM (factor ≈ 1.1).

Ways in which the environmental impact can be improved have been examined. It has been argued that, since passenger load factors are already relatively high, the most effective way to improve the *ETRW* is to carry more belly cargo. This is particularly important on short flights. Improvements in technology have also been considered. Structure weight reduction through the extensive use of carbon fibre composite could eventually deliver an 8–15% reduction in *ETRW*, whilst the use of the open-rotor propulsor could provide a 10–15% reduction on appropriate aircraft and there are some small improvements available made through better aerodynamic design. In addition, it has been shown that *ETRW* is dependent upon the design cruise Mach number, with a minimum occurring at about 0.65–0.7. The *ETRW* at this condition is about 10% less than that for an aircraft with a design cruise Mach number of 0.85 and it would cruise at altitudes of about 30 000 ft, i.e. below those associated with contrail formation.

If a new single-aisle aircraft took advantage of all these improvements, the *ETRW* would be 35% lower than the aircraft that it would replace. With better load factors and more efficient ATM, this number could be closer to 50%. Similar levels of improvement are possible for other types of aircraft.

Looking beyond 2025, further reduction in *ETRW* will require either the introduction of high-risk technologies or new configurations like the blended wing body. However, whilst aviation is dependent upon oil-derived kerosene, it will always pose a threat to the environment. Sooner or later, this dependency must be broken, either by the use of bio-fuels or through a radical change in propulsion technology. The inescapable requirement to transform aviation into a safe, clean and highly efficient transportation system growing, without constraint, to supply the needs of a sustainable global economy is the greatest challenge yet faced by the aerospace and aviation communities.

6.12   Acknowledgements

This chapter is based upon work conducted in the context of the Omega programme, which was supported by the Higher Education Funding Council Innovation Fund (HEIF 3). The author would like to acknowledge the many people who have given him their opinions, lent their expertise and spared valuable time. Special mention must go to Professors David Lee, Keith Shine, Kevin Garry, Mark Savill and Joe Morris, to Drs Andreas Schaffer, Howard Smith, Robert Jones, John Green and Raj Nangia, to Jeff Jupp and Les Hyde formerly of Airbus UK, to Keith Mans and the members of the Greener by Design Executive Committee and to Dodge Bailey, Chief Test Pilot NFLC.

6.13   References

1. Airbus. *Airbus Global Market Forecast 2007–2026*. Available at http://www.airbus.com/fileadmin/documents/gmf/PDF_dl/00-all-gmf_2007.pdf.
2. The Boeing Company. *Current Market Outlook 2008–2027*. Available at http://www.boeing.com/commercial/cmo/pdf/Boeing_Current_Market_Outlook_2008_to_2027.pdf.
3. IPCC. *Climate Change 2007, Synthesis Report*. Available at http://www.ipcc.ch/pdf/assessment-report/ar4/syr/ar4_syr.pdf.
4. Schumann, U. (1996). 'On conditions for contrail formation from aircraft'. *Meteorol. Zeitschrift, N.F.* **5**, February, 4–23.
5. Roberson, W. (2007). 'Fuel conservation strategies: cost index explained'. *Boeing Commercial Aeromagazine*, Quarter 2.
6. Hileman, J.I., Katz, J.B., Mantilla, J.G. and Fleming, G. (2008). 'Payload fuel energy efficiency as a metric for aviation environmental performance'. *26th International Congress of the Aeronautical Sciences*, September 2008, Anchorage, Alaska, USA.
7. Cavcar, M. (2006). 'Bréguet Range Equation?', *Journal of Aircraft*, **43** (5): 1542–1544.
8. US Department of Transportation (2008). *Air Carrier Summary Data, Form 41 Schedule T-2 for 1991–2008*. Department for Transportation, Washington DC, USA.
9. Reynolds, T.G. (2008). 'Analysis of lateral flight inefficiency in global air traffic management'. *8th AIAA Aviation Technology, Integration and Operations Conference*, September 2008, Anchorage, Alaska, USA.
10. Hileman, J.I., Wong, H.M., Ortiz, D., Brown, N., Maurice, L. *et al.* (2008). 'The feasibility and potential environmental benefits of alternative fuels for commercial aviation', September 2008. *26th International Congress of the Aeronautical Sciences*, Anchorage, Alaska, USA.

7

The human factors that relate to technological developments in aviation

D. HARRIS, HFI Solutions Ltd, UK

Abstract: The discipline of human factors has its roots in the aerospace industry. This chapter provides a brief overview of the development of human factors, from its birth during World War II to the present day. Flight deck design, pilot selection and training, and crew resource management (CRM) are all considered. While human factors has been primarily associated with aviation safety during the first 50 years of its life, it is argued that it is now time for the discipline to take a more pro-active role in improving the performance of airlines and reducing operating costs. This can only be achieved by taking a more integrated approach to the application of the various aspects of the discipline.

Key words: flight deck design, automation, training, crew resource management (CRM), human factors integration (HFI).

7.1 Introduction to human factors as a discipline

Human factors, as a whole, is a relatively new discipline, arguably with its nascency in the 1940s in the aviation domain. However, it is also a somewhat fragmented discipline, drawing upon basic science from psychology, sociology, physiology/medicine, engineering and management science, to name but a few.

From an overall system perspective, three generic, antagonistic parameters can be applied to evaluate system functioning: safety, performance and cost. Airworthiness authorities are concerned solely with safety aspects of aircraft design, pilot training and airline operations. However, airlines are required to balance the requirement for safety against both cost and performance considerations, but, as will be argued, the human factors discipline has until recently concentrated almost exclusively on the safety aspects of the system function troika.

7.2 Human factors in a socio-technical system context

The organisation and operation of aircraft in an airline is a socio-technical system. This can be described using the five 'M's model (Harris and Harris, 2004; see Fig. 7.1). The operation of a commercial airliner is not just about the union of pilot (huMan) and aircraft (Machine) to perform a flight (or Mission) within the constraints imposed by the context of the physical environment (Medium). The Mission of a commercial aircraft is simply to deliver passengers and cargo at the greatest possible speed (and in comfort, as required) while maintaining the

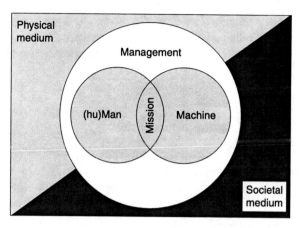

7.1 Five 'M's conceptual model of socio-technical systems. (Source: Harris and Harris, 2004.)

highest possible standards of safety and achieving all this in a cost-effective manner. However, the societal aspects of the environment (another component of the Medium) and the role of Management (both internal and external to the aircraft) are also central to the control of safety, performance and cost.

The (hu)Man aspect of the five 'M's model encompasses issues such as the basic skills and abilities of the pilot in addition to their size, strength and fuel requirements (all elements falling within the 'traditional' realms of psychology and ergonomics). From a user-centred design perspective, the (hu)Man is the ultimate design forcing function. It cannot be changed. When (hu)Man and Machine elements come together they perform a Mission. It is usually the Machine and Mission components on which developers and designers fixate.

However, aircraft designers and engineers must not only work within the bounds of the technology available, the abilities of the pilots and the physical aspects of the Medium; they must also abide by the rules and norms of society (part of the societal Medium). For example, the minimum safety performance standards for human–machine systems such as an aircraft are actually determined by societal norms (regulations), e.g. the level of redundancy required (aircraft certification) or minimum standards of user competence (flight crew licensing). Aviation is an international business and as such culture (another part of the societal Medium) also has a profound, yet largely unseen, effect on operations. Management must also work within these rules. The airline Management is the link between the (hu)Man, Machine, Mission and Medium. It performs the integrating role to ensure compliance with operating, crew licensing, maintenance and aircraft certification requirements to promote safe and efficient operations.

Regulatory objectives are specifically aimed at enhancing safety. Organisational aims, however, need to balance safety against performance, comfort and economy.

Until relatively recently most human factors research has been biased toward satisfying the regulatory/safety component of the performance troika. However, with new technology and operating concepts, human factors may also be able to make a positive contribution to the other two aspects.

7.3   A history of human factors

7.3.1   Early developments: 1940s–1960s

The roots of human factors in the aviation domain lie within the work undertaken in the UK and North America during and shortly after World War II. From the mid-1940s to the mid-1960s the discipline was essentially building its applied science base, drawing heavily from experimental and social psychology, and aerospace medicine. Early work completed after World War II in the USA by pioneers such as Alphonse Chapanis, Paul Fitts, Paul Jones and Walter Grether identified deficiencies in the design and layout of cockpit instrumentation that led to accidents, and produced recommendations for improvements.

In a book looking back at the nascency of aviation human factors, Chapanis (1999) reviewed his work at the Aero Medical Laboratory in the early 1940s where, amongst other things, he was asked to investigate why pilots sometimes retracted the gear instead of the flaps after landing in certain types of aircraft (specifically, the Boeing B-17 Flying Fortress, the North American B-25 Mitchell and the Republic P-47 Thunderbolt). He observed that that the toggle switches and the actuation levers for the gear and the flaps in these aircraft were both identical in shape and were located immediately next to each other. He also noted that the corresponding controls on the Douglas C-47 Skytrain (or Dakota) were not adjacent and their methods of actuation were quite different, and that pilots of this aircraft never retracted the gear inadvertently after landing. Chapanis's insight into human performance, especially in stressed or fatigued pilots, enabled him to understand how they might have confused the two controls. He proposed coding solutions to the problem: separate the controls (spatial coding) and/or shape the controls to represent their corresponding part (shape coding). Hence, the flap lever was shaped to resemble a flap and the landing gear lever resembled a tyre. The pilot could therefore ascertain, either by looking at it or by touching it, what function the lever controlled.

Losses involving US Army Air Corps pilots during World War II were approximately equally distributed: about one-third were lost in training crashes, a further one-third in operational accidents, and the remaining third were lost in combat (Office of Statistical Control, 1945). This suggested various deficiencies inherent in the system. In some further work investigating human-centric design deficiencies in early military cockpits, Grether (1949) described the difficulties of reading the early three-pointer altimeter. Previous work and numerous fatal and non-fatal accidents had shown that pilots frequently misread this instrument. Grether investigated the effects of different designs of altimeter on interpretation time and error rate. He asked pilots to use six different variations of the altimeter containing combinations of three, two and one pointers (both with and without inset counter displaying the altitude of the aircraft) as well as three types of vertically moving scale (similar to a digital display). The results showed that there were marked differences in the error rates for the different designs of the altimeters. They showed that traditional three-needle altimeters took slightly over seven seconds to interpret and produced over 11% errors of 1000 ft or more. The vertical, moving-scale altimeters took less than two seconds to interpret and produced fewer than 1% errors of similar magnitude. Eventually, the electro-mechanical counter-pointer altimeter replaced the three-pointer altimeter. This is generally regarded as being an excellent design. It is fast to read and produces very low error rates. The sweep of the 100s-of-feet hand provides the pilot with good rate of change information; the counter with good state information.

Both of these early examples, one of control design and one of display design, suggest that it is not 'pilot error' that causes accidents; rather it is 'designer error' (i.e. confusing system controls or the poor presentation of information). Nowadays, the notion of blaming the last person in the accident chain (the pilot) has lost credibility. The modern perspective is to take a systemic view of error (and also of cost and performance) by understanding the relationships between all the components in the system, both human and technical. However, the work by Chapanis and Grether (amongst many) was quite radical in this respect in the 1940s.

At the same time in the UK, related research was also being undertaken in the MRC Applied Psychology Unit in Cambridge, on such issues as the direction of motion relationships between controls and displays (Craik and Vince, 1945). Mackworth was developing his famous (if you are a psychology student) Mackworth clock to investigate the effects of prolonged vigilance in radar operators (Mackworth, 1944, 1948). During the latter half of the war, with the introduction of more complex, longer-range, higher-performance aircraft that placed greater demands on their pilots, fatigue became a big issue in RAF aircrew. Losses as a result of fatigue rather than combat were mounting. Perhaps the most famous piece of apparatus developed by Kenneth Craik was the Fatigue Apparatus, which universally became known as the Cambridge Cockpit (Craik, 1940). It was based upon the cockpit of a disused Supermarine Spitfire. As early as 1940, experiments using the Cambridge Cockpit established that the nature of control inputs from a fatigued pilot were considerably different from those from a fresh pilot.

The tired operator tends to make more movements than the fresh one, and he does not grade them appropriately . . . he does not make the movements at the right time. He is unable to react any longer in a smooth and economical way to a complex pattern of signals, such as those provided by the blind-flying instruments. Instead his attention becomes fixed on one of the separate instruments. . . . He becomes increasingly distractible.

Drew, G.C. (1940).

An excellent summary of this early work in aviation human factors can be found in Roscoe (1997).

7.3.2   The late twentieth century and beyond

With increasing knowledge and specialisation, the discipline of human factors began to fragment, with sub-disciplines in human-centred design, training and simulation, selection, management aspects (organisational behaviour), health and safety, and so on. Nevertheless, from the mid-1960s human factors began to make increasingly large contributions, particularly in the three areas of selection, training, and the design of flight decks.

7.3.3   Selection

The processes and methods for the selection of flight crew particularly began to develop in their degree of sophistication throughout the 1970s. This coincided with a change in the nature of work of the airline pilot. The job has changed somewhat over the past half century from that of being a 'hands on throttle and stick' flyer to one of being a flight deck manager. This change has been particularly pronounced in the last half of this period.

In the 1950s and 1960s airlines tended to rely quite heavily on the military for producing trained pilots. Even until relatively recently it was reported that, in the USA, 75% of new-hire airline pilots were recruited after commencing their flying career in military aviation (Hansen and Oster, 1997). Selection techniques tended to rely on techniques that assume candidates are already trained and competent (e.g. reference checks, background checks, interviews, often combined with a flight check – see Suirez *et al.*, 1994). Military selection procedures tend to place a great deal of emphasis on spatial and psychomotor skills in the evaluation of aircrew candidates (e.g. Hunter and Burke, 1994); however, military aviation is very different from commercial aviation, requiring a great deal more 'hands on throttle and stick' flying.

In Europe, where there has traditionally been less emphasis on recruiting pilots from a military background (a trend that has been developing as the demand for commercial pilots has increased), there has been greater emphasis on selection processes for *ab initio* trainees, where emphasis is placed upon an assessment of the candidate's potential to become a successful pilot. After pre-screening, a typical assessment centre for *ab initio* pilot candidates will involve personality assessments, tests for verbal and numerical reasoning, tests of psychomotor skills, group discussions and structured interviews. For example, Cathay Pacific used selection criteria in six main areas: technical skill and aptitude; judgement and problem solving; communications; social relationships, personality and compatibility with Cathay Pacific; leadership/subordinate style and motivation and ambition (Bartram and Baxter, 1996). The main point to note here is that flying skills *per se* formed only a relatively small component. Hörmann and Maschke (1996) analysed personality data from *circa* 300 pilots in addition to data from a simulator checkride and other biographical data (e.g. age, flight experience and command experience). Three years later it was found that pilots graded as being below-standard pilots had significantly lower scores on the interpersonal scales and higher scores on the emotional scales of the Temperament Structure Scales (Maschke, 1987), a psychometric instrument developed specifically for assessing aircrew personality. An earlier review by Chidester *et al.* (1991) also found that better-performing airline pilots scored higher on traits such as 'mastery' and 'expressivity', and lower on 'hostility and aggression'. Non-personality-related selection tests (e.g. verbal and numerical reasoning tests) have also shown a strong relationship with performance, which probably reflects the job of the airline pilot, where flight administration and planning are as important as the psychomotor skills required for flying.

Appropriate and effective selection procedures (especially for *ab initio* trainees), although expensive, can ultimately save airlines a great deal of money. They help to ensure that the pilots selected are likely to complete their training (dropouts from training are very expensive) as well as making sure that capable, safe pilots are recruited to the airline. In the selection of personnel, even relatively modest increases in the success of the selection process, especially since training costs are high, can pay back the initial investment ten-fold.

7.3.4 Training

Until relatively recently, pilot training and licensing concentrated on flight and technical skills (manoeuvring the aircraft, navigation, system management and fault diagnosis, etc.). Training was largely undertaken in the aircraft or in relatively low-fidelity (by today's standards) flight simulators. A great deal of the emphasis was placed upon technical training (e.g. how to handle the aircraft's systems or how to fly the aircraft manually) and training for emergencies resulting from a technical failure (e.g. engine failure at V_1 or performing flapless approaches). However, with increasing technical reliability, it became evident that the major cause of air accidents was human error. The failure of the flight deck crew to act in a well-coordinated manner further contributed to this end on many occasions. This resulted in a series of intra-cockpit flight crew management programmes being instigated.

The human factors discipline really began to come to prominence with the cockpit – later – crew resource management (CRM) revolution, which introduced applied social psychology and management science onto the flight deck. This placed emphasis on training to facilitate the flight deck (later whole aircraft) crew acting as a coordinated team. The Joint Airworthiness Authorities (JAA) (1998) defined CRM as 'the effective utilisation of all resources (e.g. crewmembers, aeroplane systems and supporting facilities) to achieve safe and efficient operation'. CRM evolved as a result of a series of accidents involving perfectly serviceable aircraft. The main cause of these accidents was a failure to utilise the human resources available on the flight deck in the best way possible. Many would argue that the initial stimulus from the CRM revolution was the accident involving a Lockheed L1011 (Tristar) in the Florida Everglades in 1972 (National Transportation Safety Board, 1973). At the time of the accident the aircraft had a minor technical failure (a blown light bulb on the gear status lights) but actually crashed because nobody was flying it! The crew were 'head down' on the flight deck trying to fix the problem. Other accidents also highlighted instances of the captain trying to do all the work while the other flight crew were almost completely unoccupied or happened as a result of a lack of crew cooperation and coordination.

Pariés and Amalberti (1995) suggested that CRM has progressed through four eras. First-generation CRM focussed on improving management style and interpersonal skills. Emphasis was placed upon improving communication, attitudes and leadership to enhance teamwork, although, in many airlines, only captains underwent CRM training! Second-generation CRM also included topics such as stress management, human error, decision making and group dynamics. However, CRM also began to extend beyond the flight deck door: cabin crew became part of the team, and training was extended to include whole crews together, rather than training flight deck and aircraft cabin separately. The CRM concept also began to extend into the airline organisation as a whole. Indeed, the evolution of the abbreviation of CRM itself has exemplified the change in culture in commercial aviation in the last quarter of a century. Further evolution of this approach has seen CRM extend beyond the aircraft to ramp operations and maintenance, and even beyond the airline (e.g. to air traffic control (ATC)). By the fourth generation, CRM training *per se* was beginning to disappear as the concepts were being absorbed into all aspects of flight training and the development of procedures. Helmreich *et al.* (1999) suggested that fifth-generation CRM will extend throughout the organisation and will basically involve a culture change. Early CRM approaches were predicated upon avoiding error on the flight deck. Fifth-generation approaches will assume that, whenever human beings are involved, error will be pervasive and the emphasis will change toward the error management troika: avoid error, trap errors and/or mitigate the consequences of errors.

The advent of CRM training was partly contingent upon a change in training philosophy towards line-oriented flight training (LOFT). Indeed, there is now a mandated requirement for crew training as part of the Airline Transport Pilot Licence (ATPL) syllabus. LOFT places emphasis on training as a crew and acting as a crew member. During a LOFT session (which will take place in a full-flight simulator) crews fly a complete trip (or part thereof) just as they would in normal operations. However, they will be presented with a series of in-flight problems and emergencies that require them to act as a team. During these simulated flights the instructors do not intervene, but crews' actions are recorded for later analysis. After the LOFT session, the crews' performance is reviewed with respect to how they handled flying the aircraft, the technical aspects of the problem and, most importantly, how the crew was employed to address the issue (see Foushee and Helmreich, 1988).

The effectiveness of the LOFT approach depends upon two key factors: the development and implementation of the flight scenarios within which the training takes place, and the adequacy with which crews are de-briefed by the instructors after the exercise. The role of the facilitator during the de-briefing after the LOFT exercise is central to its effectiveness. However, Dismukes *et al.* (2000) observed great variations in facilitator performance. It was noted that the mean debriefing duration of a post-LOFT simulator ride (which typically lasted approximately two hours) was only 31 minutes. One-third of this time was often spent reviewing incidents on the video and, in the remaining time, the facilitator overseeing the session often spent more time talking than did the crew being trained.

Despite the change in the training philosophy in later years, LOFT scenarios were based largely around handling technical failures. As will be seen in the section looking at the present day, this situation is beginning to change.

7.3.5   Flight deck design

Reising *et al.* (1998) have suggested that flight deck displays have progressed through three eras: the mechanical era, the electro-mechanical era and, most recently, the electro-optical (E-O) era. With the advent of the 'glass cockpit' revolution (when E-O display technology began to replace the electro-mechanical flight instrumentation) opportunities were presented for new formats for the display of information, and human factors started to play an increasingly important part in their design. However, while the new display technology represented a visible indication of progress in the third generation of airliners (e.g. Airbus A300/310; Boeing 757/767 and McDonnell-Douglas MD-80 series), the true revolution in the way in which these aircraft were being operated lay in the less visible aspects of the flight deck, specifically the increased level of automation available as a result of the advent of the flight management system (FMS) or flight management computer (FMC); see Billings (1997). The glass cockpit display systems were merely the phenotype: the digital computing systems that were being introduced represented a true change in the genotype of the commercial aircraft. The higher levels of automation on the flight deck not only allowed opportunities for reducing the number of crew on the flight deck (in many instances) from three to two (eradicating the function of the flight engineer) but also required a change in the skill set required by flight crew. Aircraft were now 'managed', not 'flown'. This trend continued further in later designs with even higher levels of automation and integration, such as the Airbus A320, Boeing 747-400 and McDonnell Douglas MD-11. For a historical perspective on flight deck design, Coombs (1990) provides an excellent overview, providing interesting insights behind some of the design solutions found in modern aircraft. Harris (2004) provides a more comprehensive, technical examination of all aspects of the human factors of commercial flight deck design.

With these increasing levels of automation on the flight deck in the 1980s and early 1990s much research was undertaken in the areas of workload measurement and prediction, and enhancing the pilot's situational awareness. Autoflight systems certainly reduced the physical workload associated with flying an aircraft. Computerised display systems also reduced the mental workload associated with routine mental computations associated with in-flight navigation (and considerably increased navigational accuracy and reduced the number of errors). However, it is wrong to say that automation reduced workload. It simply changed its nature. Wiener (1989) called the automation in these more modern aircraft 'clumsy' automation. It reduced crew workload where it was already low (e.g. in the cruise) but increased it dramatically where it was already high (e.g. in terminal manoeuvring areas). Pilots' workloads have changed to become almost exclusively mental workloads associated with the management of the aircraft's automation.

The new (at the time) breed of multifunctional displays coupled with high levels of aircraft automation also had the ability both to promote Situation Awareness through new, intuitive display formats and simultaneously to degrade it by hiding more information than ever before. In the scientific literature there are various definitions of situation awareness. For example, Smith and Hancock (1995) defined it as 'the up-to-the minute comprehension of task relevant information that enables appropriate decision making under stress'. Boy and Ferro (2004) suggested that the concept was a function of several quasi-independent situation types: the *available situation* on the flight deck, available from data originating from either the aircraft, the environment (including ATC) or other crew members; the *perceived situation* by the crew, which may be affected by various parameters such as workload, performance, noise and interruptions; the *expected situation* derived by the crew from their planning and decision-making processes, or the *inferred situation* by the crew, compiled from incomplete and/or uncertain data.

However, automation can make a system 'opaque' and hence degrade situation awareness. Dekker and Hollnagel (1999) described automated flight decks of the time as being rich in data but deficient in information; hence they did little to enhance situation awareness. Christoffersen and Woods (2000) developed guidelines to suggest ways to optimise human–automation interaction and turn it into a 'good team player'. Good team players, they opined, make their activities observable for their fellows and are easy to direct. As a result they suggested that system information should be event-based (not state-based displays, conveying the status of the machine and its mission goals). The displays must be future-oriented, allowing operators to enhance their situation awareness by being able to project ahead, and the display formats should also be form-based, enhancing the pilots' ability to detect patterns rather than having to engage in arduous calculations, integrations and extrapolations of disparate pieces of data.

While it is undoubted that the advanced-technology aircraft being introduced were safer and had a much lower accident rate (see Boeing Commercial Airplanes, 2009), they introduced a new type of accident (or perhaps they merely exaggerated the incidence of an already existing underlying problem). Dekker (2001) describes these as 'going sour' accidents, which involved these newer-generation, highly automated airliners being 'managed' into an accident. The majority of these accidents exhibited a common underlying set of circumstances: a series of human errors, miscommunications and mis-assessments of the situation. Dekker argued that the accidents occurred as a result of a number of factors pertaining to the flight deck design, the situation and the crews' knowledge/training, which conspired against their ability to coordinate their activities with the aircraft's automation. The characteristics of the automation involved factors such as autonomous actions on the part of the aircraft and limited feedback about its behaviour, making it what, in CRM parlance, could be deemed a bad team player (Sarter and Woods, 1997).

It can be seen that it is almost impossible to separate design from procedures from training if it is desired to optimise the human element in the system. One of the great problems aviation has (which can also be a great strength) is the longevity of designs and working practices. Change is never for the sake of change (it is a very conservative industry, which aids in maintaining safety as much as intensive research and development). However, this always means that, as new problems come to the surface, solutions must be found that can be applied to the many legacy designs operating in the world-wide fleet. For example, Wood and Huddlestone (2007) have proposed a new method of training pilots to use the automation in their airliners; however, this only partially overcomes the fundamental design issues concerning the way that the automation on the flight deck has been implemented. What is actually required is an integrated, systemic approach to human factors.

7.4   Recent developments and current trends

The practice of human factors in the aerospace industry today is largely an incremental development of the position described in the previous section. However, there is now universal acceptance of its importance to safe operations. Nevertheless, emphasis is still firmly on the safety aspect of the system-function troika. Perhaps the two biggest changes in the operation and design of modern aircraft involving human factors are again in the areas of training and design.

7.4.1   Training

The *raison d'être* of LOFT was to encourage effective flight deck management practices through team-based training and de-briefing performed within abnormal and emergency flight scenarios. However, the problems faced in a LOFT session are not the major threats to operational safety faced by crews on a daily basis. Line Operations Safety Audits (LOSA) were introduced into airlines during the late 1990s. LOSA data are collected by trained observers during line operations. These data form the basis of an audit process to check the everyday safety health of the operation. Data are collected in three categories: external threats to safety (e.g. ATC problems, adverse weather or system malfunctions), errors and the crew's responses to the errors committed, and non-technical skills evaluation (essentially CRM processes and techniques). The concept of the LOSA methodology can be found in Helmreich *et al.* (1999). Thomas (2003) compared threats gathered from LOSA data with problems faced by aircrew in LOFT training scenarios. He observed that almost 70% of LOFT scenarios involved an aircraft malfunction; however, this only occurred in 14% of operational threats in the LOSA audit data. The most frequent external safety threat in line operations was weather (almost 21%), but this was incorporated in only 4% of LOFT sessions. Other external threats to safety, such as operational pressures on crews, air traffic and ground handling events, did not occur in any LOFT scenarios at all. Thomas (2003) also observed that crew performance during LOFT sessions was considerably superior to that observed when flying on the line, a classic instance of training appearing effective in the simulator but failing to transfer effectively to the workplace itself.

The approach of developing training needs directly from line operational requirements reflects the latest training philosophy outlined by the Federal Aviation Administration (FAA) in the Advanced Qualification Program (AQP) and also being adopted elsewhere (e.g. by the European Aviation Safety Agency (EASA) 2008). The emphasis in the AQP has moved away from time-based training requirements to fleet-specific, proficiency-based requirements (see AC 120-54; Federal Aviation Administration, 1991). In the AQP process, the airline develops a set of proficiency objectives based upon its requirements for a specific type of operation (e.g. based upon a threat and error management process developed from a LOSA audit). The AQP is based upon a rigorous task analysis of operations but with emphasis placed firmly upon the behavioural aspects of the flight task, such as decision making or the management of the aircraft's automation. The revolutionary aspect of this process for the aviation industry is that many of the regulatory shackles are released when approving the content of the training programme. However, the complexity of the AQP process means that considerable skills and resources have to be applied to gain approval, and Maurino (1999) has suggested that only major airlines with such resources will benefit through its adoption.

It has already been noted that the pilot's task has changed considerably as a result of increasing levels of automation on the flight deck, but regulations always lag behind technological advances. While regulations now require professional pilots to undertake multi-crew cooperation courses, there has been no such corresponding advance in the training requirements for the understanding and management of advanced automation. There is a considerable discrepancy between what pilots are required to know to gain a professional licence and what they need to be able to do to act as a First Officer. One such gap is in the management of flight deck automation (Dekker, 2000). Wood and Huddlestone (2007) observed that this problem was not an issue in managing the automation interface, but was rather an issue in understanding *what* the automation was doing and *how* it was trying to control the aircraft. Without this knowledge it is difficult to 'manage' the automation effectively. Automation should not be looked upon as a separate 'add on' as it is central to the design and operation of modern airliners (Rignér and Dekker, 1999). Any inspection of the avionics architecture of a modern fourth-generation airliner (e.g. Airbus A330/340 or Boeing 777) reveals this to be the case, but this is not reflected in the training of modern commercial pilots.

7.4.2   Flight deck design

In September 2007, EASA implemented a new airworthiness rule (Certification Specification 25.1302) that mandates for the error-tolerant design of flight deck equipment on all new large commercial aircraft. The stimulus for the rule was the FAA Human Factors Team *Report on the Interfaces between Flightcrews and Modern Flight Deck Systems* (1996), which was commissioned as a result of several accidents occurring to new (at the time) technology airliners. This was the first time a rule explicitly addressing human factors issues had been implemented in the airworthiness regulations.

The roots of human error are manifold and have complex interrelationships with all aspects of the operation of a modern airliner, especially training. During the last decade 'design-induced' error has become of particular concern to the airworthiness authorities, particularly in the highly automated third (and now fourth) generations of airliners. While the high level of automation in modern airliners has doubtless contributed to considerable advances in safety, accidents have begun to occur as a direct result of the manner in which it has been instantiated (Woods and Sarter, 1998): for example, the Nagoya Airbus A300-600 (where the pilots could not disengage the go-around mode after inadvertent activation as a result of a combination of lack of understanding of the automation and poor design of the operating logic in the autoland system); the Cali Boeing 757 accident (where the poor interface on the flight management computer and a lack of logic checking resulted in a Controlled Flight Into Terrain (CFIT) accident); and the Air Inter Airbus A320 accident on Mount St Odile, near Strasbourg (where the flight crew inadvertently set an excessive rate of descent on the mode control panel instead of manipulating the flight path angle, as a result of both functions utilising a common control interface and a poor associated display). However, as noted at the beginning of this chapter, many aspects of 'pilot error' are actually 'designer error'.

As a result of such accidents the FAA commissioned an exhaustive study of the pilot–aircraft interfaces in highly automated aircraft (FAA, 1996). This report identified several major shortcomings and deficiencies in flight deck design. There were criticisms of the interfaces, such as pilots' autoflight mode awareness/ indication, energy awareness, confusing and unclear display symbology and nomenclature, and a lack of consistency in flight management systems' interfaces and conventions. The report also heavily criticised the flight deck design process itself, identifying in particular a lack of human factors expertise in design teams and placing too much emphasis on physical ergonomics and insufficient on cognitive ergonomics. Fifty-one recommendations came out of the report, including, 'The FAA should require the evaluation of flight deck designs for susceptibility to design-induced flightcrew errors and the consequences of those errors as part of the type certification process.'

Subsequently, in July 1999, the US Department of Transportation tasked the Aviation Rulemaking Advisory Committee to develop new regulatory standards and/or advisory material to address design-related flight crew performance vulnerabilities and the prevention (including the detection, tolerance and recovery) of error (US Department of Transportation, 1999). Since September 2007 the rules and advisory material developed from this process have been adopted by EASA as CS 25.1302 and AMC (Acceptable Means of Compliance) 25.1302. At the time of writing, the FAA is shortly expected to adopt the same rule in 2012. The rule requires that it must be demonstrated that flight deck equipment can be used by qualified flight-crew members to perform their tasks safely by providing them with the necessary information in a timely and appropriate format and that, if a multi-function interface is used, the information should be accessible in a manner consistent with the urgency, frequency and duration of the flight task in question. Furthermore, the flight deck interfaces must be predictable, unambiguous, and designed to enable the crew to intervene in a manner appropriate to the task if necessary. Finally, the rule requires that as far as is possible the flight deck equipment must enable the flight crew to manage errors resulting from the kinds of interaction that can be reasonably expected (see CS 25.1302 for the full wording of the airworthiness requirement).

However, one of the great challenges of devising a certification rule to address the adequacy of the human factors aspect of the flight deck is that it is essentially attempting to evaluate a hypothetical construct. The pilot/aircraft interface doesn't really exist!

Figure 7.2 (from Harris, 2011) shows why this is the case. On the output/human input side of the interface, 'images' on the aircraft displays should convey

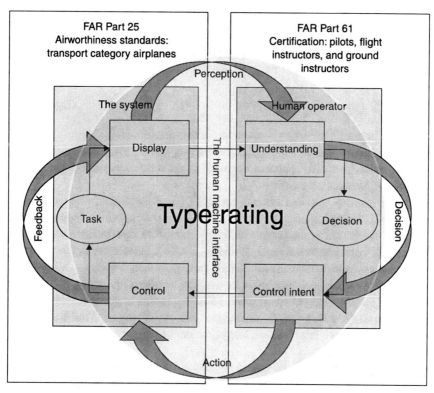

7.2 The concept of the human/machine interface superimposed over a representation of the classical 'perception-decision-action-feedback' control loop. The components where airworthiness regulations from FAR 25 apply are on the left and where FAR 61 applies are on the right.

appropriate information to the pilot. This information should be interpreted correctly (within the context of the situation and the mission goals) and be transformed into knowledge and understanding that allows management of the aircraft. This is a function of part 61 of the Federal Airworthiness Regulations, which addresses pilot training. On the human output/machine input side of the control and monitoring loop, control intent from the pilot needs to be translated into the desired aircraft output. A good control system (be it a flight control system or system management interface) will translate pilot intent into system output in the manner desired and with minimal physical or mental effort. This is a function of part 25 of the Regulations, which addresses aircraft design. A 'high-quality' pilot/aircraft interface consists of a good 'fit' between the skills, knowledge and ability of the user and the controls and displays of the machine. However, neither certification of the aircraft alone nor satisfactorily attaining the requirements for an Air Transport Pilot's Licence (ATPL) can ensure the pilot/aircraft fit. At the moment, the only way that this fit is managed is via the requirement for an aircraft type rating, which attempts to make sure that the generic training obtained as part of the ATPL can be translated into the specific requirements to manage the flight deck interfaces of one particular type of aeroplane.

7.4.3 A systemic view

To a certain degree the fragmented nature of the discipline reflects the fragmented nature of the regulations within which airlines operate. This in turn mitigates the opportunities for a truly systemic approach to the implementation of the Human Science in the aviation industry.

Taking a socio-technical systems approach based upon the 5Ms model, some very basic, high-level representations of the manner in which human factors operates in the airline industry can be proposed that incorporate many of the issues addressed so far (see Fig. 7.3) (Harris, 2007b). These representations are by no means a comprehensive model. They do not address such things as the negative effects of psychological performance shaping factors (e.g. stress, fatigue and workload) or physiological performance shaping factors (e.g. noise, vibration and temperature). However, they do attempt to make explicit the relationships between the various human factors disciplines and how they contribute to the overall function of an airline.

To start off with, the requirements of the flight task (Mission) drive the design of the Machine (and, hence, to a certain degree the design of the flight deck). The role of the flight deck is to support the four basic tasks of the flight crew (aviate, navigate, communicate and manage) and to protect the crew from the physiological stressors imposed by the Medium. These basic functions of the flight deck are pivotal to the design of the training (essentially a process of modifying the huMan). Training is all about teaching someone how to do something, with something. To train someone you need training devices (a full flight simulator approved by the regulatory authorities for training and licensing purposes, equipped with a daylight visual system and six-axis motion platform; cockpit procedures trainers; other part-task trainers; computer-based training facilities for aircraft systems and procedures; and also 'regular' classroom facilities). The training overlay is more than just a curriculum: it also dictates which training devices will best be suited for delivering which aspects of the training course. The nature of the task will also dictate what sort of person is required: what are the basic aptitudes that they need that will make them likely to successfully complete their training and become a safe and efficient pilot? It has already been seen that basic 'stick and rudder' skills are only a very small part of the makeup of the modern pilot. Cognitive and team working (flight deck Management skills) are now also essential. There is an intimate relationship between all these components, all of which help to inform the design of each other. Hopefully, as a result of this process there will be a positive, beneficial effect on the huMan under

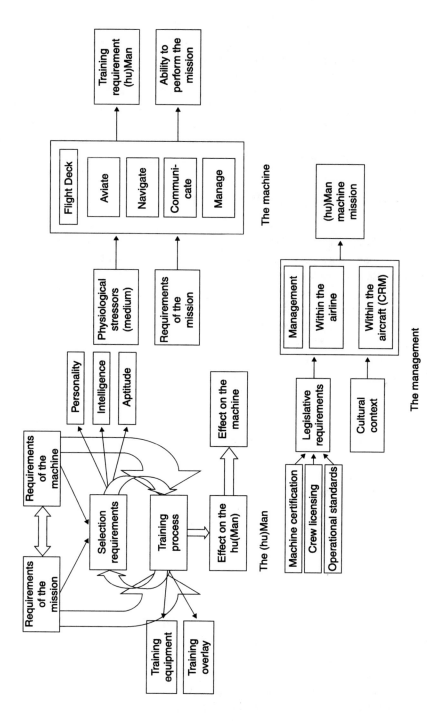

The machine

Training requirement (hu)Man

Ability to perform the mission

Flight Deck
Aviate
Navigate
Communi-cate
Manage

Physiological stressors (medium)

Requirements of the mission

Personality
Intelligence
Aptitude

Effect on the machine

Requirements of the machine

Requirements of the mission

Selection requirements

Training process

Effect on the hu(Man)

Training equipment

Training overlay

The (hu)Man

Management
Within the airline
Within the aircraft (CRM)

(hu)Man machine mission

Legislative requirements

Machine certification
Crew licensing
Operational standards

Cultural context

The management

7.3 A system-wide depiction of the interactions between the various human factors sub-disciplines.

training who, by thrashing around on the stick and throttle and jabbing at various buttons, will have the desired effect on the aircraft (i.e. use it effectively to complete the Mission tasked by the Management).

The role of airline Management, in addition to making money, is to ensure that all operations fall within the legislative requirements (e.g. for flight crew licensing and aircraft maintenance) required by society. This is essentially a safety management role that itself contains lots of human factors issues associated with confidential reporting schemes, safety culture and the analysis of accident and incident data. These issues, though, are well beyond the scope of this short chapter.

7.4.4   Human factors as a hygiene factor

The major emphasis of the work in these formative years of the discipline (particularly in commercial aviation) has been safety related. Human factors has been seen as a discipline necessary to help in avoiding error (and hence accidents). Poor human factors increase the likelihood of error. Human factors is almost regarded as a 'hygiene factor'. From a design perspective, a failure to consider the requirements of the aircraft/pilot interface will result in a product that is difficult to use and promotes error. However, from a manufacturer's perspective, providing a 'good' aircraft/pilot interface does not 'add value' to a product: a failure to provide a user-friendly interface merely detracts from its value. As a result, it is difficult to make a convincing argument to invest heavily in human factors research and for airlines to indirectly pay for such research via increased unit cost. All modern flight deck interfaces are generally very good; hence minor deficiencies become a training or selection issue to be dealt with within the airline rather than being the manufacturer's problem. But training is also a major cost. There is the cost to the airline in terms of the time and expense devoted to training and in the provision of the training equipment (simulators, computer-based training equipment, etc.). There is also the cost as a result of time lost to revenue-earning operations.

The human factors discipline needs to enter a third era, where it can begin to make a far more positive contribution to performance and cost aspects of airline functioning.

7.5   Future trends

Human factors alone cannot improve the operational efficiency of an aeroplane (Harris, 2006). A wider, 'system perspective' is required. Human factors integration (HFI), which is a sub-discipline of systems engineering, began to emerge as a concept during the 1990s. HFI provides an integrative framework for the application of human factors. HFI originally encompassed six domains that were regarded as essential for the optimum integration of the human element into a system (UK Ministry of Defence, 2001). These were:

- Staffing (how many people are required to operate and maintain the system?)
- Personnel (what are the aptitudes, experience and other human characteristics required to operate the system?)
- Training (how can the requisite knowledge, skills and abilities to operate and maintain the system be developed and maintained?)
- Human factors engineering (how can human characteristics be integrated into system design to optimise performance within the human/machine system – essentially human-centred design?)
- Health hazards (what are the short or long-term hazards to health resulting from normal operation of the system?) and
- System safety (how can the safety risks which humans might cause when operating or maintaining the system be identified and eliminated, trapped or managed?)
- Recently a seventh domain has been added, the organisational and social domain, which encompasses issues such as information sharing and interoperability.

Simultaneous with the emergence of the HFI approach has been the development of powerful and robust computing technology, the development of large, cheap, flexible displays and the advent of robust, high-speed data links. These technologies provide the support for truly human-centred design and a revolution on the flight deck (and consequently across airline operations as a whole). Previously, to a large degree human-centred design had been a technology-driven process. However, with these technologies being developed, it is possible to start designing the tools and automation that are really required to support users in their tasks, rather than simply making the interfaces more 'user friendly' and less likely to induce error. Put together, the emergence of these new technologies coupled with a system-wide HFI approach means that human factors need no longer merely be a hygiene factor. It can have positive benefits by enhancing performance and reducing both operational and through-life costs. In short, it can 'add value'.

7.5.1   Evolution of the flight deck

The trend in flight deck design has been one of progressive 'de-crewing'. The common flight deck complement is now just two. Fifty years ago, it was not uncommon for there to be five crew in the cockpit of a civil airliner (two pilots, flight engineer, navigator and radio operator). Now just two crew, with much increased levels of assistance from the aircraft, can accomplish the same tasks once undertaken by five personnel. Many of the functions once performed by flight crew are now wholly (or partially) performed by the aircraft itself. As referred to several times earlier, the emphasis in the role of the pilot has changed from one of being a 'flyer' to one of being a flight deck manager, where the aircraft and its systems are usually under supervisory, rather than manual, control. The pilots are now effectively outer-loop controllers (setters of high-level goals) and monitors of systems, rather than inner-loop ('hands-on', minute-to-minute) controllers. Emphasis is now on crew and automation management rather than flight path control *per se.*

Harris (2007a) has suggested that with the judicious use of a range of technologies developed during the past decade there are no major reasons why a single-pilot-operated aircraft is not feasible. The individual technologies required have now all reached a suitable level of maturity. The military have been operating complex, high-performance single crew aircraft for many years, and uninhabited air vehicles (UAVs) are now a regular part of operations. It is time for these technologies to be spun out into the commercial domain. The greatest obstacle to the operation of a civil single-pilot aircraft is not technological. It is combining the extant technology, designing the user interfaces and developing a new concept of operations to make such an aeroplane. The human factors requirements are the prime driver, *not* the hardware and software technologies. Such an aircraft will offer considerable cost savings. For regional aircraft of up to 50 seats it has been estimated that, over a 200 nautical mile leg, between 15 and 35% of the direct operating costs can be accounted for by crew costs (Alcock, 2004). As a result, considerable savings are possible with a reduction in the number of flight deck crew. Furthermore, with such a completely new design, there are relatively few legacy issues to overcome, thereby giving the designers a much freer hand to explore new concepts of operation.

Many assumptions about the role of the second pilot are also incorrect (or at least the received wisdom is questionable); for example, the second pilot's role as a means of reducing workload and acting as an error-checking mechanism. This is not necessarily true. Firstly, there is a workload cost associated with crew management. It costs effort to work as a team. The requirement to coordinate crew, cooperate and communicate on the flight deck itself has workload associated with it. Doubling the number of crew does not halve the workload on each member. Far from it. Furthermore, poor CRM has been implicated as a causal factor in over 17% of all fatal commercial jet aircraft accidents (Civil Aviation Authority, 1998). From the same data set the effectiveness of the second pilot as an 'error checker' can also be questioned. Omission of action or inappropriate action was implicated in nearly 37% of accidents, and a deliberate non-adherence to procedures was implicated in 12.2%. Becoming 'low and slow' (a failure to cross-monitor the flying pilot) was a factor in 19% of accidents. As a cross-check on the position of the aircraft, the pilot not flying's (PNF) effectiveness would seem to be questionable, as a lack of positional awareness was identified as a causal factor in over 41% of cases. It is accepted that these data cannot show the number of cases where the actions of the second pilot averted an accident. However, LOSA data obtained during routine operations again question the effectiveness of the second pilot. Thomas (2003) reported that 47.2% of errors committed by captains during normal line operations involved intentional non-compliance with standard operating procedures or regulations, and 38.5% were unintentional procedural non-compliance. He also reported that, in observations of line operations, crews did not demonstrate effective error detection, with more than half of all errors remaining undetected by one or both of the flight crew. It can be argued that removing the second pilot actually reduces the scope for accidents occurring as a result of miscommunication or misunderstanding between the pilots and, furthermore, it would not double the workload.

Modern commercial aircraft are already equipped with intelligent electronic checklist systems that effectively perform the cross-checking role of the second pilot. However, it would be tempting simply to adopt an approach whereby the automation in the aircraft simply undertakes the role of the second pilot. This would simply be falling into the 'electric horse' trap. The single-crew commercial aircraft provides an opportunity to fundamentally re-think the role of the pilot and to operate aircraft in a new and better way. Simply automating the functions of the second pilot would not provide a step change in operational effectiveness. New designs and new operating concepts are required. A system-wide (HFI) approach is required. For example, there is no reason why many functions need to be physically undertaken on the flight deck, for example navigation, surveillance and routine system monitoring. Harris (2007a) suggests not *replacing* many of these functions, but *displacing* them to dedicated, specialist teams on the ground, which will not necessarily be comprised of pilots. These teams can simultaneously look after many aircraft, giving economies of scale. Even with the advent of pilot incapacitation, UAV technology has become mature enough to allow the aircraft to be recovered.

Such a change in operational concept will require concomitant changes in training and organisational structures, not to mention a change in the airworthiness requirements. However, all these things are possible.

7.6   Conclusion

Human factors as a discipline has come of age. It must, however, avoid its natural inclination to rush to claim the moral high ground by marking its territory solely within the realm of aviation safety. The discipline must also coalesce once again in order that the maximum benefit from an integrated, through-life approach can be realised. To a large degree, while increasing levels of specialisation have served to develop the science, this has also simultaneously militated against its coherent application in commercial aviation. Nevertheless, the opportunity now exists to capitalise on the developments made by this relatively new discipline, which was originally born in the aviation domain just half a century ago.

7.7   References

Alcock, C. (2004). 'New turboprop push tied to rising fuel costs'. *Aviation International News*, 29 September 2004. Available at http://www.ainonline.com/Publications/era/era_04/era_newturbop18.html. (Accessed 30 January 2007.)

Bartram, D. and Baxter, P. (1996). 'Validation of the Cathay Pacific Airways Pilot Selection Program'. *International Journal of Aviation Psychology*, **6**: 149–69.

Billings, C.E. (1997). *Aviation Automation*. Lawrence Erlbaum Associates, Mahwah, NJ.

Boeing Commercial Airplanes (2009). *Statistical Summary of Commercial Jet Airplane Accidents Worldwide Operations 1959–2008*. Boeing, Seattle, WA.

Boy, G.A. and Ferro, D. (2004). 'Using cognitive function analysis to prevent controlled flight into terrain'. In D. Harris (ed.), *Human Factors for Civil Flight Deck Design*, pp. 55–68. Ashgate, Aldershot, UK.

Chapanis, A. (1999). *The Chapanis Chronicles: 50 years of Human Factors Research, Education, and Design*. Aegean Publishing Company, Santa Barbara, CA.

Chidester, T.L., Helmreich, R.L., Gregorich, S.E. and Geis, C.E. (1991). 'Pilot personality and crew coordination: Implications for training and selection'. *International Journal of Aviation Psychology*, **1**: 25–44.

Christoffersen, K. and Woods, D.D. (2000). 'How to make automated systems team players'. *Advances in Human Performance and Cognitive Engineering* Research, **2**: 1–12.

Civil Aviation Authority (1998). *Global Fatal Accident Review 1980-1996 (CAP 681)*, Civil Aviation Authority, London.

Coombs, L.F.E. (1990). *The Aircraft Cockpit*. Patrick Stephens Ltd, Wellingborough, UK.

Craik, K.J.W. (1940). 'The fatigue apparatus (Cambridge cockpit)'. *British Flying. Personnel Research Committee Memorandum No. 119*. Flying Personnel Research Committee, British Air Ministry, London.

Craik, K.J.W. and Vince, M.A. (1945). '*A Note on the Design and Manipulation of Instrument-Knobs*'. Applied Psychology Laboratory, Cambridge University Report. Cambridge University.

Dekker, S.W.A. (2000). *Cockpit Automation and* Ab Initio *Training: A European Investigation*. Centre for Human Factors in Aviation Technical Report. Linköping Institute of Technology, Linköping.

Dekker, S.W.A. (2001). 'The re-invention of human error'. *Human Factors and Aerospace Safety*, **1**(3): 247–66.

Dekker, S. and Hollnagel, E. (eds) (1999). *Coping with Computers in the Cockpit*. Ashgate, Aldershot.

Dismukes, R.K., McDonnell, L.K. and Jobe, K.K. (2000). 'Facilitating LOFT debriefings: Instructor techniques and crew participation'. *International Journal of Aviation Psychology*, **10**: 35–57.

Drew, G.C. (1940). 'An experimental study of mental fatigue'. *British Flying. Personnel Research Committee Memorandum No. 227*. Flying Personnel Research Committee, British Air Ministry, London.

European Aviation Safety Agency (2008). *Certification Specifications for Large Aeroplanes (CS- 25): Amendment 5*. Cologne: EASA. Available at http://www.easa.europa.eu/ws_prod/g/rg_certspecs.php#CS-25.

Federal Aviation Administration (1991). *Advanced Qualification Program (Advisory Circular AC 120-54)*. US Department of Transportation, Washington, DC.

Federal Aviation Administration (1996). *Report on the Interfaces Between Flightcrews and Modern Flight Deck Systems*. Federal Aviation Administration, Washington, DC.

Foushee, H.C. and Helmreich, R.L. (1988). 'Group interaction and flight crew performance'. In E. L. Weiner and D. C. Nagel (eds), *Human Factors in Aviation*, pp. 189–227. Academic Press, San Diego, CA.

Grether, W.F. (1949). 'The design of long-scale indicators for speed and accuracy of quantitative reading'. *Journal of Applied Psychology*, **33**: 363–72.

Hansen, J.S. and Oster, C.V. (1997). *Taking Flight: Education and Training in Aviation Careers*. National Academy Press, Washington, DC.

Harris, D. (2004). *Human Factors for Civil Flight Deck Design*. Ashgate, Aldershot.

Harris, D. (2006). 'The influence of human factors on operational efficiency'. *Aircraft Engineering and Aerospace Technology*, **78**(1): 20–5.

Harris, D. (2007a). 'A human-centred design agenda for the development of single crew operated commercial aircraft'. *Aircraft Engineering and Aerospace Technology*, **79**(5): 518–26.

Harris, D. (2007b). *Human Factors Integration (HFI) in civil aviation – Taking a systems perspective on Human Factors*. Invited address to the Aviation Safety Council, Taiwan (Republic of China). Taipei, 1 November 2007.

Harris, D. (2011). *Human Performance on the Flight Deck*. Ashgate, Aldershot.

Harris, D. and Harris, F.J. (2004). 'Predicting the successful transfer of technology between application areas; a critical evaluation of the human component in the system'. *Technology in Society*, **26**(4): 551–65.

Helmreich, R.L., Merritt, A.C. and Wilhelm, J.A. (1999). 'The evolution of crew resource management training in commercial aviation'. *International Journal of Aviation Psychology*, **9**: 19–32.

Hörmann, H.-J. and Maschke, P. (1996). 'On the relation between personality and job performance of airline pilots'. *International Journal of Aviation Psychology*, **6**, 171–8.

Hunter, D.R. and Burke, E.F. (1994). 'Predicting aircraft pilot-training success: A meta-analysis of published research'. *International Journal of Aviation Psychology*, **4**: 297–313.

Joint Aviation Authorities (1998). *Crew Resource Management – Flight Crew. Temporary Guidance Leaflet 5 (JAR-OPS)*. Administrative and Guidance Material (Section 4 – Operations). JAA, Hoofdorp, NL.

Mackworth, N.H. (1944). '*Notes on the clock test – A new approach to the study of prolonged visual perception to find the optimum length of watch for radar operators*'. MRC Applied Psychology Unit, Cambridge University Report. Cambridge. Subsequently published as:

Mackworth, N.H. (1948). 'The breakdown of vigilance during prolonged visual search'. *Quarterly Journal of Experimental Psychology*, **1**: 6–21.

Maschke, P. (1987). *Temperament Structure Scales (TSS)*. (Tech. Report ESA-TT-1069). Oberpfaffenhofen: European Space Agency.

Maurino, D.E. (1999). 'Crew resource management: A time for reflection'. In D. J. Garland, J. A. Wise and V. D. Hopkin (eds), *Handbook of Aviation Human Factors*, pp. 215–34. Lawrence Erlbaum, Mahwah, NJ.

National Transportation Safety Board (1973). *Aircraft Accident Report, Eastern Air Lines, Inc. Miami, Florida, December 29, 1972, L-1011, N310EA: Report Number: NTSB-AAR-73-14*. Washington, DC.

Office of Statistical Control (1945). *Army Air Forces Statistical Digest – World War II*. Air Force Historical Research Agency. Available at http://www.afhra.af.mil//.

Pariés, J. and Amalberti, R. (1995). *Recent Trends in Aviation Safety: From Individuals to Organisational Resources Management Training*, pp. 216–28. Risøe National Laboratory Systems Analysis Department Technical Report (Series 1). Risøe National Laboratory, Roskilde, Denmark.

Reising, J.M., Ligget, K.K. and Munns, R.C. (1998). 'Controls, displays and workplace design'. In D. J. Garland, J. A.Wise and V. D. Hopkin (eds), *Handbook of Aviation Human Factors*, pp. 327–54. Lawrence Erlbaum Associates, Mahwah, NJ.

Rignér, J. and Dekker, S.W.A. (1999). 'Modern flight training: Managing automation or learning to fly?' In S. W. A. Dekker and E. Hollnagel (eds), *Coping With Computers in the Cockpit*, pp. 145–52. Ashgate, Aldershot.

Roscoe, S.N. (1997). *The Adolescence of Engineering Psychology*. Human Factors and Ergonomics Society, Santa Monica, CA.

Sarter, N.B. and Woods, D.D. (1997). 'Teamplay with a powerful and independent agent: Operational experiences and automation surprises on the Airbus A-320'. *Human Factors*, **39**(4): 553–69.

Smith, K. and Hancock, P.A. (1995). 'Situation awareness is adaptive, externally-directed consciousness'. *Human Factors*, **37**: 137–48.

Suirez, J., Barborek, S., Nikore, V. and Hunter, D.R. (1994). *Current Trends in Pilot Hiring and Selection (AAM-240-94-1)*. Federal Aviation Administration, Washington, DC.

Thomas, M.J.W. (2003). 'Improving organisational safety through the integrated evaluation of operational and training performance: An adaptation on the line operations safety audit (LOSA) methodology'. *Human Factors and Aerospace Safety*, **3**: 25–45.

UK Ministry of Defence (2001). *Human Factors Integration (HFI): Practical Guidance for IPTs*. UK Ministry of Defence, London.

US Department of Transportation (1999). Aviation Rulemaking Advisory Committee; Transport Airplane and Engine: Notice of new task assignment for the Aviation Rulemaking Advisory Committee (ARAC). *Federal Register*, **64**, No. 140: 22 July 1999.

Wiener, E.L. (1989). *Human Factors of Advanced Technology ('Glass Cockpit') Transport Aircraft*. NASA Contractor Report 177528. NASA Ames Research Center, Moffett Field, CA.

Wood, S.J. and Huddlestone, J.A. (2007). 'Requirements for a revised syllabus to train pilots in the use of advanced flight deck automation'. *Human Factors and Aerospace Safety*, **6**(4): 359–70.

Woods, D.D. and Sarter, N. (1998). *Learning from Automation Surprises and Going Sour Accidents*. Institute for Ergonomics, Report ERGO-CSEL-98-02. National Aeronautics and Space Administration, Ames, CA.

8

Innovation in supersonic passenger air travel

H. SMITH, Cranfield University, UK

Abstract: Key issues relating to the supersonic business jet (SBJ) concept are reviewed, assessing the readiness of enabling technologies and the concept itself. The chapter overviews the market, environmental issues, the sonic boom and solutions, technological issues, including prediction methods, flight testing, systems, certification, and interested aerospace companies and design organisations. Reducing the sonic boom signature is vital if the vehicle is to be operated over land and economically viable. Investment is aimed at de-risking enabling technologies and raising readiness levels. Technologies are moving beyond theoretical and numerical analysis into the experimental and flight test domains. Collaboration between civil and military sectors is increasing. Concerns for the environment are balanced against the 'value of time' benefits offered by the SBJ concept.

Key words: aircraft design, supersonic business jet (SBJ), market, sonic boom, shock wave physics, aerodynamics, manufacturers, environmental issues.

8.1   Introduction

The demand for low ticket prices has been a very visible aspect of air travel in recent years. It has resulted in the rise of the low-cost airline and the availability of air travel to a wider range of customers than ever before. Nevertheless, there still exists a customer for whom the value of time remains a dominant driver. Despite great advances in communications technologies, there still remains a need for face-to-face meetings in the business world. The business jet enables quicker ground times than scheduled flights, flexibility in departure times and wider choice in airfield use due to superior field performance.

This ability to save time can be considerably enhanced by increasing the aircraft's flight speed into the supersonic regime. A supersonic business jet (SBJ) could offer considerable time savings; thus a B777 flying from New York to London would take about 11 hours door-to-door (7 hours in the air plus a further 4 hours on the ground) compared with 5 hours for an SBJ (4 plus 1). Compare this with Concorde's 7 hours (3.5 air plus 4). If we factor in the greater ability to fly point-to-point then we achieve even greater time savings.

An attractive design would still need the usual attributes of safety, security, comfort, reliability, performance and operational flexibility. Furthermore, a practical design would need to make a significant attempt at sonic boom reduction. This aspect would require a high priority as, at best, it could prevent over-land operation or, at worst, prevent operation entirely.

8.2   Historical background

The birth of serious interest in supersonic civil transport aircraft can be traced back to the 1950s. Technology had developed to a point where engineers were exploring the possibility of civil air travel at speeds of about Mach 1.2 or even as high as 1.8–2.0. The military application of these technologies was, as is usual, leading the way. In fact, a major milestone was a flight from New York to Paris of a Convair B-58 Hustler bomber. This non-stop flight was made at a supersonic speed of Mach 2 in 3 hours and 20 minutes. Not all the technical challenges had been beaten, as evident by the need for two decelerations and descents to air–air refuel.

The UK Supersonic Transport Aircraft Committee was set up in 1956 with a technical subcommittee[1] that included the Air Registration Board, the Aircraft Research Association, the National Physical Laboratory and the Cranfield College of Aeronautics. Thus was set in motion the research, design and development programme that led to the production of the Concorde supersonic transport (SST) airliner (Fig. 8.1).

Whilst US companies had been working on SST designs since the late 1950s, it was not until the Anglo-French agreement was signed in 1962 that President Kennedy set up a committee to evaluate the situation. In June 1963 the US formally started their own SST programme. Boeing and Lockheed developed concepts that by 1969 had evolved into a 234 passenger, 3575 nm, Mach 3 aircraft. Whilst the programme was eventually cancelled in 1971 on the grounds of economic feasibility, sonic booms and environmental effects, the progress made was notable. It is not widely realised that, by the time of the cancellation, the American taxpayer had paid about as much as the total cost of Concorde up to the granting of its full Certificate of Airworthiness.

Further to the original US programme, two further programmes were commenced (Supersonic Cruise Aircraft Research (SCAR) and High Speed Civil Transport (HSCT))[2] but failed to reach completion. NASA's HSR programme commenced in 1990 with in-depth studies in the environmental issues and continued with technology studies from 1995 including Tu-144 test flights (1996–8).

8.1 Concorde. (Source: NASA C-1975-00768.)

It would require considerable political will to embark on a new large SST aircraft programme. Major airframe manufacturers would need to commit significant resources to the project. With manufacturers currently committed to either large or environmentally benign aircraft, it may be some time before large, high-speed transport comes to the fore. However, it is believed that a significant market may exist for a supersonic business jet. Whilst significant technical, economic and environmental issues remain to be addressed, this goal could be a feasible venture for an airframe manufacturer.

8.3   Operational issues

8.3.1   Market

Gulfstream[2] have presented the results of a number of market studies based on a $50–100 million unit price offering a large cabin business aircraft. The study attempted to assess the number of customers who would, given the availability, wish to upgrade from subsonic business jets to SBJs. The results indicated a minimum potential market of 180 aircraft (compare this with a typical Gulfstream production run of 200 units) up to a possible 350 aircraft over a 10-year period. An independent market survey by Meridian International Research for small $50–100 million SBJ indicated a market of 250–450 units over a 10–20-year period.

Additional issues concluded by the Gulfstream studies include requirements for >4000 nm range and supersonic 'overland' flight capability. The latter issue is cited as being fundamental in achieving the economic viability of the concept. Data presented indicate that a random sample of Gulfstream jets operating over a one-year period only achieved 25% of their flight hours over water. It is not clear whether the Gulfstream studies accounted for the difference in research, development, test and evaluation (RDT and E) costs between a low-boom SBJ and a concept with restricted performance overland. Chudoba *et al.*[3,4] present an extensive analysis of market trends and drivers and their implications for the SBJ. Operational aspects are also discussed.

8.3.2   Environment

It is usual to subdivide environmental issues into four domains:

- Global environmental impact issues relate to the effect of aircraft operating emissions on the Earth's climate,
- Local environmental issues include emissions, noise and other pollution in close proximity to the airfield,
- Cabin environmental issues relate to the health and comfort of the passengers and crew,
- Sustainability relates to life cycle issues.

Sonic boom, being an issue special to this concept, will be discussed separately.

Global environment

Considerable effort has been invested into the potential effect of SBJ emissions and the global environment. Definitive conclusions are not yet available for three main reasons. Firstly, the precise characteristics of a future SBJ are not yet known in terms of payload range, engine size and performance, including cruise altitude. In fact, some of these parameters may ultimately be constrained to improve environmental impact. Secondly, the potential market, in terms of numbers, routes and other operational considerations, is also uncertain. Finally, the considerable effort that is currently focused on atmospheric chemistry, physics and global climate modelling indicates that our understanding is still not complete. However, an aircraft[5] that burns 30 tonnes of fuel to carry six passengers 4800 nm will have a greater CO_2 and water vapour impact than its subsonic counterpart.

The potential of SBJs causing damage to stratospheric and tropospheric ozone chemistry has been studied by NASA.[6,7] Here a number of SBJ emission scenarios, developed for NASA's Ultra-Efficient Engine Technology (UEET) Programme, were selected to include fuel burn, cruise altitude and emission index of NO_x (fuel burn being a function of fleet size, utilisation and engine characteristics). The atmospheric modelling was performed using the UIUC 2-D chemical radiative-transport model[8] of the chemistry and physics of the global atmosphere. This zonally averaged model is used to analyse artificial and natural forces on the global atmosphere from pole to pole and sea level to 84 km altitude. For the most probable scenario (18 M lb/day fuel burn, 20 g/kg $EI(NO_x)$ at 15–17 km altitude) the maximum local ozone depletion is 0.038%. The analysis indicates that the impact on the total column ozone change reduces with altitude, with point of inflexion at about 14.5 km (47 000 ft), below which it becomes positive.

Local environment

Whilst a 6–12 passenger business jet would usually be a fairly light aircraft, the SBJ will be of a similar mass to, say, the RJ100 regional jet. As airframe noise is related to aircraft mass, it would be anticipated that this aspect of the local environment would require due attention. Engine noise, the second main source of local noise, is referred to in Section 8.5.3.

Cabin environment: pressurisation/radiation

Under the category of comfort it should be noted that, for SBJs, cabin diameter comes at greater cost than for subsonic jets of similar cabin volume. Aerodynamic wave drag considerations impose a strong constraint on the fuselage fineness ratio. Thus an increase in cabin diameter will result in a proportionate increase in fuselage length or a significant increase in propulsion demands. This characteristic, when taken in conjunction with all the design constraints (including sonic boom), ultimately limits the cabin diameter to slightly less than that generally available to business jet passengers (Fig. 8.2, 8.3)

Atmospheric ionising radiation is caused by galactic and solar cosmic rays (high-energy subatomic particles) colliding with atoms in the upper atmosphere, thus generating a shower of further subatomic particles. For life on the planet surface the atmosphere provides effective shielding. At the operating altitude of typical subsonic jet aircraft the level of shielding drops to about 0.2 of the sea-level value. At typical SBJ operating altitudes this drops to 0.05 of the sea level value. The result of this is that the total effective dose rate (μSv h^{-1}) at 30000 ft is 90 times the sea-level value.[9] This increases to 180 times the sea level value at 40000 ft and then to 360 times at 65000 ft. Passengers flying at twice the speed of a subsonic vehicle will be exposed for half the time.

The dose rate can increase during solar flares (or decrease due to the accompanying disturbances to the magnetic field). It is not yet possible to predict the dose rate increase at commercial aircraft operating altitudes. These phenomena

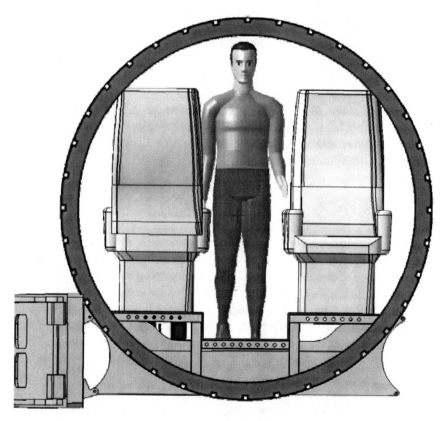

8.2 Typical SBJ cabin cross-section. (Source: Smith.[10])

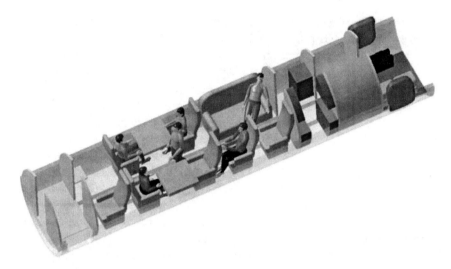

8.3 Typical SBJ cabin layout. (Source: Smith.[10])

can, over a short period of time, accrue an exposure equivalent to one month of typical exposure. These events are relatively rare, occurring approximately once per year. However, during one of the most intense observed occurrences, Concorde's on-board monitoring equipment did not measure sufficient levels to trigger its alarm (0.5 mSv h^{-1}). Nevertheless, an accurate physics-based atmospheric ionising radiation model is still required to evaluate potential design features of future aircraft to improve safety. NASA Langley Research Center continues to work[11] towards this aim.

Sustainability: materials, fossil fuels

Research into sustainable materials selection and environmentally benign manufacturing processes continues and is not specific to the SBJ concept. It is likely that, to ensure reasonable mission performance, the airframe will have a relatively high percentage of composite materials. Perhaps the most outstanding sustainability issue will be that of fuel burn per passenger mile. One study[5] gives some typical SBJ performance figures; six passengers plus baggage over 4800 nm requires 30.3 tonnes of fuel. This equates to about 1 kg per passenger nm, perhaps 2.5 times that of a business jet and some 20 times that of a large subsonic civil airliner.

Sonic boom

The effects of transport noise, in general, on the public can be categorised into sleep interference, annoyance, speech interference, task interference and hearing loss. Thresholds at which continuous noise causes these effects are known,[12,13] to some degree (sleep ~45 dB, annoyance ~55 dB, speech ~60 dB and hearing loss ~115–140 dB).

Due to their transient nature, sonic booms are thought to have more limited concerns for the public. Although only limited evidence exists, the impact is thought to relate to sleep interference, annoyance and, possibly, task interference, primarily due to the sonic boom 'startle' effect.

Concerns have been expressed over the impact of sonic booms on marine mammals (US policies: Marine Mammal Protection Act and Endangered Species Act). Evidence suggests that loud noises used to deter wildlife from potential food sources are soon ignored and, in some cases, eventually alert wildlife to the presence of the food.[14] Studies indicate that sonic boom frequency components above about 500 Hz will dissipate within 15 ft from the water surface.

Sonic booms are also known to cause physical damage to structures.[15] Here the effects are related to the peak over-pressure, some examples of which are given below:

- 0.5–2 psf: fine cracks in plaster, extension of existing cracks. Glass rarely shattered; cracks either partial or extension of existing. Slippage of existing loose roof tiles/slates. Bric-a-brac carefully balanced or on edges can fall; and fine glass, e.g. large goblets, can fall and break. Dust falls in chimneys.
- 2–4 psf: failures in glass, plaster and roof dependent on their existing localised condition.
- 4–10 psf: regular glass failures of well-installed glass; industrial as well as domestic greenhouses. Partial plaster ceiling collapse of good plaster; complete collapse of very new, incompletely cured, or very old plaster. High probability rate of failure of roofs in nominally good state. Old outside free-standing walls, in fairly good condition, can collapse. Interior walls known to move at 10 psf.
- Greater than 10 psf: some good glass would fail regularly to sonic booms from the same direction. Glass with existing faults could shatter and fly. Large window frames move. Most plaster affected. Domestic chimneys dislodged if not in good condition. Interior walls can move even if carrying fittings such as hand basins or taps; secondary damage from water leakage. Some nominally secure bric-a-brac can fall; e.g., large pictures, especially if fixed to party walls.
- Greater than 720 psf:[16] damage to human eardrums. (Sonic booms created by supersonic aircraft flying at very low altitude that have generated over-pressure of between 20 and 144 psf have been experienced by observers without causing injury.)
- Greater than 2160 psf: damage to human lungs.

In addition to studies into acceptable over-pressure limits, studies have also explored the effect of sonic boom wave form on human tolerance. The Japanese

Advanced Aircraft Technology Development Centre has performed human factors trials[17] on 61 N-type and ramp-type wave forms in a simulator chamber.

8.4   Technological issues: sonic boom

8.4.1   Physics

In steady supersonic flight, an aircraft will cause the formation of conical shock waves, not just at the bow and rear of the vehicle, but at every discontinuity along its length. Of these shock waves, those with a pressure greater than ambient have a tendency to propagate at a higher speed and hence coalesce at the front of the disturbance field (Fig. 8.4).

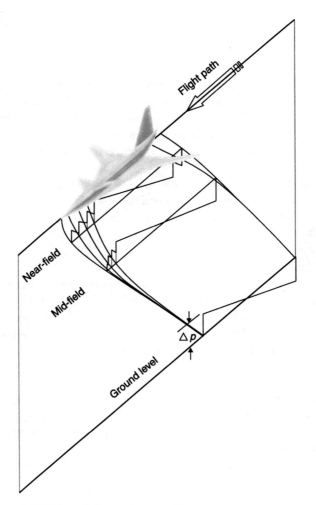

8.4 SBJ sonic boom propagation.

Those shocks with lower than ambient pressure propagate more slowly and hence will coalesce at the rear of the disturbance field, the result being that at some distance from the aircraft the shock waves will have coalesced into two stable conical waves, one to the front and one to the rear of the aircraft, as depicted in Fig. 8.4. Between the mid-field and ground level the pressure disturbance sharpens, under the influence of this effect, to form a rapid pressure rise followed by a gradual reduction terminated by a second rapid rise back to ambient pressure. This gives rise to the characteristic far-field N wave (Fig. 8.5). The initial pressure rise is often referred to as the maximum over-pressure. As this pressure field passes over an observer at ground level, the two sudden pressure increases are heard as a double 'crack' referred to as the sonic boom. As the aural response level is directly related to the magnitude of the pressure rise, the initial maximum over-pressure acts as a good metric for the perceived boom loudness. For more detailed discussion see, for example, Maglieri and Plotkin.[18]

The magnitude of the over-pressure, Δp, is typically 1–3 lb/ft^2 for an aircraft at cruise altitude and has a duration of 0.1–0.3 seconds (Table 8.1).

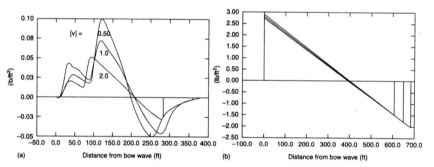

8.5 Typical near and far-field signatures. (a) Near-field fine grid analysis; (b) Ground signature. Signatures extrapolated from $v = 0.5$, 1.0 and 2.0. (Source: McCurdy.[19])

Table 8.1 Typical over-pressure of various aircraft types

Aircraft	Mach	Altitude (feet)	Δp (lb/ ft^2)	Reference
Concorde	2.0	52000	1.94	NASA[20]
SR-71	3.0	80000	0.9	NASA[20]
F-104	1.93	48000	0.8	NASA[20]
Space Shuttle	1.5	60000	1.25	NASA[20]
HSCT			2.5	
'Fighter'	1.12	60	~100	Carlson[21]
	1.06	590	~30	Carlson[21]
	1.18	5540	~6	Carlson[21]
	1.93	48000	0.7	Carlson[21]
F-18			1.3	Howe[22]
DARPA QSP			0.3	Howe[22]

The pressure field propagates three-dimensionally, in a complex conical manner, giving rise to primary and secondary boom areas (Fig. 8.6). In addition, areas of focused booms occur where the aircraft accelerates or manoeuvres (Fig. 8.7), referred to as 'super-booms'.

8.4.2   Sonic boom studies

Plotkin[16] traces some of the recent history of sonic boom studies from the 1960s through to the present day. Early (1960s) studies between the National Sonic Boom Evaluation Office, USAF, NASA and FAA generated some 100 000 sonic boom recordings, physiological and psychoacoustic studies, structural studies and animal effects research. The 1970s through to the mid-1980s saw ongoing NASA studies into boom minimisation, with some focus on the space shuttle launch and entry boom effects, in addition to US National Environmental Policy Act-driven studies for military supersonic airspace. Early optimism in the late 1980s towards large SSTs encouraged further research into human factors issues relating to sonic booms and sonic boom propagation modelling with further work on effects[23] on animals and structures. Recent work now focuses more on supporting the development of supersonic business jets, as large SSTs now seem further into the future.

8.4.3   Prediction methods

Whilst the existence and nature of shock waves from supersonic artillery shells, referred to as ballistic waves, have been known since the nineteenth century, it was some time after the first supersonic aircraft flight that the connection was made

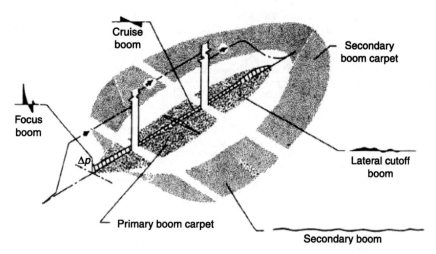

8.6 Primary and secondary boom carpets. (Source: Maglieri *et al.*[18])

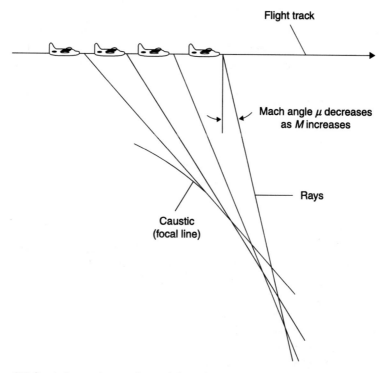

8.7 Sonic boom (super-boom) focusing due to acceleration. (Source: Maglieri *et al.*[18])

with sonic booms. Whitham's[24] theory describes the wave steepening mechanism. Compared with experimental data, Whitham tended to under-predict the strength of the sonic boom by as much as 40%.[25] Whitham's theory did not attempt to model aircraft lift or the atmospheric gradients typical of high-altitude flight.

Where Whitham's theory relates primarily to the projectile's longitudinal cross-sectional area distribution (more strictly, normal projections of sections taken along Mach planes), Walkden[26] extends the theory through the introduction of an equivalent cross-sectional area term based on the longitudinal lift distribution. George,[27] with a higher order form of analysis, was able to start to modify configurations to minimise the sonic boom.

Whilst most attention tends to be directed towards the sonic boom generated by supersonic aircraft while in the cruise, considerable perturbations from this can be induced during aircraft manoeuvres. In particular, while turning or accelerating a focused boom or 'superboom' may result with three or more times the magnitude of a steady boom.

As advances in computer power progressed, more sophisticated models[28] became feasible and were able to include the effect of manoeuvring as well as wind. Further development[29] resulted in a computer model known as ARAP. Turbulence-related variations, giving rise to significant boom amplitude variations and shock rise times, were analysed by Crow.[30] Extending the earlier work of George,[24] George and Seebass[31] developed the work to produce one of the most important tools for the minimisation of sonic booms for supersonic aircraft. Analysis[32] was eventually developed to encompass the superboom phenomenon.

Airbus France coordinated a three-year study, the Sonic Boom European Research program (SOBER),[33] to develop numerical models (and experimental models; see below) to predict some of the more complex phenomena associated with sonic boom generation and propagation. Here, Euler simulations of the near field are coupled to atmospheric propagation modelling and the Tricomi algorithm.[34,35] The resulting software model, referred to as BANGV, can predict booms for manoeuvring aircraft, including focusing and off-track boom, allowing for wind and temperature stratification, air humidity and ground impedance. It is unclear whether the approach has been utilised to confirm the propagation of a low-boom signature from the near field down to ground level.

Further development of the focus-boom version of the computer model developed by Thomas[36] led to a widely used model, PCBOOM4. The application of molecular relaxation to the development of the sonic boom has led to the ZEPHYRUS computer model, produced as part of a PhD thesis.[37] An inverse design[38] approach was explored under Defense Advanced Research Projects Agency (DARPA) Quiet Supersonic Platform (QSP) funding based on a set of adjoint equations and their corresponding boundary conditions. The formulation and application to 3-D of the inverse design of a biconvex wing and a business jet wing body configuration were explored. See also Nadarajah et al.[39]

Howe[22] has developed an approach that, utilising a 3-D non-linear, non-isentropic Euler computer model, more rigorously examines benefits in modifying the longitudinal distribution of cross-sectional area in a non-axisymmetric manner. The key element here is that higher surface curvatures and steeper surface angles to the freestream flow give rise to stronger pressure disturbances and, similarly, lower surface curvatures and surface angles give rise to weaker pressure disturbances. This additional degree of freedom can be used to shape the near field signature and, if it persists to the far field, reduce the sonic boom at ground level.

8.4.4   Flight testing

There have been many flight test programmes over the years in which sonic boom measurements have been made, for example Elmer and Joshi.[40] By 1989 there was sufficient interest in the design of aircraft for low boom signature and the validation of the prediction that a low boom signature would persist to the far field that the possibility of flight testing a modified UAV (Firebee BQM-34E) was explored[41] (Fig. 8.8). Flight tests were also proposed involving a modified

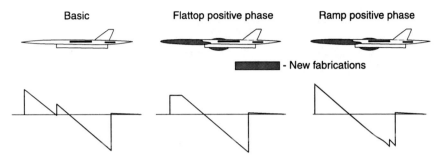

8.8 Firebee UAVs and signatures. (Source: Maglieri *et al.*[41])

SR-71.[19] Unfortunately, the reduction in funding for civil SST research at this time prevented these programmes from being pursued.

However, DARPA had retained an interest in small supersonic aircraft with a view to application in a small supersonic civil transport or supersonic bomber. In this light, a programme was initiated in 2000 to investigate design issues for this class of aircraft, referred to as the Quiet Supersonic Platform (QSP). A prime driver of this programme was to achieve an over-pressure of no more than 0.3 lb/ft². By 2001 DARPA had initiated its Shaped Sonic Boom Demonstration programme. Here the F-5 was modified, particularly in the forebody region, to produce a low-boom configuration, F-5 SSBD-24B (Fig. 8.9). The shaping analysis utilised a range of computer models including lower-order methods such as PCBOOM and more recently developed 3-D full potential propagation methods.

By 2003, the programme[42–50] had successfully demonstrated the concept of sonic boom reduction through aircraft shaping and generated significant quantities of data suitable for methodology and tool validation (Fig. 8.10).

Gulfstream have developed a technique[51] that offers the possibility of correctly scaling a sub-scale low-altitude demonstrator to simulate the far-field boom signature perceived at ground level of a full-scale aircraft flying at high altitude. The method will be of particular use in the determination of acceptability criteria for civil airworthiness requirements. The approach appears straightforward for N-wave sonic booms and, by trading low altitude and low Mach number, can also produce useful results for low boom signatures (Fig. 8.11).

8.4.5 Experimental simulation

Due to the scale issues relating to far-field distance relative to the aircraft length, the derivation of far-field over-pressure data from direct wind tunnel testing is difficult. It is more usual to measure the near-field pressures then propagate these to ground level via a sonic boom propagation methodology, for example.[52,53]

8.9 (a) F-5); (b) F-5 SSBD. (Source: Haering.[45])

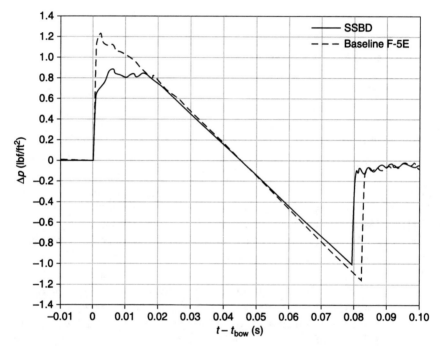

8.10 Baseline and low boom signature comparison. (Source: Haering.[45])

Whilst full-scale flight testing should provide the best source of data for rigorous validation of numerical or analytical prediction methods, in practice it usually contains some degree of uncertainty. Parameters that are difficult to determine include wind and temperature vertical distribution and atmospheric turbulence. Laboratory-scale simulation of sonic boom focusing has been attempted[36] using electronically controlled piezoelectric transducers (representing the aircraft) in a water tank. The non-linear sound propagation in water is perfectly

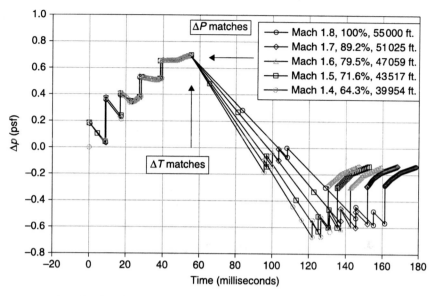

8.11 Scaled simulations of full scale signature. (Source: Howe.[51])

analogous to that of air, permitting, in this experimental set-up, a 1:100 000 scaling. The approach was used to validate the numerical simulation of boom focusing due to acceleration.

The NASA Ames Research Centre aeroballistic range[54] has been exploring the possibility of using the facility to provide test data, beyond the near field, to assess configuration potential for sonic boom reduction. Results indicate that the concept is viable and could progress to the free flight testing of models with wings and fins.

The National Aerospace Laboratory (NAL) of Japan has (2005) successfully demonstrated the technique of flying a scaled demonstrator[55] at Mach 2 and 12 km. Whilst, in this case, the programme was aimed at supporting the design of a 300 passenger SST, it would present a useful test facility for the development of any supersonic business jet.

Under the QSP programme, experimental work[56] is underway to validate the model predictions for off-board energy addition and to explore, in real time, near-field interactions of the energy with a simple shock. The Mach 2.4 wind tunnel setup enables localised energy addition through laser-induced air breakdown.

8.4.6 Low boom solutions

A number of possible methods are proposed that can help reduce the boom signature at ground level. Figure 8.12 illustrates some of these.

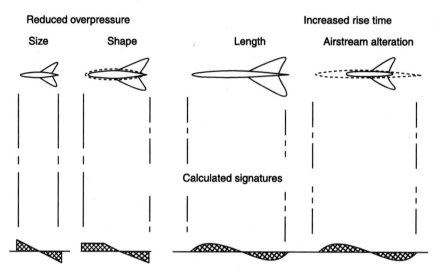

8.12 Sonic boom minimisation solutions. (Source: Maglieri *et al.*[18])

Low boom aircraft shaping

As discussed earlier, the pressure disturbance created by a supersonic aircraft propagates through the atmosphere in such a way as to create the worst possible signature by the time it reaches the ground. The resulting N wave gives the maximum aural response, i.e. it is perceived by human subjects as being loud. As the magnitude of the boom is a function of both the volume and the mass of the aircraft, the smaller the aircraft the less intense the boom. Thus, all other things being equal, the SBJ will be quieter than a 100-passenger SST.

Beyond this, it is necessary to carefully shape the aircraft in an attempt to prevent the coalescing of the pressure disturbances. A large compression at the aircraft nose and an expansion at the tail help stretch the ends of the signature more than the middle pressures, hence suppressing the N wave tendency.

Active lift control

Active lift control[57] attempts to modify the longitudinal lift distribution without affecting the trim state of the aircraft by pumping fuel forward or aft.

Active thickness and camber control

A low boom signature can be achieved by deploying a device from the lower surface of the aircraft extending in the longitudinal direction to generate expansions ahead of compressions produced in off-design conditions. SAI hold a patent[58] for an aircraft thickness/camber control device that operates in this way, and enables the maintenance of a low boom signature.

Extendable nose spike

The concept[59] proposed here is that the development of the nose shocks be controlled through a deployable spike that extends significantly beyond the nose of the aircraft (Fig. 8.13, 8.14).

As the nose shocks are no longer controlled solely by the nose of the aircraft, the forward fuselage can be tailored to the payload and structural requirements. Wind tunnel testing and CFD show the potential of the approach. Further work[60] investigates the engineering aspects of integrating the nose spike. Work reports on proposed extension and retraction mechanisms and drive systems, prototype hardware and ground tests. Flight testing[61] will establish the structural and structural dynamic behaviour in addition to gaining operational experience of the deployment mechanism (Fig. 8.15).

Off-body energy addition

Off-body air heating has been considered as a means of suppressing the coalescence of bow, lifting and other shockwaves at the far field. Approaches include combustion, exploding projectiles, hot jets, plasma and electromagnetic interactions. Miles *et al.*[62] proposed an approach to energy addition whereby the energy addition

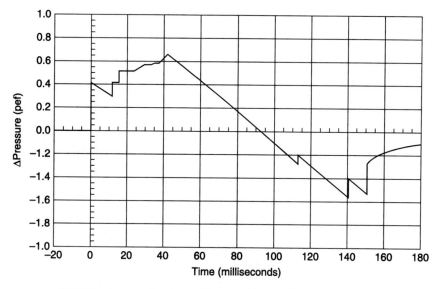

8.13 Low boom signature. (Source: Howe.[59])

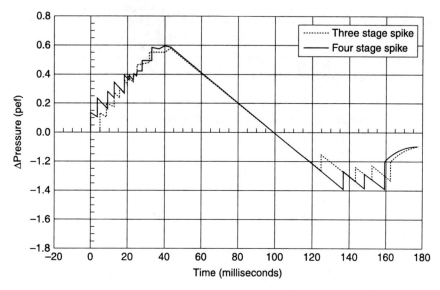

8.14 Low boom signature with nose spike. (Source: Howe.[59])

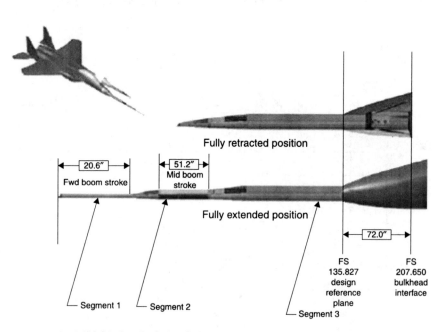

8.15 NASA F-15B 'Quiet Spike' test configuration. (Source: Simmons.[61])

process increases the apparent length of the airplane and reduces the extent of shock wave coalescence. Here an initial rise of 0.8 lbs/ft^2 reduced to 0.2 lbs/ft^2 with a peak rise of 0.6 lbs/ft^2.

8.5   Technological issues: aerodynamics

8.5.1   Cruise efficiency

The aerodynamic cruise efficiency at a given Mach number, as characterised by L/D, is important for all aircraft as it has a direct impact on the achievable range. L/D reduces transonically dependent upon the specific aircraft configuration. At Mach 2, Concorde achieves an L/D of 7.5 (compare this with typical subsonic civil transport 18–20) and at low speed barely gets up to 9. The latter could prove significant if a supersonic business jet is required to operate subsonically for extended periods, such as over land if the boom is unacceptable. Aerodynamic optimisation[56, 62–66] continues to attract considerable interest with a variety of techniques, both high and low order methods, including sonic boom propagation with atmospheric variations, thermo-viscous absorption, molecular relaxation and diffusion phenomena.[67]

8.5.2   Laminar flow control

Supersonic civil transport aircraft capable of any reasonable range tend to have a high fuel fraction. This makes the concept very sensitive to parameters such as the specific fuel consumption and the aircraft lift-to-drag ratio. As such, small changes to the aerodynamic efficiency of the vehicle can give rise to marked improvements in range and/or payload capacity.

NASA successfully achieved laminar flow at supersonic speeds over the period from 1988 to 1996 in a flight research programme utilising two F-16XL aircraft[68] (Fig. 8.16). The aim of the work was to achieve laminar flow over 50–60% of the wing chord on a swept wing configuration (70° inboard and 50° outboard). At Mach 2 and 53 000 ft, a laminar flow extent of 46% was achieved (Reynolds number of 22.7×10^6). The scheme used an active flow-control glove made from perforated titanium. Marshall[69] discusses variables that directly affect the ability to obtain laminar flow. Whilst natural laminar flow can be achieved through aerofoil profile design in certain situations, the sweep-induced cross-flow disturbances associated with the turbulent boundary layer over a highly swept wing need to be controlled through, for example, an active suction system.

Integration of such a system requires careful attention to potential shockwave generators that could disturb the flow. Even after the incorporation of a 20″ vertical shock-fence to block an inlet shock from the wing leading edge, the F-16XL still required a 1.5° sideslip angle to prevent shockwaves from interfering with the laminar boundary layer. The system, whilst beneficial in terms of skin

8.16 F-16XL with SLFC 'glove'. (Source: NASA EC96-43548-8.)

friction drag reduction, does incur a mass penalty in addition to propulsion off-take demands. Wing structure, leading-edge high-lift devices and de-icing systems can also be constrained in addition to implications for fuel reserves and pre-flight operating procedures.

Boeing[70] has explored, amongst other technologies, the possibility of utilising both natural and hybrid laminar flow control under the DARPA QSP programme. Skin friction reductions of about 24% could be achieved, leading to a total drag reduction of about 8%. (This accounts for system mass increases and power demands.) In the case of passive laminar flow control, distributed leading-edge roughness has been shown[71] to limit the growth of naturally dominant instabilities that would lead to early transition.

8.5.3 Propulsion

As with all aspects of supersonic aircraft design, the propulsion system is heavily constrained by numerous requirements. High thermal efficiency is, of course, of high importance to improve the payload fraction and will also contribute to a reduction in carbon footprint. Other emissions such as combustion-driven NO_x are also high on the environmental agenda, both in cruise and on take-off.

Maintaining this cruise efficiency, whilst achieving low take-off noise, is a notable challenge.

Rolls-Royce[72] has studied a number of possible propulsion system concepts as applied to a range of supersonic aircraft from 300-seat SSTs to eight-passenger supersonic business jets. Engines include variable cycle conventional turbofan (CTF), mixed nozzle ejector (MNE) and mid-tandem fan (MTF) configurations.

NASA[73] investigated six potential propulsion systems for use on supersonic civil aircraft: turbojet, turbine bypass, mixed flow turbofan, variable cycle, Flade (Fan-on-blade is a modified turbofan engine that introduces an additional airstream up front in the engine fan) and the inverting flow valve engine. These were evaluated for performance, weight and size. Effects of take-off noise, cruise emissions and cycle design rules were examined. The primary concept selected was the mixed flow turbofan for its good performance, low weight and low risk and complexity (Fig. 8.17).

NASA's Ultra-Efficient Engine Technology Program (UEETP) aims to enhance the performance of future aircraft propulsion systems by developing a number of key technologies. These technologies include Highly Loaded Turbomachinery, Emissions Reduction, Materials and Structures, Controls and Propulsion-Airframe Integration. The top level aims are to reduce landing and takeoff NO_x by 70% and mission CO_2 by 8%. The technologies are aimed at a number of concept aircraft, including the supersonic business jet. A study[74] performed under this programme resulted in the design of an efficient (23% fuel burn reduction), reduced emission (81% NO_x reduction) mixed-flow turbofan engine with a fan pressure ratio of 3.0 and a throttle ratio of 1.143.

Numerous other studies have investigated engine cycle analysis[75,76] for efficient and quiet use over the wider operating envelope characteristic of supersonic business jets. Gulfstream and Rolls-Royce[77] have shown beneficial results from the application of a highly variable cycle (HVC) engine in terms of reduced full burn.

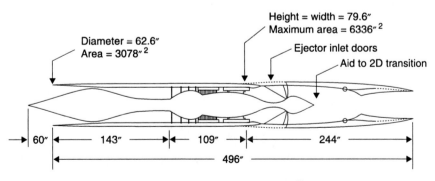

8.17 Mixed flow turbo fan. (Source: McCurdy.[19])

Advanced technology solutions have been explored to address the problem of engine noise reduction for supersonic business jets, such as that caused by the mixing of two streams at different velocities, the interaction of turbulence with shocks, and radiation turbulence convected at velocities that are supersonic relative to the speed of sound in the surrounding medium. Areas that show promise[78] are mixer ejectors and inverted velocity profile approaches. Another technology being investigated is jet noise reduction through plasma formation. The effect of plasma formation created by glow discharge, pulse repetitive discharge and HF discharge has been demonstrated[79] to achieve noise reductions of 4–6 dB (for 4–100 kHz) in laboratory tests.

Beyond optimal aerodynamic and propulsive performance, engine airframe integration now takes on the additional function of contributing to the minimisation of boom signature. A one-year DARPA-funded study[80] using Computational Fluid Dynamics (CFD) investigated the impact of engine placement on sonic boom generation. Axial positions were analysed for over-wing and under-wing arrangements. The over-wing arrangement proved beneficial. Desktop Aeronautics[81] have investigated the propulsion/airframe integration issue with a view to optimising the Aerion SBJ configuration performance. The problem of designing an engine intake compatible with both high-speed cruise conditions and low-speed field performance conditions has been considered by Aerion.[82] Here a translating cowl slides 2 ft forward from 0.75 fan diameters ahead of the engine face to achieve the required mass flow at take-off.

8.5.4 Structures

The high fuel fraction required for the SBJ concept to achieve the desired range demands minimum empty mass and thus a low structural mass (Fig. 8.18). This increases the need for optimal design and materials selection.

The supersonic business jet gives rise to many additional structural issues that need to be considered in the design phase, such as thermal effects, supersonic panel flutter and acoustic fatigue. Many of these considerations have been encountered with military aircraft, but not always with the proposed materials or with the intended endurance.

NASA's High Speed Research (HSR) programme partially achieved all of its sonic fatigue aims.[83] These included:

- To be able to predict inlet, exhaust and boundary layer acoustic loads,
- To measure high cycle fatigue data for materials developed during the HSR programme,
- To develop advanced sonic fatigue calculation methods to reduce required conservatism in airframe designs,
- To develop damping techniques for sonic fatigue reduction where weight-effective,
- To develop wing and fuselage sonic fatigue design requirements,

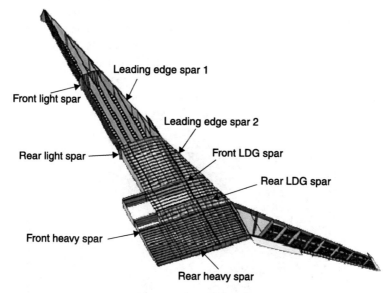

Front light spar

Leading edge spar 1

Leading edge spar 2

Rear light spar

Front LDG spar

Rear LDG spar

Front heavy spar

Rear heavy spar

8.18 Typical SBJ wing structure. (Source: Smith.[10])

- To perform sonic fatigue analyses on HSCT structural concepts to provide guidance to design teams.

With the variety of aero-structural challenges needing to be addressed in the supersonic business jet concept, it is no surprise that aspects of aero-structural design, analysis and sensitivity have attracted much effort.[84]

The impact of aeroelastic vibration on the vehicle's flight dynamics, flight control and flying qualities has been investigated.[85] The study highlights examples of biodynamic coupling and shows that structural stiffening and visual display compensation were of little benefit, whilst increased damping and elimination of control effector excitation of low-frequency modes offered improvements. With a view to reducing the internal cabin noise component induced by structural acoustic vibrations (10–1000 Hz), analysis models have been developed[86] to assist in the engine-airframe mounting design.

8.5.6   Systems

Pilot vision during the take-off and landing phases can be restricted by the slender nose cones of supersonic transport aircraft. Concorde circumvents this difficulty by lowering the nosecone when required. It has been suggested[10] that the application of synthetic vision systems (SVS) could provide a viable alternative solution.

Further benefits include mass reduction and enhanced pilot situation awareness in addition to greater aircraft packaging flexibility. NASA has developed an SVS concept[87] that promises to offer pilots an enhanced intuitive view of the flight environment, hazard and obstacle detection and display, and precision navigation guidance for all phases of flight: from ground operations, departure and en route to arrival irrespective of low-visibility conditions due to meteorology (Fig. 8.19).

The high fuel fraction characteristic of the SBJ makes the design of the fuel system challenging from both a management and packaging point of view (Fig. 8.20). Again, synthetic vision offers some flexibility here.

8.19 Synthetic vision system. (Source: NASA EL-1998-00169.)

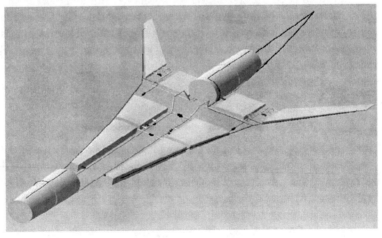

8.20 Typical fuel tank arrangement. (Source: Smith.[10])

8.5.7   Design and integration

Various conceptual design studies have been performed to investigate the viability of the supersonic business jet and the various enabling technologies that are required to be integrated. The Langley Research Centre[88] developed a Mach 2, 4000 nm, ten-passenger, 0.5 lbf/ft^2 vehicle. The study was augmented by a wind tunnel test programme.[5] A number of observations were made that are generally consistent with most other studies. Whilst the reduction in mass resulting from reducing the passenger load from 300, typical of large SST studies, to ~10 assists greatly with the sonic boom issue, this is partly offset by the reduction in length. The low boom wing planform greatly constrains the aspect ratio. Passenger comfort considerations need careful trade against drag (and hence range) and sonic boom. Fuel and landing gear packaging required suboptimal aerofoil thickness (Fig. 8.21).

High fuselage volume requirements, largely for passengers and fuel, pushed up fuselage length but assisted with low boom. Aft fuselage-mounted engines

8.21 Critical wing packaging area. (Source: Smith.[10])

resulted in numerous technical challenges, although with additional resources this may be resolved. The canard location was challenging due to the constraints from low-speed controllability and low boom characteristics.

Raytheon performed a NASA-funded study[89] to quantify the engineering 'cost' of the low boom constraints. Two supersonic business jet designs were produced to meet identical requirements except that one was limited to an initial sonic boom over-pressure of 0.4 lb/ft^2. The results showed that the boom-constrained concept was 25% longer and 18% heavier (Fig. 8.22, 8.23).

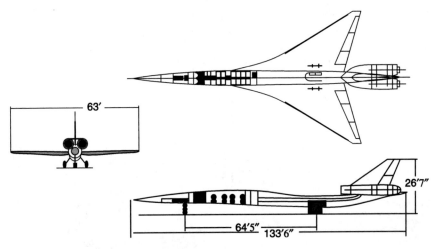

8.22 Unconstrained boom configuration. (Source: Aronstein and Schueler.[89])

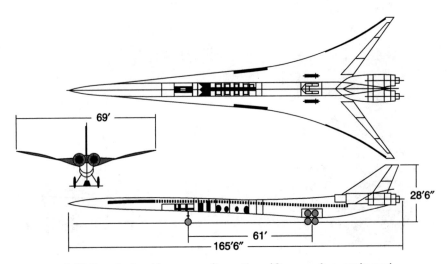

8.23 Constrained boom configuration. (Source: Aronstein and Schueler.[89])

Under the DARPA QSP programme, Northrop Grumman and Raytheon investigated the design issues relevant to long-range supersonic aircraft within the context of a 'dual-relevant' approach.[90] This would aim to form the basis of both a military strike aircraft and a civilian business jet. That is to say, whilst the actual aircraft would be different they would be based largely on common technologies. Consistent with, or slightly more demanding than, many business jet studies, the concept called for a vehicle capable of Mach 2.4, 0.3 lb/ft^2 sonic boom, 6000 nm and take-off mass ~100000 lb. The wide-ranging study investigated thin wings, multiple wings, delta and arrow wings, boom shaping and other exotic concepts.

Although there is considerable experience of military swing-wing aircraft, the proposed Boeing SST is the only civil example that has attracted significant investigation. A small study[91] has been performed into the performance benefits achieved through the application of a swing wing to the supersonic business jet concept. The study showed small improvements in both subsonic and supersonic performance, sonic boom and field performance, but with greater complexity and mass and possible certification issues.

A more extensive study by Gulfstream[92] investigated some of the structural design challenges, pivot location issues, aeroelastic concerns and materials choices (Fig. 8.24). The systems aspects were also investigated, including the actuation systems, wing fuel interconnect, flight control systems and reliability issues. Certification issues are discussed and a way forward is presented.

Design issues associated with hybrid laminar flow control, which could offer 8% drag reduction, include increased system complexity, mass and power off-takes. A Boeing study[73] indicated an additional 360 lbs systems weight and 90 hp power demand.

8.24 Swing wing structural analysis. (Source: Simmons and Freund.[92])

The implementation of off-body air heating requires the consideration of two design aspects: the generation of sufficient energy through some appropriate secondary power system and the projection of that energy to the required location. Miles[62] proposes a solution to the latter in the form of four 2 m by 2 m multi-element arrays focussing 49 MW of pulsed microwave energy at 3 GHz over a 1 m^2 spot some 46 m below the aircraft. Typical secondary power requirements[5] for a SBJ are likely to be of the order of 100 kW. The heavy constraints and strongly coupled design issues have been explored by numerous optimisation approaches, such as genetic algorithms[93] and response surface methods.[94]

The oblique flying wing, whilst offering drag benefits, is unlikely to be compatible with the requirements for a small business jet due to anthropomorphic constraints, e.g. wing thickness driven by passenger height. At a larger scale, studies[95] indicate excessive sonic boom problems.

8.6   Technological issues: airworthiness

FAR 91.817[96] states:

'(a) No person may operate a civil aircraft in the United States at a true flight Mach number greater than 1 except in compliance with conditions and limitations in an authorization to exceed Mach 1 issued to the operator under appendix B of this part.
(b) In addition, no person may operate a civil aircraft for which the maximum operating limit speed ^{M}MO exceeds a Mach number of 1, to or from an airport in the United States, unless –

(1) Information available to the flight crew includes flight limitations that ensure that flights entering or leaving the United States will not cause a sonic boom to reach the surface within the United States.'

Progress needs to be made in the area of certification if supersonic flight overland is to be realised. Certainly, at this time, there is uncertainty relating to specific sonic boom requirements. Research[97] has started to develop methods to support decision making for engine architecture selection in the light of uncertainty in regulation.

However, more general design requirements pertaining to structural and systems design, as contained in current FAR and CS handbooks, will be inadequate in ensuring comparable levels of safety in supersonic aircraft. British TSS standards,[98] to which Concorde was designed, and the US equivalent, to which Boeing's early SST was being designed, can act as guidelines for initial studies. However, effort will be required to develop requirements applicable to this class of aircraft incorporating twenty-first-century technologies. These early requirements were contemporary with 1960s BCARs.

8.7   Manufacturers and design organisations

8.7.1   Sukhoi-Gulfstream

After the Cold War, Sukhoi teamed with Gulfstream to build S-21 SBJs. Expected orders never materialised.

8.7.2   Gulfstream

Gulfstream is researching the feasibility of developing a supersonic business jet. The Quiet Supersonic Jet (QSJ), which would likely cost $80–100 million, is a swing-wing, aft tail configuration. Its most notable feature is a nose spike device. The company[99] has employed NASA F-15 flight trials of its telescopic 'Quiet Spike' installation. The trials were intended to verify the structural integrity of the system, the F-15 not being a suitable configuration for boom suppression measurement trials. The spike, weighing 470 lbs, extends from 14 ft subsonically to 24 ft at supersonic flight speeds.

Gulfstream/Northrop Grumman[100] is one of four industry teams awarded $1 m by NASA to investigate the feasibility technologies that would enable supersonic flight over land without significant disturbance to populated areas. A supersonic acoustic signature simulator has been developed to demonstrate the difference in perceived sonic boom between a supersonic jet and a low-boom supersonic jet.

8.7.3   Aerion Corporation

Aerion Corporation,[101] a company based in Reno, Nevada, has plans to build a supersonic business jet that could enter the market as early as 2011. The concept utilises the already certified P&W JT8D-219 powerplant, thus minimising the development risk, and will incorporate a natural laminar-flow wing section. The latter, combined with the wing design, should permit efficient flight at both supersonic and subsonic flight speeds. This aspect mitigates the risk of airworthiness authorities restricting supersonic flight to over-water operations only. The design should be scalable to carry larger numbers of passengers. Development costs are placed between $1.2 and 1.5 billion.

Development work includes:[102] in-flight testing of wing section on a NASA F-15 test bed, CFD airframe analysis, FE analysis, engine–airframe integration studies, systems studies and high-speed wind tunnel tests. Pratt and Whitney indicates that jet noise level should be within Stage4/Chapter 4 limits. Work with several vendors continues to develop system architectures (fuel, secondary power, FCS, ECS, anti-icing). Aeroelastic analysis continues. There has been rocket sled testing (zero to 100 mph in 1.8 s) up to Mach 1.5[103] and flight deck mock-ups have been produced.

8.7.4   Supersonic Aerospace International

Las Vegas-based Supersonic Aerospace International[104] plans a jet that would cruise at Mach 1.8 over land, hitting the market in 2014. The company, led by the son of Gulfstream founder Allen Paulson, is pushing a concept being developed by Lockheed Martin under a $25 m contract from SAI.

The distinctive features[105] of the concept include an inverted V-tail to ensure a mass-efficient support structure for the aft-mounted engines and a heavily boom-constrained configuration resulting from an optimum longitudinal lift distribution (Fig. 8.25).

The main design driver here is to reduce the sonic boom to a level that would be deemed acceptable for supersonic operation over land. Reflecting the design complexity involved in ensuring a low boom signature, SAI's projected development costs are of the order of $2.5–3 billion. The required 33–35 000 lbf thrust engines will be part of this development.

SAI claim that the design should be scalable and envision a 30-seat version being feasible.

8.7.5   Dassault

Dassault is developing a 1990s design by taking the lead in a European/Russian 'Environmentally Friendly High Speed Aircraft' research programme (HiSAC)[106], an EU funded (14 m Euro),[107] 37-member, 48-month consortium of industry and academia including airframe makers Alenia, Dassault and Sukhoi[108] (Fig. 8.27, 8.28, 8.29).

8.25 SAI SBJ concept. (Source: SAI.[107])

8.26 SAI Cabin Interior Concept (SAI[107]).

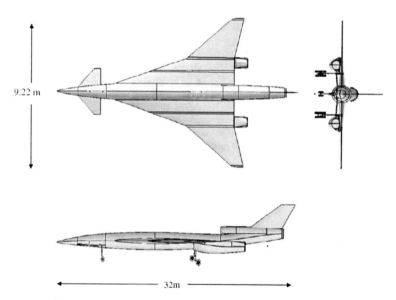

8.27 HiSAC Low Noise Family (Pietremont).[109]

Sukhoi's low sonic boom adaptation[110,111] was selected as the programme's reference configuration and will be assessed by all the other partners (Fig. 8.29). Wind tunnel evaluation at one-tenth scale will commence shortly. Snecma is developing a noise-reducing mixer-ejector engine. The programme was funded until 2009.

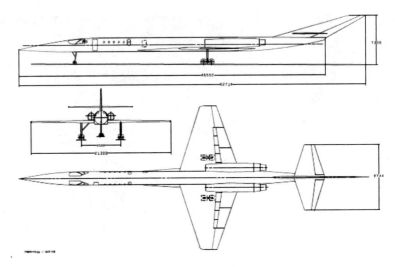

8.28 HiSAC Long Range Family (Pietremont).[109]

8.29 HiSAC Low Boom Family (Pietremont).[109]

8.7.6 DARPA

DARPA, whilst not an airframe manufacturer, has maintained an interest in this class of vehicle as many of the technologies would be common to a small military supersonic strike vehicle.

Its Quiet Supersonic Platform (QSP)[112] programme envisions a Mach 2–2.4, 6000 nm, 100 000 lb take-off mass of which 20% would be payload. The sonic boom Δp would be less than 0.3 lb/ft^2. Lockheed Martin and Northrop Grumman are involved in the systems studies segment of the programme (Fig. 8.30, 8.31).

8.30 Lockheed Martin QSP Concept (DARPA).[112]

8.31 Northrop Grumman QSP Concept (DARPA).[112]

8.7.7   Tupolev Design Bureau

The Tupolev Design Bureau[113] has turned its 40 years of supersonic design experience to the challenge of supersonic business jets. The Tu-444 appears to be, at this time, an internally funded conceptual design study. The concept appears consistent with other proposed concepts, although with a slightly higher Mach number of 2 (Fig. 8.32). No information is presented on sonic boom issues.

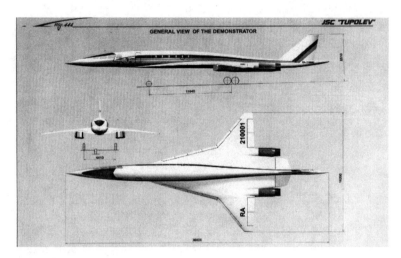

8.32 Tupolev SBJ Concept (Tupolev).[113]

8.7.8   Mitsubishi Heavy Industries

Whilst Japanese interest has primarily been focused on large civil SSTs, some designs[109] have been produced in the supersonic business jet class. Low boom/low drag designs have predicted over-pressures of 1.14 lb/ft^2.

8.7.9   Supersonic Cruise Industry Alliance

The Supersonic Cruise Industry Alliance ('Super 10') is a group of airframe and engine manufacturers with the common interest of enabling quiet, safe, affordable, environmentally acceptable supersonic flight over land in the next ten years. The group includes Boeing, Cessna, Gulfstream, Lockheed Martin, Northrop Grumman, Raytheon, General Electric, Pratt and Whitney, Rolls-Royce and Netjets. With NASA funding it is hoped that a small low-boom X-plane might fly in 2008.

Table 8.2 summarises the characteristics of some of the concepts discussed in Section 8.7 taken from references.[104–113]

Table 8.2 Key characteristics of some industry concepts

Company	Mach	Range (nautical miles)	Passenger seats (PAX)	Mass (lb)	Price (US $m)
Aerion	1.6	4000		90000	80
SAI/L-M	1.8	4600	12	153000	80
Gulfstream	1.8		15		
Tupolev Tu-444	2	4660	6–10	90400	

8.8   Conclusion

History shows that public acceptance is important in the introduction of new technologies; more so today than ever, with the environmental lobby building strength. It is apparent that the need to reduce the sonic boom signature is vital if the vehicle is to be permitted to operate over land and hence have a good chance of being economically viable.

This single design constraint permeates all aspects of the design, from planform and fuselage geometry, empennage and internal layout, through propulsion integration, fuel and other systems, to the application of advanced systems such as nose spikes and off-body-energy addition. The analysis leading to almost any characteristic of the SBJ will at some point be underpinned by the target far-field sonic boom over-pressure. If that target moves, all aspects of the design, including risk, development and operational costs, change. It is clear that sonic boom acceptability requirements must be set if resources are to be effectively focused and designs are to converge. This challenge must be grasped. Theoretical studies and laboratory simulations will not convince the public. A possible approach is the development of a flight demonstrator scaled to produce a representative reduced boom waveform signature and to operate it, with public consultation, over a populated area for a short duration to enable human factors specialists to gather data. Whilst this would require investment, it would significantly reduce future development costs by limiting the design space.

Despite this challenge, considerable investment is aimed at de-risking many of the enabling technologies and raising readiness levels. Many technologies are moving beyond theoretical and numerical analysis into the experimental and flight test domains. Collaboration between the civil and military sectors is increasing. The possible 'dual use' nature of many of the SBJ technologies implies that their development funding can be equally 'dual sourced'.

Clearly, supersonic air travel is not an efficient means of personal conveyance; however, concerns for the environment are difficult to balance against the 'value of time' benefits offered by the SBJ concept. Currently, other than the sonic boom issue, no existing legislation is likely to preclude the development or operation of the SBJ. The possibility of further environmental laws will need to be monitored. Air travel, of which this is only a specialised form, is, and will continue to be, important to the global economy.

In summary, some of the proposed technical solutions to the many SBJ design issues appear to be founded on physics that is sufficiently well understood to permit engineering application. These solutions are starting to be validated and/or de-risked through experiment and some flight testing. No 'show-stoppers' have, as yet, been encountered. Many fundamental challenges are to be found in the non-engineering domain and include issues such as human factors, customer demand, and certification and requirements. The derivation of data to support decisions in these areas should clearly be given priority.

8.9   Acknowledgement

This chapter is based on the Royal Aeronautical Society paper: Smith H., 'A review of supersonic business jet design issues', first published in *The Aeronautical Journal*, December 2007, **111**, no. 1126.

8.10   References

1. Owen K. (2001). *Concorde: Story of a Supersonic Pioneer*, NMSI Trading Ltd, London.
2. Henne P.A. (2003). The case for small supersonic civil aircraft. AIAA Paper 2003–2555. In *AIAA/ICAS International Air and Space Symposium and Exposition*, 14–17 July 2003, Dayton, Ohio.
3. Chudoba B., Coleman G., Roberts K., Mixon B., Mixon B. *et al.* (2007). What price supersonic speed? – A design anatomy of supersonic transportation – Part 1. AIAA 2007-851. In *45th AIAA Aerospace Sciences Meeting and Exhibit*, 8–11 January 2007, Reno, Nevada.
4. Chudoba B., Coleman G., Roberts K., Mixon B., Mixon B. *et al.* (2007). What price supersonic speed? – An applied market research case study – Part 2. AIAA 2007–848. In *45th AIAA Aerospace Sciences Meeting and Exhibit*, 8–11 January 2007, Reno, Nevada.
5. Mack R.J. (2003). *A Supersonic Business-jet Concept Designed for Low Sonic Boom*, NASA/TM-2003-212435 (Corrected copy 3/22/04).
6. Baugheum S.L. (2002). *First order ozone impact of a fleet of supersonic business jets (SSBJs)*. NASA/CR-2002-211898.
7. Dutta M., Patten K. and Wuebbles D. (2004). *Parametric analysis of potential effects on stratospheric and tropospheric ozone chemistry by a fleet of supersonic business jets projected in a 2020 atmosphere*. NASA/CR-2004-213306.
8. Wei, C.F., Larson S.O., Patten K.O. and Wuebbles, D.J. (2001). Modelling of ozone reaction on aircraft-related soot in the upper troposphere and lower stratosphere. *Atmospheric Environment*, **35**: 6167–80.
9. Bagshaw M. (2009). *Cosmic Radiation in Commercial Aviation*. King's College London. Available at http://www.iaasm.org/documents/Cosmic_Radiation.pdf.
10. Smith, H. (2006). *E-5 Supersonic business jet: Design specification, DES 0500*, Cranfield University, Cranfield.
11. Wilson J.W., Goldhagen P., Rafnsson V., Clem J.M., De Angelis G. *et al.* (2003). Overview of atmospheric ionizing radiation (air) research: SST-present. *Advances in Space Research*, **32**(1): 3–16.
12. Finegold L.S., Harris C.S. and Von Gierke H.E. (1994). Community annoyance and sleep disturbance: Updated criteria for assessing the impacts of general transportation noise on people. *Noise Control Engineering Journal*, **42**: 1, January–February 1994.
13. Federal Interagency Committee on Noise (1992). *Federal Agency Review of Selected Airport Noise Analysis Issues*. Federal Interagency Committee on Noise, Burlington, MA.
14. Darder, C.M. (1998). *An Overview of NASA's HSR Program: Environmental Issues and Economic Concerns*. Available at http://techreports.larc.nasa.gov/ltrs/PDF/1998/mtg/NASA-98-eccmas-cmd.pdf.
15. United States Air Force (2000). *Supplemental Environmental Impact Statement for the Evolved Expendable Launch Vehicle Program*. Available at http://fas.org/spp/military/program/launch/eelv-eis2000/.

16. Plotkin K.J. (2003). *Recent (More or Less) Government Sonic Boom Programs*. Wyle Laboratories, El Segundo, California. Available at http://www.faa.gov/about/office_org/headquarters_offices/apl/noise_emissions/supersonic_aircraft_noise/media/4-Panel1-Plotkin-Wyle.pdf. (Accessed 14 May 2007.)

17. Kubota H. (2003). Sonic boom research in Japan. AIAA 2003-3578. In *33rd AIAA Fluid Dynamics Conference and Exhibit*, 23–26 June 2003, Orlando, Florida.

18. Maglieri D.J. and Plotkin K.J. (1991). Sonic boom. In H. H. Hubbard (ed.), *Aeroacoustics of Flight Vehicles*, Vol. 1, pp. 519–561, NASA RP 1258.

19. McCurdy D.A. (ed.) (1999). *High-Speed Research: 1994 Sonic Boom Workshop. Configuration Design, Analysis and Testing*, NASA CP-1999-209699.

20. NASA (2007). *Sonic Booms*. NASA Facts. Available at www.nasa.gov/centers/dryden/pdf/120274main_FS-016-DFRC.pdf. (Accessed 6 May 2007.)

21. Carlson H.W. (1967). Experimental and analytic research on sonic boom generation at NASA. In *Sonic Boom Research Conference*, 12 April 1967, Washington DC.

22. Howe D.C. (2003). Sonic boom reduction through the use of non-axisymmetric configuration shaping. AIAA 2003-929. In *41st Aerospace Sciences Meeting and Exhibit*, 8–9 January 2003, Reno, Nevada.

23. Kull R. (2007). *Sonic Booms and their Effects on Wildlife*. NSBIT. Available at http://www.iemr.org/Wrkgrps/uploads/Sonic_boom_Lecture.pdf. (Accessed 14 May 2007.)

24. Whitham G.B. (1952). The flow pattern of a supersonic projectile. *Communications on Pure and Applied Mathematics*, **5**: 301–48.

25. Plotkin K.J. and Maglieri D.J. (2003). Sonic boom research: History and future. In *AIAA 2003-3575, 33rd AIAA Fluid Dynamics Conference*, 23 June 2003, Orlando, Florida.

26. Walkden F. (1958). The shock pattern of a wing-body combination, far from the flight path. *Aeronautical Quarterly*, **IX**(2): 164–94.

27. George A.R. (1969). Reduction of sonic boom by azimuthal redistribution of overpressure. *AIAA Journal*, 7: 68–159.

28. Friedman M.P., Kane E.J. and Sigalla A. (1963). Effects of atmosphere and aircraft motion on the location and intensity of a sonic boom. *AIAA Journal*, **1**(6): 1327–35.

29. Hayes W.D., Haefeli R.C. and Kulsrund H.E. (1969). *Sonic Boom Propagation in a Stratified Atmosphere, With Computer Program*, NASA CR-1299.

30. Crow S.C. (1969). Distortion of sonic bangs by atmospheric turbulence. *Journal of Fluid Mechanics*, **37**: 529–63.

31. Seebass R. and George A.R. (1972). Sonic boom minimization. *Journal of the Acoustic Society of America*, **51**, No. 2 (prt 3), 686–94.

32. Seebass R. (1970). *Nonlinear Acoustic Behaviour at a Caustic*. NASA SP-255, 87–120.

33. Coulouvrat F. (2003). SOnic Boom European Research program (SOBER): Numerical and laboratory-scale experimental simulation. In *7th CEAS-ASC Workshop, Aeroacoustics of Supersonic Transport*, 13–14 November 2003, CTU-FEE Prague, Czech Republic.

34. Gill P.M. and Seebass A.R. (1975). Non-linear acoustic behaviour at a caustic: An approximate solution. In H. J. T. Nagamatsu (ed.), *AIAA Progress in Astronautics and Aeronautics*. MIT Press, Cambridge.

35. Marchiano R., Coulouvrat F. and Grenon R. (2003). Numerical simulation of shock wave focusing at fold caustics, with application to sonic boom. *The Journal of the Acoustical Society of America*, **114**(4): 1758–71.

36. Thomas C.L. (1972). *Extrapolation of Sonic Boom Pressure Signatures by the Waveform Parameter Method.* NASA TN D-6832.
37. Robinson L.D. (1991). *Sonic Boom Propagation Through an Inhomogeneous, Windy Atmosphere*, Thesis (PhD). The University of Texas at Austin.
38. Nadarajah S.K., Jameson A. and Alonso J.J. (2005). Adjoint-based sonic boom reduction for wing-body configurations in supersonic flow. *Canadian Aeronautics and Space Journal*, **51**(4).
39. Nadarajah S.K., Soucy O. and Balloch C. (2007). Sonic boom reduction via remote inverse adjoint approach. AIAA 2007–56. In *45th AIAA Aerospace Sciences Meeting and Exhibit*, 8–11 January 2007, Reno, Nevada.
40. Elmer K.R. and Joshi M.C. (1994). *Variability of Measured Sonic Boom Signatures*, Vol 1. NASA TR 191483.
41. Maglieri D.J., Sothcott V.E. and Keefer T.N. (1993). *Feasibility Study on Conducting Overflight Measurements of Shaped Sonic Boom Signatures Using the Firebee BQM-34E RPV.* NASA CR 189715.
42. Pawlowski J.W., Graham D.H., Boccadoro C.H., Coen P.G. and Maglieri D.J. (2005). Origins and overview of the Shaped Sonic Boom Demonstration Program. AIAA 2005-5. In *43rd AIAA Aerospace Sciences Meeting and Exhibit*, 10–13 January 2005, Reno, Nevada.
43. Meredith K., Dahlin J., Graham D., Haering E., Malone M. *et al.* (2005). CFD development and measurement of inlet spillage of an F-5E in supersonic flight. AIAA 2005-6. In *43rd AIAA Aerospace Sciences Meeting and Exhibit*, 10–13 January 2005, Reno, Nevada.
44. Graham D., Dahlin J., Meredith K. and Vadnais J. (2005). Aerodynamic design of shaped sonic boom demonstration aircraft, Northrop Grumman Corporation, El Segundo, CA. AIAA-2005-8. In *43rd AIAA Aerospace Sciences Meeting and Exhibit*, 10–13 January 2005, Reno, Nevada.
45. Haering A.E., Murray J.E., Purifoy D.D., Graham D.H., Meredith K.B. *et al.* (2005). Airborne shaped sonic boom demonstration pressure measurements with computational fluid dynamics comparisons. AIAA 2005-9. In *43rd AIAA Aerospace Sciences Meeting and Exhibit*, 10–13 January 2005, Reno, Nevada.
46. Graham D., Dahlin J., Page J., Plotkin K., and Coen P. (2005). Wind tunnel validation of shaped sonic boom demonstration aircraft design. AIAA-2005-7. In *43rd AIAA Aerospace Sciences Meeting and Exhibit*, 10–13 January 2005, Reno, Nevada.
47. Plotkin K., Haering E., Murray J., Maglieri D., Salamone J. *et al.* (2005). Ground data collection of shaped sonic boom experiment aircraft pressure signatures. AIAA-2005-10. In *43rd AIAA Aerospace Sciences Meeting and Exhibit*, 10–13 January 2005, Reno, Nevada.
48. Plotkin K., Martin L., Maglieri D., Haering E. and Murray J. (2005). Pushover focus booms from the shaped sonic boom demonstrator. AIAA-2005-11. In *43rd AIAA Aerospace Sciences Meeting and Exhibit*, 10–13 January 2005, Reno, Nevada.
49. Morgenstern J., Arslan A., Lyman V. and Vadyak J. (2005). F-5 shaped sonic boom demonstrator's persistence of boom shaping reduction through turbulence. AIAA-2005-12. In *43rd AIAA Aerospace Sciences Meeting and Exhibit*, 10–13 January 2005, Reno, Nevada.
50. Kandil O., Ozcer I., Zheng X. and Bobbitt P. (2005). Comparison of full-potential propagation-code computations with the F-5E 'Shaped Sonic Boom Experiment' program. In *AIAA-2005-13, 43rd AIAA Aerospace Sciences Meeting and Exhibit*, 10–13 January 2005, Reno, Nevada.

51. Howe D.C. and Henne P.A. (2006). Improved sonic boom scaling algorithm. AIAA 2006-27. In *44th AIAA Aerospace Sciences Meeting and Exhibit*, 9–12 January 2006, Reno, Nevada.

52. Morgenstern J.M. (2004). Wind tunnel testing of a sonic boom minimised tail-braced wing transport configuration. In *10th AIAA/ISSMO Multidisciplinary Analysis and Optimization Conference*, Albany, 30–31 August 2004, New York.

53. Lyman V. and Morgenstern J.M. (2004). Calculated and measured pressure fields for an aircraft designed for sonic-boom alleviation. AIAA 2004-4846. In *22nd Applied Aerodynamics Conference and Exhibit*, 16–19 August 2004, Providence, Rhode Island.

54. Tam T., Ruffin S., Yates L., Gage P., Bogdanoff D. *et al.* (2000). Sonic boom testing of Artificially Blunted Leading Edge (ABLE) concepts in the NASA Ames Aeroballistic Range. AIAA 2000-1011. In *38th Aerospace Sciences Meeting and Exhibit*, 10–13 January 2000, Reno, Nevada.

55. Makino Y. and Noguchi M. (2003). Sonic boom research activities on unmanned scaled supersonic experimental airplane. AIAA 2003-3574. In *33rd AIAA Fluid Dynamics Conference and Exhibit*, 23–26 June 2003, Orlando, Florida.

56. Chung H-S. and Alonso J.J. (2002). Design of a low-boom supersonic business jet using Cokriging approximation models. AIAA 2002-5598. In *9th AIAA/ISSMO Symposium on Multidisciplinary Analysis and Optimization*, 4–6 September 2002, Atlanta, Georgia.

57. Lee H., Morganstern J.M. and Aminpour H. (2005). *Aircraft with active centre of gravity control*. United States Patent 6913228.

58. Morgenstern J.M. and Arslan A.E. (2006). *Aircraft thickness/camber control device for low sonic boom*. United States Patent 7070146.

59. Howe D.C. (2005). Improved sonic boom minimization with extendable nose spike. AIAA 2005-1014. In *43rd AIAA Aerospace Sciences Meeting and Exhibit*, 10–13 January 2005, Reno, Nevada.

60. Simmons F. and Freund D. (2005). Morphing concept for quiet supersonic jet boom mitigation. AIAA 2005-1015. In *43rd AIAA Aerospace Sciences Meeting and Exhibit*, 10–13 January 2005, Reno, Nevada.

61. Simmons F. and Spivey N.D. (2007). Quiet spike: The design and validation of an extendable nose boom prototype. AIAA 2007-1774. In *48th AIAA/ASME/ASCE/ASC Structures, Structural Dynamics and Materials Conference*, 23–26 April 2007, Waikiki, Hawaii.

62. Miles R.B., Martinelli L., Macheret S.O., Shneider M., Girgis I.G. *et al.* (2002). Suppression of sonic boom by dynamic off-body energy addition and shape optimisation. AIAA 2002-0150. In *40th AIAA Aerospace Sciences Meeting and Exhibit*, 14–17 January 2002, Reno, Nevada.

63. Jameson A., Sriram, Martinelli L., Cliff S. and Thomas S. (2005). Aerodynamic shape optimisation of transonic and supersonic aircraft configurations. AIAA 2005-1013. In *43rd Aerospace Sciences Meeting*, 10–13 January 2005, Reno, Nevada.

64. Chung H-S. and Choi S. (2003). Supersonic business jet design using a knowledge-based genetic algorithm with an adaptive, unstructured grid methodology. AIAA-2003-3791. In *21st AIAA Applied Aerodynamics Conference*, 23–26 June 2003, Orlando, Florida.

65. Alauzet F. and Mohammadi B. (2003). *Optimisation 3D du nez d'un Supersonic Business Jet basee sur l'adaptation de Maillages. Application a la reduction du bang sonique*. INRIA, Report No. 5053.

66. Daumas, L., Heron, N., Johan, Z., Roge, G. and Vigneron, S. (2006). Aerodynamic design process of a supersonic business jet. AIAA 2006-3459. In *24th Applied Aerodynamics Conference*, 5–8 June 2006, San Francisco.

67. Rallabhandi, S.K. and Mavris N. (2006). Design and analysis of supersonic business jet concepts. AIAA 2006-7702. In *6th AIAA Aviation Technology, Integration and Operations Conference (ATIO)*, 25–27 September 2006, Wichita, Kansas.

68. NASA (2007). *NASA Technology Facts, F-XL Supersonic Laminar Flow*. Available at http://www.nasa.gov/centers/dryden/about/Organizations/Technology/Facts/TF-2004-12-DFRC.html. (Accessed 6 May 2007.)

69. Marshall L.A. (2000). Summary of transition results from the F-16XL-2 supersonic laminar flow control experiment. AIAA 2000-4418. In *AIAA August 2000 Conferences*, 14–17 August 2000, Denver, Colorado.

70. Hartwich P.M., Burroughs B.A., Herzberg J.S. and Wiler C.D. (2003). Design development strategies and technology integration for supersonic aircraft of low perceived sonic boom. AIAA 2003-0556. In *41st Aerospace Sciences Meeting and Exhibit*, 6–9 January 2003, Reno, Nevada.

71. Choudhari M., Chang C-L., Streett C. and Balakumar P. (2003). Integrated transition prediction: A case study in supersonic laminar flow control. AIAA 2004-0973. In *41st Aerospace Sciences Meeting and Exhibit*, 6–9 January 2003, Reno, Nevada.

72. Whurr J. (2003). Propulsion system concepts and technology requirements for quiet supersonic transports. In *7th CEAS-ASC Workshop, Aeroacoustics of Supersonic Transport*, 13–14 November 2003, CTU-FEE Prague, Czech Republic.

73. Berton J.J., Haller W.J., Senick P.F., Jones S.M. and Seidel J.A. (2005). *A comparative propulsion system analysis for the high-speed civil transport*. NASA/TM-2005-213414.

74. Bruckner R.J. (2002). Conceptual design of a supersonic business jet propulsion system. AIAA 2002-3919. In *38th AIAA Joint Propulsion Conference*, 7–10 July 2002, Indianapolis, Illinois.

75. Debiasi M. (2001). Cycle analysis for quieter supersonic turbofan engines. AIAA 2001-3749. In *37th AIAA Joint Propulsion Conference and Exhibit*, 8–11 July 2001, Salt Lake City, Utah.

76. Richter H. and Cosner A.A. (2003). Meeting future fuel burn and emission requirements for corporate and regional aircraft by effective technologies. AIAA 2003-2563. In *AIAA International Air and Space Symposium and Exposition*, 14–17 July 2003, Dayton, Ohio.

77. Conners T.R., Howe D.C. and Whurr J.R. (2005). Impact of engine cycle selection on propulsion system integration and vehicle performance for a quiet supersonic aircraft. AIAA 2005-1016. In *43rd AIAA Aerospace Sciences Meeting and Exhibit*, 10–13 January 2005, Reno, Nevada.

78. Stone J.R., Krejsa E.A., Halliwell I. and Clark B.J. (2000). Noise suppression nozzles for a supersonic business jet. AIAA 2000-3194. In *36th AIAA/ASME/SAE/ASEE Joint Propulsion Conference and Exhibit*, 16–19 July 2000, Huntsville, AL.

79. Klimov A., Bityurin V. and Mironov A. (2003). Jet noise reduction by plasma formation. In *7th CEAS-ASC Workshop, Aeroacoustics of Supersonic Transport*, 13–14 November 2003, CTU-FEE Prague, Czech Republic.

80. Howe D. (2002). Engine placement for sonic boom mitigation. AIAA 2002-0148. In *40th AIAA Aerospace Sciences Meeting and Exhibit*, 14–17 January 2002, Reno, Nevada.

81. Rodriguez D.L. (2007). Propulsion/airframe integration and optimisation on a supersonic business jet. AIAA 2007-1048. In *45th AIAA Aerospace Sciences Meeting and Exhibit*, 8–11 January 2007, Reno, Nevada.

82. Garzon G.A. (2007). Use of a translating cowl on a SSBJ for improved takeoff performance. AIAA 2007-25. In *45th AIAA Aerospace Sciences Meeting and Exhibit*, 8–11 January 2007, Reno, Nevada.

83. Beier T.H. and Heaton P. (2005). *High Speed Research Program Sonic Fatigue Summary Report*, NASA/CR-2005-213742.

84. Martins J.R.R.A., Alonso J.J. and Reuther J.J. (2003). High-fidelity aero-structural design optimization of a supersonic business jet. AIAA 2002-1483. In *43rd AIAA Structures, Structural Dynamics and Materials Conference*, 22–25 April 2002, Denver, Colorado.

85. Choudhari M., Chang C.-L., Streett C. and Balakumar P. (2003). Integrated transition prediction: A case study in supersonic laminar flow control. AIAA 2004-0973. In *41st Aerospace Sciences Meeting and Exhibit*, 6–9 January 2003, Reno, Nevada.

86. Baklanov V. and Pastnov S. (2003). New technologies for required vibroacoustical characteristics in pressurized cabin of supersonic aircraft and execution of norms of noise on land (at take-off). In *7th CEAS-ASC Workshop, Aeroacoustics of Supersonic Transport*, 13–14 November 2003, CTU-FEE Prague, Czech Republic.

87. Williams, D.M., Waller, M.C., Koelling J.H., Burdette D.W., Capron W.R. *et al.* (2001). *Concept of Operations for Commercial and Business Aircraft Synthetic Vision Systems*, NASA/TM-2001-211058.

88. Mack R.J. (2003). *A Supersonic Business-jet Concept Designed for Low Sonic Boom*, NASA/TM-2003-212435. (Corrected copy 3/22/04.)

89. Aronstein D.C. and Schueler K.L. (2004). Conceptual design of a sonic boom constrained supersonic business aircraft. AIAA 2004-697. In *42nd AIAA Aerospace Sciences Meeting and Exhibit*, 5–8 January 2004, Reno, Nevada.

90. Komadina S., Drake A. and Bruner S. (2002). Development of a quiet supersonic aircraft with technology applications to military and civil aircraft. AIAA 2002-0519. In *40th Aerospace Sciences Meeting and Exhibit*, 14–17 January 2002, Reno, Nevada.

91. Phan L.L., Yamaoka Y. and Mavris D.N. (2003). Implementation and benefits of variable geometry wings for a supersonic business jet. AIAA 2003-6812. In *AIAA 3rd Annual Aviation Technology, Integration and Operations Technology*, 17–19 November 2003, Denver, Colorado.

92. Simmons F. and Freund D. (2005). Wing morphing for quiet supersonic jet performance – variable geometry design challenges for business jet utilization. AIAA 2005-1017. In *43rd AIAA Aerospace Sciences Meeting and Exhibit*, 10–13 January 2005, Reno, Nevada.

93. Buonanno M.A. and Mavris D.N. (2004). Small supersonic transport concept evaluation using interactive evolutionary algorithms. AIAA 2004-6301. In *AIAA 4th Aviation Technology, Integration and Operations Forum*, 20–22 September 2004, Chicago, Illinois.

94. Briceño S.I, Buonanno M.A., Fernández I. and Mavris D.N. (2002). A parametric exploration of supersonic business jet concepts utilizing response surfaces. AIAA 2002-5828. In *AIAA Aircraft Technology, Integration, and Operations (ATIO) 2002 Technical Forum* 1–3 October 2002, Los Angeles, California.

95. Li P., Seebass R. and Sobieczky H. (1995). The sonic boom of an oblique flying wing SST, CEAS/AIAA-95-107 *1st Joint Aeroacoustics Conference*, Munich.

96. Federal Aviation Administration (1963). *FAR 91.817 Part 91 General operating and flight rules subpart I – Operating noise limits*. Federal Airworthiness Administration, Department of Transportation.

97. Nam T., Shih K. and Mavris D.N. (2003). Assessment of environmental and regulatory uncertainty impacts on propulsion system design. AIAA 2003-6805. In *AIAA 3rd Annual Aviation Technology, Integration and Operations Technical Forum*, 17–19 November 2003, Denver, Colorado.

98. The Air Registration Board (1963). *Supersonic Transport Aircraft TSS Standard*, No. 0 issue 1, 5th November 1963.

99. AINonline (2007). *Gulfstream Demonstrates 'Quiet Spike' for Potential Supersonic Bizjet*. Available at http://ain.gcnpublishing.com/content/news/single-news-page/article/gulfstream-demonstrates-quiet-spike-for-potentialsupersonic-bizjet/?no_cache=1&cHash=cbde58a2c5. (Accessed 6 May 2007.)

100. Gulfstream News (2007). *Gulfstream Continues Research Efforts in Sonic Boom Suppression*. Available at http://www.gulfstream.com/news/releases/2005/051108d.htm. (Accessed 6 May 2007.)

101. Aerion Corporation (2007). Available at http://www.aerioncorp.com/. (Accessed May 2007.)

102. AINonline (2005). *Two Companies Reveal Supersonic Bizjet Plans*. Available at http://www.ainonline.com/issues/11_04/11_04_twocompaniesp1.html. (Accessed 24 May 2005.)

103. Aerion Corporation (2007). *News Update*. Available at http://www.aerioncorp.com/about_aerion-update-report.html. (Accessed 6 May 2007.)

104. Supersonic Aerospace International (2007). Available at http://www.saiqsst.com/. (Accessed May 2007.)

105. Hagerman E. (2007). All sonic, no boom. *Popular Science*, March 2007.

106. HISAC (2007). Available at http://www.hisacproject.com/index.html. (Accessed 6 May 2007.)

107. CORDIS Search (2007). Available at http://cordis.europa.eu/search/index.cfm?fuseaction=proj.simpledocumentlucene&HD_ID=7977959&CFID=475029&CFTOKEN=39551818. (Accessed 6 May 2007.)

108. AINonline (2007). *Aerion Expects Partners on SSBJ by NBAA 2007*. Available at http://ain.gcnpublishing.com/content/news/single-news-page/article/aerion-expects-partners-on-ssbj-by-nbaa-2007/?no_cache=1&cHash=c3aae389b1. (Accessed 6 May 2007.)

109. Yoshimoto M. and Uchiyami N. (2003). Optimization of canard surface positioning of supersonic business jet for low boom and low drag design. AIAA 2003-3576. In *33rd AIAA Fluid Dynamics Conference and Exhibit*, 23–26 June 2003, Orlando, Florida.

110. Wall R. (2007). Gaining Speed. *Aviation Week & Space Technology*, 12 March 2007.

111. Pietremont N. and Deremaux Y. (2005). *Executive Public Summary of the Three Preliminary Aircraft Configuration Families*. 10 November 2005. HISAC-T-5-1-1. Available at http://www.hisacproject.com/publications.html.

112. DARPA (2002). *Quiet Supersonic Platform Phase II Contractors Selected*. Available at http://www.darpa.mil/body/news/2002/qspph2.pdf. (Accessed 6 May 2007.)

113. Tupolev (2007). Available at http://www.tupolev.ru/. (Accessed on 6 May 2007.)

Part II
Change

9

The process of innovation in aeronautics

M. HENSHAW, Loughborough University, UK

Abstract: It is argued that, within organisations, innovation is a cultural attribute, rather than something that is based on process. Nevertheless, one can conceive an informal process through which change is introduced from its first conception through to realisation as comprising the following activities: scan, focus, resource, implement, learn (Bessant, 2003). Successful innovation is the property of organisations that take an holistic approach to design and that pay significant attention to the last activity of the notional process, i.e. learning. Innovation concerns not only invention, but the vision and ability to take an idea through to application (which may be commercial success or delivery of societal benefit). In many respects, it is the tacit knowledge that an organisation has at its disposal that determines its levels of creativity and innovation. An historical perspective on innovation in aeronautics is also provided.

Key words: holistic design, innovation culture, learning organisation.

9.1 Introduction

The advances of aeronautics during the last 100 years or so have been staggering; technology capable of a short flight at about 7 mph has developed into machines capable of supersonic flight, regular flight carrying more than 800 passengers, space flight, etc. Furthermore, these astounding innovations have turned a novelty into a powerhouse of the world economy. But many would assert (e.g. Young, 2007; Kroo, 2004) that the first 50 of those years were resplendent with significant innovations, whereas the next 50 were characterised by incremental, evolutionary development of established concepts. However, if the nature of the innovation process has changed in terms of the product itself, it is still worthwhile considering the innovation process as applied to the means of production and use of aeroplane technology. Rothwell (1992) has noted that the speed of development became an important consideration in the 1980s, driven, one might conjecture, by the economic circumstances of the day, and that this required innovation in terms of the organisations engaged in product development. But what do we mean by innovation process, and is there a process of innovation? Through consideration of innovation across a number of industrial sectors and the changes that have taken place in the challenges faced by aviation, we shall explore the concept of an innovation process and determine some of the features that enable effective innovation.

9.2   Definitions and sources of confusion

Before attempting to define and explain the concept of an innovation process, it is important to first arrive at a definition of innovation itself. Unfortunately, there is confusion about the term. Bessant (2003) has noted that many confuse invention with innovation, regarding innovation as the eureka moment (a notional event we shall refute later) and others that it is simply about science and technology. There is also disagreement about whether innovation concerns radical changes or whether it can include incremental steps. The word innovate comes from the Latin *innouare*, which the concise dictionary of English etymology interprets as to renew or make new (Skeat, 1993). Hence, to innovate is to introduce something new. With this definition, we shall understand innovation to imply the introduction of a change that may be major or minor, and assert that it must include the realisation of the change, not just the idea from which the change may come. Figure 9.1, adapted from Tidd and Bessant (2009), shows the range of types of innovation.

There is further confusion in the meaning of the term 'innovation process'. In a good deal of the literature, the term innovation process is used synonymously with product development; this is an unhelpful confusion, because the quality of innovation then becomes measurable only through efficiency (reduced time or costs) in bringing products to market. These values are not those that are most applicable to the commonly accepted examples of the greatest innovators. A more subtle, but nonetheless important, distinction that we shall make is between the 'process of innovation' and the 'business processes that support innovation.' Technology readiness levels (TRLs) (Young, 2007; Mankins, 1995) are a means

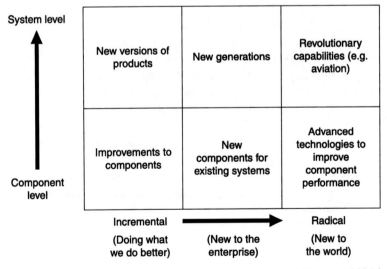

9.1 Dimensions of innovation. (Adapted from Tidd and Bessant, 2009.)

of assessing maturity in the process of product development and provide a framework for managing project risk; they are not, however, related to innovation beyond influencing decision makers in the level of change they may be prepared to accept. The use of TRLs generally encourages organisations towards a more incremental approach to product development and perhaps encourages an incremental innovation environment. To be clear, the process of innovation concerns the manner in which a change is introduced from its first conception through to realisation, whereas the processes that support innovation are components of environments in which innovation can flourish.

Societal evaluation of an innovation, i.e. whether it is regarded as innovative or not, depends upon the perspective from which it is viewed. Incremental innovation is almost always measured in terms of cost, time or risk reduction, but probably directly affects a limited population (e.g. the company within which it occurs). Historically, radical innovation affects large proportions of society but very often does not seem to result in financial gain for the innovator. Rothwell (1992) has defined the term 'industrial innovation' to mean the commercialisation of technological change. Similarly, Dodgson *et al.* (2002) have defined innovation '. . . as the productive use of knowledge manifested in the successful development and introduction of new products, processes and/or services'. They go on to provide a description of the fifth-generation innovation process (Rothwell, 1992) through a set of features concerned with organisation, creativity, strategy, knowledge-based competition, lean production and computer-integrated manufacturing. However, this appears to be a set of environmental factors to support accelerated exploitation, rather than a process for innovation itself.

In all contexts (business or otherwise) innovation should be regarded as an important characteristic of agility. In this sense, innovation is the process through which an idea of use is turned from an idea into reality to gain market advantage or individual prestige, secure an escape or effect a capture. In this most fundamental consideration, innovation is simply a process of turning an idea into a tangible benefit. Indeed, the Department of Trade and Industry (DTI, 2005) offer a very straightforward definition of innovation as 'the successful exploitation of new ideas'. This works very well, provided one keeps in mind that success can be measured in many different ways and not all of them are financial. The Confederation of British Industry, recognising the diversity of advances that might be termed innovations, remark that: 'About the only thing all successful innovations have in common is that they fill a need and so have value. But quite often that need, and that value are not perceived until the innovation exists – at least as a concept' (Townsend, 1990).

One of the greatest innovators, Thomas Edison, recognised that the real challenge of innovation was not coming up with new ideas, but making those inventions work technically and commercially. Noting this, Tidd and Bessant (2009) remark that innovation is more than coming up with good ideas; it is the process of growing them to practical use. That is to say, innovation is, itself, a process. Our task, then, is to define what that process might be. The process must have a clear and focused direction, according to Tidd and Bessant (2009, p. 19), and they assert that a critical requirement is the organisational conditions that allow focused creativity.

9.3   How to measure innovation

If you type the words 'Isambard Kingdom Brunel Financial Success' into Google, the first thing that strikes you about the results is that, mostly, Brunel's innovations were not financially successful. That is to say, the man often regarded as the greatest British innovator, the impact of whose innovations is still much in evidence 150 years after his death, did not achieve commercial success with many of his innovations. From the world of aviation we can reflect that, whilst Concorde was hugely innovative in a technical sense, it did not achieve the same success commercially. Berkum (2007), in his excellent debunking of the myths of innovation, has remarked that 'an ideas man is not motivated by wealth, but by the desire to succeed at something (technical)'. Such a man does not measure innovation by a scale other than the technical novelty. The difference between an inventor and an innovator is that the latter not only has an idea, but also the vision and drive to take it through to completion. The question, then, of whether something is innovative or not, or whether it is more or less innovative than something else, cannot be answered through a generalised measurement, because the metrics vary both according to context and according to the interests and passions of the beholder. Some comparison between innovations is, however, possible in a subjective fashion. Tidd and Bessant (2009) have described some aspects of innovation; these are:

- Degree of novelty – whether the innovation is incremental, i.e. part of continuous improvement of business processes, or radical.
- Platforms and families of innovation – which essentially implies that a base innovation can be stretched to create a family of products that derive from the base.
- Discontinuous innovation – this covers innovations that completely change the rules of the game. Sometimes this type of change is referred to as a disruptive technology, although, strictly it is not the technology that is generally disruptive, but the use to which it is put.
- Level of innovation – referring to Fig. 9.1, this distinguishes between component and architecture level.
- Timing – this refers to the innovation lifecycle. For instance, at the beginning it may be the product itself that is innovative, but later in the lifecycle innovation may be applied to the way that the product is developed, or marketed.

Despite the difficulty in measuring innovation itself, there are some measurable parameters associated with innovation that are important from the point of considering the innovation process.

One such parameter is risk, without which the notion of innovation would probably be fairly meaningless. If the start of the innovation process is an idea (of how a system might work), then there is risk associated with the development of that idea into the realised system; which might not work! The risk is often financial, but it could be the risk to reputation, or to personal or personnel safety, etc. The processes through which the innovation is developed will for the most part be concerned with turning uncertainty into risk, and then reducing risk as investment increases until the idea is realised. Many organisations that deal with complex systems – such as aircraft – have adopted the concept of technology readiness levels (TRLs) to characterise the maturity of the innovative system under development. These were developed by NASA for complex projects many years ago and summarised in a white paper by Mankins (1995); there are slight variations in the definition of the TRLs by different organisations, but those of NASA are shown in Table 9.1.

The TRLs represent a gradual reduction in technical risk as the system moves from one level to the next; the financial risk, however, is not monotonic. It may be relatively modest at the initial TRLs, but as the development enters TRLs 4–7 the required investment generally increases significantly, broadly following the S-curve typical of several aspects of innovation, especially take up of new products by customers (Bass, 1969). This range of TRLs (4–7) is the region in which many projects fail – the so-called TRL valley of death – which implies that an important part of any innovation process that industry might choose to adopt must be designed to maximise the chances of pulling an idea of a system through that phase of the development. TRLs are sometimes equated to the innovation

Table 9.1 Technology readiness levels

TRL	Definition
TRL 1	Basic principles observed and reported
TRL 2	Technology concept and/or application formulated
TRL 3	Analytical and experimental critical function and/or characteristic proof of concept
TRL 4	Component and/or breadboard validation in laboratory environment
TRL 5	Component and/or breadboard validation in relevant environment
TRL 6	System/subsystem model or prototype demonstration in a relevant environment (ground or space)
TRL 7	System prototype demonstration in a space environment
TRL 8	Actual system completed and flight qualified through test and demonstration (ground or space)
TRL 9	Actual system flight proven through successful mission operations

(Source: Mankins, 1995)

process, but they are really related to the product development process and the management of risk, and this does not map exactly to innovation.

The appetite for all sorts of risk is a significant factor in innovation. During the development of jet flight and the bid to break the sound barrier, the advances in aviation were stupendous, but during periods in the 1950s test pilots were being killed at a rate of about one per week (see, for example, Wolfe, 1979). World-changing innovation involves risk-taking, courage, and leadership.

The CBI asserted that innovation is disproportionately due to small companies (as opposed to large companies) because they are prepared to take greater risks (Townsend, 1990), and the imaginative aircraft designer, John McMasters, is said to have rued the consolidation of the aerospace industry into fewer, larger companies because of its negative impact on risk-taking and, hence, innovation (Kroo, 2009). Risk, then, is a critical factor in the innovation process for better or for worse.

9.4   The innovation process

'How do you systemise innovation?' 'You don't,' replied Steve Jobs, chairman and CEO of Apple Computers (Jobs, 2004). 'You hire good people who will challenge each other every day to make the best products possible ... Our corporate culture is simple.' This is a consistent message among innovative companies: innovation is a cultural attribute rather than something based on process. Nevertheless, researchers have analysed the innovation process and reported it broadly in terms of implicit (perhaps informal) processes as well as in terms of explicit processes. We shall start by considering the implicit stages of an innovation process.

The traditional view of innovation is as a process that begins with an idea (often supposedly driven by a particular need or necessity). This is referred to as the creative part of the innovation process. There then follows a period during which the idea is developed into a product or benefit that can be marketed so as to realise a financial gain (Fig. 9.2). In his insightful text, Berkum (2007) explains that this creative moment, the eureka moment, is not the sudden emergence of an idea, but rather the fitting of the last piece of a jigsaw that shows the inventor how a change may be achieved. By this, Berkum implies that an inventor may spend years mulling over a problem, gradually building up the different parts of the solution, until at some point the last piece of the problem skips into place and the invention is ready to be developed. The development phase is shaped by the same prevailing environment that probably shaped the idea in the first place. Considerable creativity will be required as the idea is shaped into a product; matching it to the perceived needs or desires of customers or society in general. Thus, we suggest that innovation is a process of research/investigation in the domain(s) of interest, the emergence of an idea for change, development of the idea to fit the environment in which it must be realised, and then realisation. Creativity is present throughout

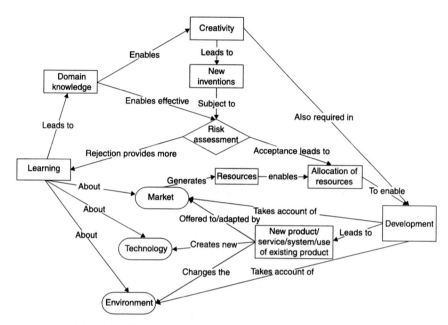

9.2 Notional innovation process, implicit in learning/innovative organisations.

(Fig. 9.2). Innovation could take place by using a technology (for instance) in a new environment or for a new purpose.

Bessant and Tidd (2007) suggest that there are three basic stages to the innovation process: generate new ideas, select the good ones, implement them. Simple. However, this is overly simple because it implies a eureka moment, whereas the organisation or individual must develop sufficient domain knowledge to recognise when the last piece of the jigsaw fits into place. These three stages might be sufficient to be innovative once (lucky), but the challenge is to continue to be innovative. This requires at least two additional stages: resourcing and learning. Thus, as noted by Bessant (2003) and many others, the process of innovation to support continued innovation in an organisation is:

- Scan – this may include horizon scanning internally and externally, or may be equivalent to having an idea,
- Focus – this means that potential innovations are filtered to determine the most likely for success,
- Resource – having decided on the ideas to develop, the development process must be adequately resourced (financially, physically, and intellectually),
- Implement – put the idea into effect; this may include experimentation to refine the implementation plan, manufacture of prototypes, user trials, design, build, and market,

- Learn – capture and manage the knowledge generated by the rest of the process to enable future innovation.

There are many more formalised definitions of the innovation process. Within the aerospace sector, Truman and De Graaf (2006) describe the innovation process as five stages comprising:

1 Technology watch and awareness,
2 Ideas creation, organisation and ingestion,
3 Ideas incubation,
4 Research programmes,
5 Implementation, production and deployment.

The focus of this process is clearly the front end part of idea generation and is perhaps rather light on the tremendously challenging processes for implementation (i.e. through the TRL valley of death). Hamel (2002) suggests: imagine, design, experiment, assess, scale, as the process for innovation; and the CBI (Townsend, 1990) the more detailed: understand company strengths, identify opportunity, screen (go/no-go), concept development – design, assess the potential, manage the technology, evaluate (go/no-go), launch. Both of these appear to regard the innovation process as pretty well synonymous with the product development process. This is not entirely surprising, nor incorrect, but these definitions imply a rather sequential, mundane view of innovation.

Freeman (1974) described innovation as a process that includes 'The technical, design, manufacturing, management and commercial activities involved in the marketing of a new (or improved) product or the first use of a new (or improved) manufacturing process or equipment.'

This is important because it specifically notes that changes must occur in organisational, management, production and commercial aspects of the firm. Hendry *et al.* (2002) develop these ideas further by first noting the importance of the whole supply chain, not just the specific firm, a point we shall return to below, and second, that a pragmatic method of technology development involves a hybrid approach in which an incremental development path checks the newly conceived technology against its fit with existing technologies and the expectations of the market. This approach, they suggest, allows for radical ideas generated through good knowledge management and careful risk management in development.

Although the basic activities of innovation discussed above are pretty consistent from age to age, the economic and business environment in which innovation takes place has tended to change the philosophical approach to it over the years. Rothwell (2002) has described five generations of innovation processes. The first generation (1950s and early 1960s) is characterised by technology push. This was an era of significant investment in research and development so that new ideas were generated and these stimulated demand. The innovation process was sequential – similar to those described above. This period coincided with many of the significant advances in aeronautical and astronautic engineering. Throughout human history, war has been a significant driver of innovation, and the Cold War (which produced the space race) was a major factor in this innovative period in the history of aviation.

Second-generation innovation processes (mid-1960s and 1970s) were characterised by market pull, in which inventiveness is stimulated by rather more specific needs. This reflected the need for strong corporate growth, and many of the new technologies in this period were extensions of existing technologies rather than radical new ideas. The innovation process itself remained sequential. It is worth pausing briefly to consider these two approaches (generations): in the first, wild and wacky ideas can flourish, stimulating entrepreneurs to say, 'gosh, with this new technology I could do such and such.' Whereas, in the case of market pull, the entrepreneur says, 'I need something that will do this better, can we not improve this or that technology somehow.'

The high inflation and concomitant pressure on resources during the 1970s and mid-1980s led to third-generation innovation processes, in which research and development was much more tightly coupled to marketing. Rothwell (2002) comments that the innovation process remained sequential but with feedback loops. Thus technology development is tied much more closely to the longer-term commercial strategy, there is little opportunity for wild and wacky ideas, and a much more complex organisational structure is required around the innovation process. To some extent, this could be seen as stifling the creativity part of innovation, with a much greater focus on incremental development and risk reduction. This coincided with a good deal of pressure on the aviation sector with, for instance, UK aerospace being nationalised to secure its competitiveness against the US private companies.

Fourth-generation innovation processes are described as being an integrated mode, in which producers and users are more closely linked, and development takes place in parallel. The aerospace sector was still contracting, with the US companies consolidating from 17 down to just three. Nevertheless, considerable innovation took place, but focused rather on the manufacturing processes and an explosion in the use of IT (and later networked IT) systems. This was strongly influenced by the production efficiency achieved in Japanese manufacturing, which enabled non-sequential approaches to be taken. To what extent the non-sequential nature is associated with the production processes, rather than the innovation process, is not clear, but it is certain that the timescales for production were considerably reduced. This period occupies the 1980s and early 1990s. The fifth-generation innovation process begins from about 1995 and is characterised by systems integration and networking.

Croom and Batchelor (1997) consider two aspects of industrial capability: those of the resources possessed by an individual company, and those associated with relational capabilities that are realised through the interaction of the organisations within the supply chain. The latter is important because it recognises that competitive advantage comes through the success of the relationships in the supply chain and, more particularly, through the knowledge that is generated thereby. We consider innovation as an aspect of management of knowledge below.

What one might term 'organised innovation' has been a feature of aerospace and other sector strategies in recent years; see, for example, BAE Systems (2009) and Merrill (2006), in which explicit reference is made to the need to pull innovation through from every part of the supply chain, and mechanisms are put in place to achieve this. Leifer *et al.* (2000) focused on the value chain as the mechanism through which innovation will be achieved. They argue that all of the value chain must benefit in order for the innovation to be realised, and that for this to occur the technical and business models must be bound together. The company under consideration, wherever it sits in the overall supply chain, must consider what its role will be in the new value chain created by the innovation. For aerospace this must be a critical consideration because of the dependencies between the many stakeholders in the value chain. The innovation process must underpin the process through which value is realised in the entire value chain. Leifer *et al.* further emphasise, as do many other authors, the critical role of leadership in innovation; unless the chief executive officer and his/her fellow directors drive an innovation culture, then the organisation cannot thrive. The components of such a culture are discussed in the next section.

9.5   Innovation environments

The preceding discussion on the innovation process provides an indication of the type of environment in which innovation will flourish. Clearly, it must be an environment in which staff are encouraged to express ideas, are sufficiently well informed to understand the possibilities for innovation, and aware of the business and technical environment through scanning for new technologies and market opportunities. The organisation must have a sufficient appetite for risk to pursue opportunities and be sufficiently agile to reconfigure (change itself) to enable the development of new offerings to the market. Needless to say, it must be sufficiently well capitalised to invest in new developments. Generally, freedom seems to be the most common attribute of organisations regarded as innovative (Kao, 1996); freedom to think, freedom to express ideas, freedom to be wrong. Dyson (2002) has described the environment and culture of his company in these terms: everyone who starts at Dyson works on a product from their first day, engineering and design are not separate, everyone is treated equally, remuneration is good, the physical working environment is designed to be pleasing. It is worth noting the result of a recent study by Sinclair *et al.* (2010) that the working environment has a much greater impact on the likelihood of team success than do the individual competences of the team members.

However, the key attribute with regard to Dyson's description is holism. Dyson take an holistic approach to design. Similarly, Bessant (2003) remarks that 'successful innovation management is not about doing one thing well but rather organising and managing a variety of different elements in an integrated and strategically coherent fashion.' The environment must enable big-picture thinking and must enable teamwork. Jim McNerney (Chairman and CEO of the Boeing Company in 2006) described innovation in aviation as 'a team sport, not a solo sport' (McNerney, 2006). In the environments espoused by Kao (1996), the team members should be free to interact in many and various ways to drive creativity; Atkinson and Moffat (2005) assert that enabling self-organisation can be a source of innovation. But for aerospace companies such freedom poses significant challenges, since engineering governance demands rather rigid structures.

The motivators for innovation include:

- Adversity and competition: the Cold War provided adversity and motivated a great deal of innovation in aerospace; competition is not adversity (Aghion *et al.*, 2002), but it drives innovation through survival needs.
- Rewards and prizes are significant motivators, especially for technical people (Young, 2007).
- The working environment, as noted above. Young (2007) has also noted that access to sophisticated design tools has a motivational effect.

The inhibitors of innovation could be stated as the opposites of the above, but a significant inhibitor is a culture of unwarranted criticism, blame and secrecy. This tends to create an atmosphere in which employees are fearful of expressing their ideas and are not full participants in the innovation process. Jones and Beckinsale (2001) have drawn particular attention to the negative effect of micro-politics, in which people work on their own careers, which can provide either encouragement or hostility towards ideas. Within a single organisation this can be a significant inhibitor, but the likelihood of it being a factor is increased by the enterprise nature of aerospace development, in which there is the risk of tribalism between the staff of the various organisations that make up the enterprise.

Another, rather curious, effect is that innovation success itself may lead, over time, to a decrease in innovativeness. This is because organisations become locked into particular products that discourage change in the future.

9.6   Innovation viewed as a management of knowledge problem

'What's wrong with bullet lists?' ask Shaw *et al.* (2002), 'Bullet lists encourage us to be intellectually lazy . . .' they respond to their own question. In an interesting article about organisational culture, Shaw *et al.* (2002) insist on the importance of detailed and specific knowledge about the organisation and its products as a key factor in innovation. There are probably few aerospace companies that do not succumb to the temptation of planning through bullet lists. But this article highlights an important aspect of innovation. Innovation is about the management of knowledge.

To be clear, this is not a matter of knowledge management, which tends to be information and communication technology (ICT) focused and deals with explicit knowledge, but rather the management of knowledge, which deals with both tacit and explicit knowledge. Indeed, in many respects it is the tacit knowledge that an organisation has at its disposal that determines its levels of creativity and innovation. Blackler (1995) describes an organisation's knowledge as being hosted in five forms:

- Embrained knowledge that depends on the conceptual skills of the employees,
- Embodied knowledge that is action-oriented (e.g. a craftsman's skills),
- Encultured knowledge that is 'the way we do things' (in a similar way to the stories told by 3M in the discussion by Shaw *et al.*, 2002),
- Embedded knowledge that exists in the routines and procedures of the organisation, and
- Encoded knowledge (held in books, computer systems, etc.).

These forms can be related to the innovation process as follows: creativity relies principally on embrained and embodied knowledge; the environment in which ideas can be expressed and selected relies on encultured and embrained knowledge; development relies on embrained, embedded and encoded knowledge; implementation relies on all five. An innovative organisation, then, should have knowledge balanced appropriately across these types. Hendry *et al.* (2002) have described new product development through the innovation process as a spiralling process of interaction between explicit and tacit knowledge. However, they draw an important distinction between the relative value of explicit and tacit knowledge, arguing that for local markets tacit knowledge is needed, whereas for global markets there is a need to make knowledge explicit wherever possible.

In many aerospace organisations, there has been significant innovation in the tools used for design (e.g. computational fluid dynamics (CFD), computer-aided design (CAD), etc.); the development of such tools essentially represents the transfer of knowledge from embrained to encoded, according to Blackler's designation. It is interesting to note the recent concern among aerospace companies that their engineers are relying on such tools and, as Dodgson *et al.* (2002) have remarked, 'simulators can lead to a lack of understanding of fundamental processes underlying the models themselves. As individuals become more reliant on models to conduct routine or basic calculations, there may be a tendency for younger generations of practitioners to fail to appreciate the full nature of the properties being tested.' Ironically, the drive for faster, cheaper solutions may be driving innovation out of aerospace design.

As Bessant and Tidd (2007) have remarked, the innovative organisation must pay attention to, what they term, the optional stage of learning in the innovation process. In fact, to remain innovative, learning cannot be optional. The organisation must be a learning organisation (Senge, 2006), which relies on shared vision, reflective conversation, and systems thinking.

Innovation is largely a matter of managing knowledge, rather than managing process, in an organisation.

9.7   Whole systems view of innovation

New developments rely on increased knowledge in the key disciplines and the ability to integrate increasingly complex systems (Young, 2007); essential attributes of concurrent development.

The development of the approaches to the innovation process (Rothwell, 2002) over the last few decades has indicated the need for increasingly complex organisational structures to deal with increasingly complex innovation situations. The effects of globalisation on the aerospace industry, and, in particular, the increasingly connected supply chain through which aerospace products are delivered, demand that would-be innovators take a whole-systems view of the situation. Maybe this has always been true of innovation, but the complexity of the environment in which innovation is required has continued to increase. As Senge (2006) suggests, we must see circles of causality and understand innovation in the context of the interrelations of complex systems. Similarly, the innovator must view the knowledge at his/her disposal and the delivery of innovation from an enterprise, by which we mean multi-organisational, perspective. There is now a need for innovative companies to be peopled by systems people, i.e. a workforce skilled in taking a whole systems view of the environment in which they must innovate and trade.

9.8   Conclusion: innovation processes of the future

We began by remarking on the history of aeronautical innovation, asserting that the first 50 years were a period of rapid and radical innovation and the next 50 a period of incremental change, which largely further developed tried and tested ideas. The notable innovations of the first 50 years were mainly concerned with the vehicles themselves and the technologies that gave them new capabilities. The challenges facing the aviation sector in the next 50 years will concern the environmental impact of aviation, the demands of increasing passenger travel, the continued need to drive down costs, and the increasing use of autonomous systems. Maybe the innovations required are not just improved technology, but changes across the whole aviation sector (legal, social, technological, environmental, financial, etc.). The innovation processes associated with aeronautics must in the future be the processes that support innovation of the whole aviation system. Perhaps it is time for a new paradigm; a new generation of innovation.

Innovation is frequently confused with invention. In fact, innovation is the set of activities through which change is introduced from its first conception through to realisation in some tangible form such as commercial success or achievement of societal benefit. Although one can conceive an innovation process to include scan, focus, resource, implement and learn, successful innovation is dependent more upon environment and organisational culture than on process. Key features of innovative organisations are a focus on learning as part of the notional innovation process and the ability to take a holistic approach to design. Innovation is often a response to challenge, and the environmental challenges for the aviation sector in the future may require a truly holistic approach encompassing not just the technologies, but the legal, social, financial and societal aspects as well.

9.9   References

Aghion, P., Bloom, N., Blundell, R., Griffith, R., and Howitt, P. (2002). *Competition and Innovation: An Inverted U Relationship*. WP02/04. London, The Inst. for Fiscal Studies.

Atkinson, S.R., and Moffat, J. (2005). *The Agile Organization – From Informal Networks to Complex Effects and Agility*. Washington DC, CCRP.

BAE Systems. (2009). *Memorandum submitted by BAE Systems to UK Parliament Business, Innovation and Skills Committee, 22 September 2009*. House of Commons, London, UK Government, available from http://www.publications.parliament.uk/pa/cm200910/cmselect/cmbis/173/173we10.htm.

Bass, F.M. (1969). A new product growth for model consumer durables. *Management Science*, 5(5): 215–27.

Berkum, S. (2007). *The Myths of Innovation*. Canada, O'Reilly Media Inc.

Bessant, J. (2003). *Managing Innovation – Moving Beyond the Steady State*. Inaugural Lecture. Cranfield School of Management, Cranfield University.

Bessant, J., and Tidd, J. (2007). *Innovation and Entrepreneurship*. Chichester, John Wiley & Sons.

Blackler, F. (1995). Knowledge, knowledge work and organizations: An overview and interpretation. *Organization Studies*, 16(6): 1021–46.

Croom, S., and Batchelor, J. (1997). The development of strategic capabilities – an interaction view. *Integrated Manufacturing Systems*, 8(5): 299–312.

Dodgson, M., Gann, D.M., and Salter, A.J. (2002). The intensification of innovation. *Int. J. Innovation Management*, 6(1): 53–83.

DTI (2005). *Creativity, Design, and Business Performance*. DTI Economics Paper No. 15. Department of Trade and Industry. Available at www.dti.gov.uk, DTI/Pub 8054/0.8k/11/05/NP. URN 05/1676.

Dyson, J. (2002). A new philosophy of business. In J. Henry, and D. Mayle, *Managing Innovation and Change*, pp. 145–54. London, SAGE Publications Ltd.

Freeman, C. (1974). *The Economics of Industrial Innovation*. London, Penguin Books.

Hamel, G. (2002). *Leading the Revolution*. Boston, Harvard Business School Press.

Hendry, C., Brown, J., Ganter, H.-D., and Hilland, S. (2002). *Understanding Innovation: How Firms Innovate and What Governments Can Do to Help – Wales and Thuringa Compared*. London, Anglo-German Foundation for the Study of Industrial Society.

Jobs, S. (2004). Voices of Innovation. *Bloomberg BusinessWeek*, 11 October 2004.

Jones, O., and Beckinsale, M.J.J. (2001). Micropolitics and Network Mapping: Innovation Management in a Mature Firm. In O. Jones, S. Conway and F. Steward (eds), *Social interaction and Organisational Change: Aston Perspectives on Innovation Networks*, pp. 41–79. London, Imperial College Press.

Kao, J. (1996). *Jamming: The Art and Discipline of Business Creativity*. New York, HarperCollins.

Kroo, I. (2004). Conference keynote. 2004 AIAA Dryden Lecture. In *Innovations in Aeronautics. 42nd AIAA Aerospace Sciences Conf.* Reno, Nevada.

Kroo, I. (2009). Adventures in aircraft design with John McMasters. In *47th AIAA Aerospace Sciences Meeting*. AIAA 2009-868. 5–8 January, Orlando, Florida.

Leifer, R., McDermot, C.M., O'Connor, G.C., Peters, L.S., Rice, M.P., *et al.* (2000). *Radical Innovation – How Mature Companies can Outsmart Upstarts*. Boston, MA., Harvard Business School Press.

Mankins, J.C. (1995). *Technology Readiness Levels*. Advanced Concepts Office, NASA.

McNerney, J. (2006). Innovation and the Global Economy. In *Distinguished Lecture Series in International Business*. St Louis University, Missouri, US. Available at http://www.boeing.com/news/speeches/2006/mcnerney_060516.html.

Merrill, S.A. (2006). *Aeronautics Innovation: NASA's Challenges and Opportunities*. Committee on Innovation Models for Aeronautics Technologies, National Research Council. Washington, DC, The National Academies Press.

Rothwell, R. (1992). Successful industrial innovation: Critical factors for the 1990s. *R&D Management*, **22**(3): 221–39.

Rothwell, R. (2002). Towards the fifth-generation innovation process. In J. Henry, and D. Mayle, *Managing Innovation and Change*, pp. 115–35. London, SAGE Publications Ltd.

Senge, P. (2006). *The Fifth Discipline – The Art and Practice of the Learning Organisation*. London, Random House.

Shaw, G., Brown, R., and Bromily, P. (2002). Strategic stories: How 3M is rewriting business planning. In J. Henry, and D. Mayle, *Managing Innovation and Change*, pp. 155–64. London, SAGE Publications Ltd.

Sinclair, M.A., Siemieniuch, C.E., Haslam, R.A., Henshaw, M.J.D.C., and Evans, L. (2010). PEAT – a tool to predict team performance in systems. In *IEEE 5th Int. Conf. on System of Systems Eng.*, 10 June, Loughborough. IEEE xplore.

Skeat, W.W. (1993). *The Concise Dictionary of English Entymology*. Wordsworth Reference.

Tidd, J., and Bessant, J. (2009). *Managing Innovation – Integrating Technological, Market and Organizational Change*. 4th ed. Chichester, John Wiley & Sons.

Townsend, D.M. (ed.) (1990). *Managing Innovation*. London, Confederation of British Industry.

Truman, T., and De Graaf, A. (2006). *Out of the Box – Ideas About the Future of Air Transport*. EU, ACARE. Brussels, European Commission.

Wolfe, T. (1979). *The Right Stuff*. New York, Picador.

Young, T.M. (2007). Aircraft design innovation: Creating an environment for creativity. *Proc. IMechE. Part G (Aero)*, **221**: 165–74.

10

Managing innovative technology development in aeronautics: technology assessment (TA) techniques

R. HENKE, RWTH Aachen, Germany

Abstract: Technology assessment (TA) is used to analyse the repercussions of the introduction of a new technology. The effect of a new technology is compared with an existing scenario, and a decision can be made whether the new technology is beneficial. Limitations of the method, application and knowledge are discussed. Examples of applications to technological and economic assessments are given. The role of technology assessment as a support tool is discussed.

Key words: technology assessment, repercussions, management decision making, scenario techniques, assumptions, product lifetime, reference comparison.

10.1 Introduction

10.1.1 The need for TA

The implementation of a technology, either as a modification of an existing aircraft or as a feature of a new aircraft, will have repercussions for both the manufacturer and the operator. For example, new production technologies may help to reduce the manufacturer's costs, or a new wing design may reduce the operator's fuel cost. On the other hand, implementing a technology will result in additional costs such as investment, either by the manufacturer, e.g. for any new tools needed, and/or by the operator, e.g. for maintenance and/or spare parts. Some technologies may lead to an advantage right from their implementation as they save weight and thus raw materials; others may pay off only in the future, based on fees, fuel price, labour costs, etc.; all of which influence the operating cost of an aircraft.

Technology assessment (TA) analyses the repercussions of the introduction of a new technology; this is always in comparison with an aircraft without that technology, which is called the reference aircraft. These repercussions will be calculated at full aircraft level, addressing penalties and benefits during the whole product lifetime, taking the non-recurring cost (NRC) as well as the recurring costs (RC) into account. Having carried out this calculation under a certain set of assumptions, the net value of that technology should be available in order to come to the decision whether or not to continue work on that technology. This could mean installing a project looking at technical feasibility, or including it in the design of a new product.

214

TA may serve as a very important tool, in technical as well as economic terms. Today, almost every company is using some sort of TA. Every technologist will be affected by it, and any new idea, e.g. in aerodynamics, will have to pass a TA at some point. However, its use asks for a critical approach, as described in this chapter.

10.1.2 History and process

In the past, benefits of technologies have mainly been estimated by brute force calculation of the primary repercussions, e.g. drag reduction by a shock bump, directly calculated as a fuel burn reduction. In the last 15 years, more sophisticated approaches have come into use, taking into account as many repercussions as possible, including snowball effects. High-fidelity tools improve structures, aerodynamics, aeroelastics, etc., but also improve the general aircraft design tools, which allows TA to become more reliable. Large companies such as Airbus have their own departments that deal with TA, and base decisions on the outcome of assessments.

Describing the initial process, specialists in a specific department, e.g. aerodynamicists, may propose a new technology; for the research and technology portfolio, or for a particular product application. Aircraft design departments such as the future project office (FPO) will then calculate benefits and penalties of this technology, still in technical terms of the overall aircraft design (OAD). The results will be used to calculate benefits and penalties in economic terms, i.e. on the internal rate of return (IRR), or the net present value (NPV), for the manufacturer and for the customer, i.e. an airline. In recent years the technical values will also be used to calculate the technology's ecological repercussions, e.g. to estimate emissions and/or noise, which may have an influence on future fees or emissions trading. Taking all data into account, TA will be used to support the management's decision on the development of the proposed technology.

During the development process of a new aircraft, or in the process of an aircraft design modification, TA will be carried out by the FPO, supported by a market analysis. Those carrying out TA will communicate important results to future customers, but also to the technologists by pointing out sensitivities of the technologies at an aircraft level, thus giving hints for improvements of specific aspects.

TA is quite often used before setting up costly technology projects. This means that TA would be the first step of such a project; with the technology department providing data, which then will be used by the FPO to evaluate the technology at full aircraft level. As TA is based on technical information this will be rather vague, because the technology programme being challenged cannot initially provide sufficient reliable data (see Section 10.2.3).

10.2   **Methods and limitations**

10.2.1 Scenario techniques

Using modern scenario techniques, all of the repercussions of a technology in a virtual environment will be listed in a cross-impact matrix. The data will be weighed by specific functions and then reduced to just a few macro scenarios. The benefits and penalties of the technology can then be calculated on the basis of each scenario, such as a 'green' scenario or a 'low-cost' scenario. This technique helps in understanding the outcome of modifications at an aircraft level, even with future uncertainties.

A technology for drag reduction and, thus, fuel reduction might not be beneficial at the time of the analysis due to the current fuel price. A simple statistical extrapolation of the fuel price will show when it would be worth introducing that technology. This is the simplest form of economic TA. Additional, more sophisticated, repercussions may be included, such as expected fuel consumption-related fees, which will also affect the direct operating cost (DOC), plus snowball effects, e.g. taking into account the decreased aircraft weight due to reduced fuel quantities. Such scenario techniques will help in understanding the technology's repercussions under different macro scenarios.

10.2.2 Technology vector

To define the performance of an aircraft, making a comparison with a reference aircraft that can be plotted at the origin of a coordinate system, the repercussions of a technology can be drawn as a vector, with the manufacturer's IRR on the x axis and the operator's IRR on the y axis (Fig. 10.1). Depending on the position or quadrant, it can immediately be seen whether the new aircraft or the aircraft with a new technology is beneficial for the manufacturer, the operator, both, or neither. It may be that the implementation of a technology will result in an additional effort on the manufacturer's side, e.g. due to a longer manufacturing period, but be of full benefit on the operator's side, e.g. from reduced fuel consumption.

Taking the price sensitivity of the aircraft as a slope, a line can be drawn, increasing the benefits on the manufacturer's side by increasing the aircraft price, hence reducing the benefits on the operator's side, until a win–win situation is reached. If this is not possible, the benefit of that specific technology needs to be challenged.

In Fig. 10.1 the technology vector is within the quadrant where there is a manufacturer's penalty and an operator's benefit. Two types of price sensitivity lines are shown. The line s_p has a slope that is small enough that, at the point where it intercepts the y axis, a positive value for the operator remains. The slope of s_q is so steep that there is no chance of making that technology beneficial at all.

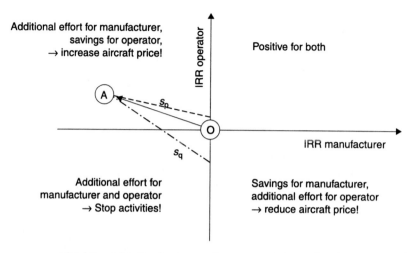

10.1 Internal rate of return – diagram and the technology vector. O, origin = reference aircraft; A = repercussion of technology in terms of cost; s = price sensitivity (s_p and s_q, see text).

10.2.3 Limitations

There are three major limitations in technology assessment:

- Limitations of the method.
  Within technical TA, the values for repercussions at full aircraft scale are bound to FPO design methods. Almost regardless of the absolute accuracy, there will be a scatter in the rough order of magnitude (ROM) of some 1–10%. Taking the operational empty weight (OEW) of an A380, 5% would mean an uncertainty in the results of the order of 104 kg. Adding a technology to such an aircraft that would result in a weight gain of, say, 100 kg, which in absolute values is a lot, will be hardly assessable by FPO methods. Of course there will be a result; but, from a scientific view, this will be within the statistical scattering within the results.
- Limitation of the application and integration.
  Any new technology must be integrated into the aircraft's architecture. However, before a detailed design is available, no such architecture exists, so the integration at an aircraft level is based on assumptions, generally based on statistical methods. At first, the uncertainty must be taken into account as addressed in the previous point. In addition, the statistical methods are all based on architectures that by nature do not include the new technology. Therefore, as well as the initial statistical scattering, there is another uncertainty, as the repercussions of the system integration is based on, for example, first- to third-generation aircraft architecture, while

results are needed for a new generation. Consequently, as the new technology's integration architecture is not known at the beginning, the repercussions of that architecture on other architectures not directly linked to the new technology are also unknown. Especially in today's aircraft, with electronic systems implemented up to the level of the electronic flight control system (EFCS), this plays an increasingly important role. As an example, consider the introduction of fuel cells to replace the auxiliary power unit (APU). This affects not only the power supply from these two different devices, but the whole aircraft's architecture. As there are no statistical data available, a complete new architecture would be needed for a reliable TA, which is impossible at the time when the decision 'fuel cell – yes or no' will be taken.

- Limitation of knowledge of the future
  This sounds trivial enough, but also needs to be looked at scientifically. The technology in today's aircraft was not originally analysed using TA methods. In order to sum up all repercussions in economic terms to calculate direct operating cost (DOC), assumptions are needed on all issues related to the DOC: what will the interest rates be in ten years' time, what fees will have to be paid, what will be the insurance rate, what will be the salaries of cockpit and cabin crews, what value can be estimated for the worldwide fuel price, what will be the labour cost for maintenance, what will be the cost for one kilogram of any material such as aluminium, titanium or CFRP? All these figures are required for any new, conventional aircraft, but, in order to point out the value of a new technology, even more details not available at that moment in time would be needed. As described in Section 10.2.1, this results in a very high uncertainty of the economic TA.

The first two topics in this section were on the subject of technical limitations, which are subject to statistical scatter; while the last topic, on the economic limitations of TA, is even worse. As a result, TA of any new technology can be assessed positively or negatively just by adjusting the assumptions for any of the above values. There are many examples in recent industrial history, never officially cited, where technologies have been adopted even after a negative TA, and others dropped even after a positive TA. A well-known example is the debate on the application of hybrid laminar flow control (HLFC) in transport aircraft. In the past 20 years, there were almost as many 'believers' in HLFC as there were 'non-believers', both arguing on the basis of 'their' TA, but both made under different assumptions, leading to different results.

It can be concluded that TA can be used for qualitative decision support, but not at present as a quantitative base for a decision, even though TA is of course a quantitative method. Much engineering knowledge is needed when a decision for or against a new technology needs to be taken; the desire for TA to give a clear and distinct answer is unrealistic.

10.3   Approach and example

10.3.1 Reference aircraft and technology description

As an example of carrying out a TA, consider an aircraft modification that results in reduced approach noise. As the very first step, a reference aircraft needs to be defined as the base against which the benefits and penalties of the technology can be rated. The technology status of the reference aircraft as well as that of the new technology aircraft must be defined, in order to make a 'fair' comparison and not to take the general evolution of performance from an old aircraft to today's aircraft as an outcome of the specific technology.

As the next step, the technology must be described in detail, with the objective of its application. This helps the inventors to structure their approach, and the FPO to address the technology in the correct way.

10.3.2 Technical technology assessment

As a second step, all impacts need to be listed:

* Benefits and penalties on the manufacturer's side. This includes positive and negative performance impacts of the modification to the aircraft, e.g. changes in component weight, drag and lift changes, system complexity, etc. Manufacturing aspects must be listed: materials to be supplied, time needed for installation, special jigs and tools, etc.
* Benefits and penalties on the operator's side. This includes direct operational issues such as fuel savings, noise fees, engine lifetime, etc., plus maintenance and spare parts issues, including the need to carry out work in a hangar instead of during an outside check.

This finishes the technical TA.

10.3.3 Economic technology assessment

The next step will then be to calculate repercussions on cost. On the manufacturer's side, there are non-recurring costs (NRC), e.g. for tools, and recurring costs (RC), e.g. for materials. On the operator's side, there are, e.g., maintenance costs or spare parts. In parallel, any ecological repercussions need to be added, e.g. noise reduction during aircraft operation. These will then be used to calculate noise fees, using either statistics or existing forecasts.

With this data available, the economic TA can be carried out in a pragmatic way. If available, a scenario process may also be added in order to see the amount of change under different boundary conditions.

10.3.4 Examples

Two examples in this chapter outline the different issues of TA. The examples consider a single-aisle medium-range aircraft, such as the Airbus A320. Both aim at flow separation suppression on the main flaps, thus prolonging the linear part of the $C_L A$ curve. This can be used for different purposes: to increase payload or to decrease stall speed v_s. The first increases the direct income of the airline, the second decreases cost in terms of reduced braking power and reduced fees because of less noise, and may allow landings on shorter runways, thus increasing the activity radius of the aircraft and/or utilisation. The reduction of approach speed may even increase safety, as there is more time for pilots to react, but this aspect is not considered here. For TA, there must be a clear technology application description, and already the simple objective 'increase C_{Lmax}' shows a variety of opportunities that may result in different figures for TA. In the examples used here, a decrease of v_s to reduce cost should be the main aim.

- Technology A is a simple add-on technology, called sub-boundary vortex generators (SBVG), to suppress separation on the flaps. These are small devices, a few millimetres in height, situated at some 20% flap chord. The flap can be deflected and retracted with the SBVGs on. The height is constant; therefore they are always present and not adjustable.

A 1: Manufacturer's efforts

A 11: Penalties

A 111: NRC

To develop this technology, some 40 man-hours plus two days of wind tunnel tests, altogether costing 10 000 €. The installation can be done during flap assembly; one day should be sufficient. Therefore, no influence on the total development time of the new aircraft is expected.

A 112: RC

The material plus machining of 2 × 300 SBVG costs approximately 1000 €. The manufacturing work plus checkout tests adds up to approximately 2000 € for each aircraft. But, if that single day of installation holds up the whole production process, the figure will change dramatically. Taking the development cost for the aircraft programme as approximately 4 billion €, which is financed at 3% interest rate before that product will produce a profit, and with a development time of five years, just the interest costs for a single day would be about € 65 000!

Considering another example, calculations indicated that installing riblet foils for turbulent drag reduction would take too long, and thus cost too much, taking up all of the benefits of some 1% drag possibly gained by that technology even over the whole operational life of the aircraft.

A 12: Benefits

There are no benefits, such as reduced development time or decreased manufacturer's efforts, linked to this technology.

A 2: Operator's efforts

A 21: Penalties

A 211: Penalties en route

A penalty for increased drag can be estimated as $DC_d = 0.01$. Actually, this is a 'DDC_d', as the drag is less than that for a separated flap; however, it is higher than for a normal attached boundary layer, which is the configuration under investigation.

A 212: Penalties during maintenance/inspection/repair

The SBVG maintenance can be done visually, and occasionally a few may need to be replaced. Without considering replacements, inspection times of about 0.1 h/day can be estimated as the input for DMC calculation.

A 22: Benefits

A 221: Benefits in operation

Due to the limited height and fixed position, the benefit can be estimated at $DC_{Lmax} = 0.1$, with repercussions on velocities such as $v_1, v_2, v_S, \ldots$

A reduction in noise fees can be assumed; the estimate would be a reduction of approximately 2% of approach noise. As each airport in the world has its own noise fee calculation procedure, a fleet calculation has to be carried out to evaluate the benefits for a specific airline. For example, this may be more in Europe and less in Africa due to different noise curfews.

A 222: Benefits in maintenance, repair and overhaul (MRO)

The reduced v_s braking results in a longer life for the brakes; 1% seems to be a reasonable estimate. About half of that, i.e. 0.5%, can be assumed for the engine, as less reverse thrust is needed.

In summary, the NRC can almost be neglected when compared with the overall aircraft development cost, while there is a measurable benefit behind the SBVGs. This simple arrangement of figures under the different cost sections already shows the sensitivities and thus can give hints to the technologists about where to improve upon costs.

- Technology B is a number of tiny loudspeakers at 40% chord of the flap, and is also aimed at the suppression of boundary layer separation. The loudspeaker output can be adjusted, based on the flap deflection angle.

To develop this technology, some more man-hours and wind tunnel tests are needed in engineering, plus design drawings in structures, plus cabling and power supply in systems. Apart from the materials, they need to be tested in advance during the power-on tests. Already a first assumption must be made: if the aircraft's performance cannot be guaranteed without these devices functioning, a certain reliability must be defined, asking for either a redundancy or a fail-safe concept.

Maintenance requires visual inspection of the surface plus functional tests of the devices. In the case of repairs, the flap needs to be opened, which cannot be done on the airfield but must be done in a hangar.

In contrast to the SBVGs, the system described here is an active one. For this, the repercussions on all ATA chapters must be listed. As a first, rough estimate, ATA chapters 6, 7, 12, 24, 27, 45, 49, 57 are affected. For each, a separate calculation must be carried out.

The aerodynamic drag penalty can be estimated as some $DC_d = 0.001$, based on the flow around the small orifices comparable to drag from any other projections.

The power required is approximately 1 kW during the final approach; no benefit is seen during take-off, climb, and initial approach.

With the chance of adjusting flap angle and loudspeaker output, the benefit can be estimated at $DC_{Lmax} = 0.2$.

In a manner similar to that for the SBVGs, a quantitative analysis must be prepared, listing all values for manufacturer's and operator's benefits and penalties, differentiating between NRC, RC, and DMC. As so many ATA chapters are involved, this would overly extend the scope of this book, and presents an exercise for the interested reader.

Figure 10.2 shows qualitative results of a TA for technology A and B; the slopes of the aircraft price-sensitivity lines are parallel for both. While A (SBVG) has a smaller benefit, the smaller effort required still makes its installation attractive, while B (loudspeakers) gives higher benefit for the operator, but is not attractive enough for the manufacturer.

Taking one further step, one could imagine the TA for two other technologies just by thinking about all of the repercussions as described above:

- The introduction of a CFRP fuselage. This asks for a huge TA effort, up to certification issues, which have a large influence, but can hardly be estimated before any CFRP fuselage has been introduced to the market.
- The introduction of HLFC. Again, many ATA chapters are affected, the maintenance effort is huge, safety is affected, route planning needs attention,

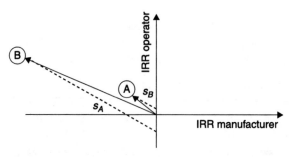

10.2 Internal rate of return – diagram for technologies A and B. s_A s_B, price sensitivity.

during design the behaviour that the laminar profile should show under turbulent conditions must be decided, which will limit the benefits, etc.

Taking the two specific examples and the two more general ones, the effort needed to carry out a TA becomes clear. The repercussions of any aerodynamics technology also ask for a detailed analysis beyond the aerodynamics, even more so in the case of active flow control.

10.4   Conclusion

At first glance, TA may be seen as a very powerful tool to rate new technologies. On the other hand, as TA always addresses future developments, there is a chance that, if it is taken literally or its results are taken just as a single digit instead of as sensitivities, it might stop a valuable technology development which would have a chance under different boundary conditions. For example, when boundary-layer suction devices for laminar flow control were under investigation in the early 1990s, laser beam drilling was the only technology available at that time. Today the tiny holes required can also be drilled by electron beam drilling, with major implications for cost and quality. Therefore, TA must be seen to be advice, rather than a decision. It can be used to support the funding of a technology and can be used for sensitivity analysis, but the owner must be aware of the limited reliability when it comes to future forecasts.

At present, and on purpose, the scenario technique is rather a story-telling. In coming years, the scenario process-based TA needs to be improved in terms of quantitative results. However, aircraft design is not a deterministic exercise, and therefore TA should play an increasing role, but should still be seen as a supporting tool.

10.5   Abbreviations

APU	Auxiliary power unit
ATA	Air Transport Alliance
C_L	Lift coefficient
C_{Lmax}	Maximum lift coefficient
C_d	Drag coefficient
CFRP	Carbon fibre reinforced plastics
DMC	Direct maintenance cost
DOC	Direct operating cost
FPO	Future project office
HLFC	Hybrid laminar flow control
IRR	Internal rate of return
MRO	Maintenance, repair and overhaul
NPV	Net present value

NRC	Non-recurring cost, e.g. development cost
OAD	Overall aircraft design
R&T	Research and technology
RC	Recurring cost, e.g. material
ROM	Rough order of magnitude
SBVG	Sub-boundary layer vortex generators
TA	Technology assessment
v_1	Decision velocity (take-off)
v_s	Stall speed

11

Mining the 'far side' of technology to develop revolutionary aircraft prototypes: the Defense Advanced Research Projects Agency (DARPA) approach

J. R. WILSON, Freelance writer, USA

Abstract: The US Defense Advanced Research Projects Agency (DARPA) was created in response to the Soviet Union's surprise launch of Sputnik I, tasked with helping America catch up with and then surpass the USSR in space. As that and other early missions shifted to NASA and other agencies, DARPA became the means not only to prevent future technology surprises, but to create such breakthroughs itself – with a growing emphasis through the years on aerospace and pushing the aeronautics envelope, often beyond what their military customers thought they wanted. 'Impossible' was redefined as DARPA-hard, and 'innovation' as turning 'far side' concepts into combat-ready reality, all from a unique 'freedom to fail' research and development model.

Key words: Defense Advanced Research Projects Agency (DARPA), stealth, unmanned aerial vehicle (UAV), hypersonic, experimental evaluation of major innovative technologies (EEMIT) aircraft, programmable matter, X-plane, in-flight shape morphing.

11.1 Introduction

When the Defense Advanced Research Projects Agency (DARPA) was set up within the US Department of Defense on 7 February 1958, it had three primary purposes: first, to get ahead of the lead the Soviet Union had grabbed in space with the recent launch of Sputnik 1, and, second, to develop technologies to detect Soviet nuclear tests, regardless of size, type or location. Both of those responsibilities soon passed on to other agencies, such as the also newly created NASA.

It was the third mission that was retained and that led to DARPA becoming the world's premier developer of cutting-edge technologies; and in so doing created one of the most unusual, perhaps unique, laboratories in the world. Essentially, it was tasked with never again allowing the US to be taken by technological surprise, and to do this by becoming the source of all future military technological surprises in the world. It was not long before the advantages of this approach crossed over to civilian applications, especially in aerospace.

225

11.2   Defense Advanced Research Projects Agency's (DARPA) philosophy and structure

In order to accomplish its goals, DARPA was given a rare freedom (one that commercially aligned industry cannot support), and that was the freedom to fail. It also was given an unheard-of structure. It had no dedicated laboratories; there were no permanent research employees, nor were there permanent programs. It was not required to be 'revenue positive'. Directors were, and still are, appointed for terms of two to four years, program managers for three to four, after which all such appointees leave DARPA to pursue careers elsewhere in government, industry or academia.

Stephen Welby, former director of DARPA's Tactical Technology Office, said in an interview,

> 'They (project managers) are all temporary hires. They are here to get something done. The clock is ticking and there is personal pressure to advance the state of the art on very aggressive timelines. You can change rapidly and move quickly when 25% of your people change out every year. You would never want to run a business this way, but for preventing technological surprise and being the engine of innovation – it's perfect.
>
> 'We start a lot of things, but we also ruthlessly kill them. It's acceptable to fail as long as we have learned from it. If it succeeds, we get the data; if it fails, we find out it's a path not to take. A project is something with a defined start, defined finish and clearly defined objectives. Something that can be written on a single piece of paper. [The goal is to] prove the feasibility of a concept and take the specific technology risk off the table. . . I think almost everything we do transitions. Even those that fail leave behind an industrial base or an aerodynamic database that can be used.'[1]

Each new director is given total freedom to choose the concepts they wish to pursue and hire program managers to do so. While any programs created by a predecessor are allowed to continue until the project manager's tenure ends, a new director has no responsibility to continue that line of research. They can, however, reach back and resurrect previous DARPA programs, usually in the event that something has changed, such as new materials, new technologies or new theories, to advance it beyond what had been hitherto possible.

Neither DARPA nor its directors and program managers are under any mandate to pursue specific or even broad areas of research, but are free to decide if they want to concentrate on space systems, aeronautics, communications, weapons, directed energy or something seemingly pulled from the pages of science fiction. As a result, during its more than half a century of work, DARPA has been the originator or primary developer of stealth aircraft, unmanned aerial systems and hypersonic flight, to name but a few.

11.3   DARPA and innovation in aviation

'In aviation, it's easy to talk about most of the achievements of the past, but a lot harder to talk about the dreams of the future, because we do a lot of black programs,' noted former DARPA director Dr Tony Tether (2001–9). 'We were created to prevent technological surprise, but we've also become the place that creates technological surprise for other people.'[2]

Most directors have had at least a few programs designed to advance some area of aeronautics. In typical DARPA fashion, however, few end-game advancements were completed by the same director under whom they began. In some cases, such as stealth, the research advanced through multiple programs, often widely spaced in time.

Tether was the lone exception to the rule on short directorship appointments, heading DARPA throughout President George W. Bush's two terms, from 18 June 2001 until 20 February 2009. It was Tether who began calling DARPA's efforts 'Far Side' research, another way of saying 'DARPA-hard'.

Tether's special assistant for space, Richard McCormick, summed up the concept at DARPATech 2005, an annual gathering of DARPA program managers and interested stakeholders from across government, industry and academia. 'First, is it DARPA-hard? That is, does it present a high chance of failure?' he said of how DARPA looks at a new research proposal.

'Forget the sure bets; that's not our business. If it's not DARPA-hard, let someone else do it.

'Conventional wisdom leads so many space program managers to be risk adverse. In DARPA, we welcome risk; without it, we will not provide the warfighter with the options and capabilities needed in this uncertain world. The problem isn't when you try and fail. The problem is when you fail to try.'[3]

While his remarks came from a space research perspective, they apply to DARPA across the board. And, often, some of the most difficult programs to pursue – technologies the presumed beneficiaries often dismissed as unnecessary, impractical or simply impossible – were aeronautical issues that led to some of DARPA's greatest successes. Tether said in an interview for this book:

'DARPA's real role is to show something can be done. I've found that is a very powerful thing – once you show people who don't believe it can be done that it can, it's amazing how much progress can be made. It removes the restraints of thinking "why waste money on this, it can't be done".'

11.4   Examples of DARPA innovation in aviation

'Since the 1980s we've put a lot of effort into vehicles that can take off and land vertically, but fly at high speeds, sort of morphing between a helicopter and a more fixed wing approach during the cruise phase. We also took on very low radar cross-sections – stealth – not just radar, but all sensors. That started in the early 1970s and has been going since. In engine technology, we have been involved with developing engines that were good through many different stages, from subsonic to supersonic to hypersonic. Our dream has always been to have an airplane that can take off horizontally, fly into space and return.

'DARPA's been the leader in aviation in the sense of having really developed the UAV business, starting in the Sixties and continuing ever since, progressing UAV capabilities from vehicles that had small range and limited endurance to vehicles with extremely high range and endurance today. Global Hawk and Predator are probably the most well known. We're still pushing that envelope further, to develop vehicles that can stay up for five years. Our latest effort is an unmanned vehicle that would be used like a satellite – you launch it and it never comes back. When you're finished with it after five years, you just dump it in the ocean, as you would a satellite.' (Wilson, citing, Tether)

Although its early emphasis was on space, nuclear test detection and ballistic missile defense, DARPA's wide-open mandate also allowed its directors to focus some of their research toward other technologies. Chief among those was flight – manned and unmanned, fixed wing and rotor, across a broad range of capabilities and potential missions. And, while it obviously is focused primarily on military requirements, much of what comes out of DARPA has significantly changed civilian aviation, just as the military ARPAnet morphed into the global civilian internet.

'Over the years, from director to director, some have thought unmanned was more important than stealth and vice versa, but I don't think anybody has ever come in and said aeronautics as a whole is not important,' Tether said.

Former DARPA Director George Heilmeier (1975–7) is perhaps best known as the inventor of the liquid crystal display (LCD) while working for RCA Laboratories in 1964. However, Heilmeier has said he considered the work he did on stealth technology while at DARPA to have been the highlight of his career. The program was called Have Blue, a proof-of-concept aircraft built for DARPA by Lockheed and made more difficult by an astonishingly tight two-year deadline.

Designed solely to demonstrate stealth capabilities and with no concern for aerodynamic performance beyond being able to fly, Have Blue met its deadline with its maiden flight in 1977. Only two years later the first operational aircraft to fully incorporate stealth technology made its maiden flight – the then super-secret F-117 Nighthawk Stealth Fighter. Included among those technologies, which provided the foundation for subsequent stealth programs, were the use of a multifaceted surface, radar-absorbent materials, infrared shielding, heat dissipation, low probability-of-intercept radar, active signature cancellation, shielded inlets and exhausts (along with exhaust cooling) and special windshield coatings.

Robert Fossum, DARPA's director at the time (1977–81), noted it was not only the prospective user of such new technologies the agency had to convince, but also the contractors the agency used to create their new realities. Most of the traditional concerns in the creation of an aircraft – varying according to the focus technology, but including engines, structure, flight dynamics and even the inclusion of safety engineers in the design process – can actually make the process too complicated to succeed.[2]

In the case of stealth, for example, Fossum ordered his contractors to ignore everything not directly related to his narrow focus on radar cross-section. As a result, Have Blue was a heavily compromised, aerodynamically inefficient and unsophisticated aircraft – but one that achieved its DARPA-hard goals in record time. It was a significant demonstration of Fossum's belief that successfully developing advanced technologies – DARPA-style – means single-minded focus on one goal, while giving all other concerns only the minimal attention required.

'Stealth is another DARPA creation, because we had some people who believed they could design and fly an airplane using Maxwell's equations and not Navier-Stokes, which basically would minimize reflection back to the source and still have enough aerodynamic features that it could fly,'

Tether explained.

'DARPA did that over the opposition of the Air Force, who understood the value of stealth, but it didn't have the performance of airplanes in the 1970s.
  'It was a hard cultural shift — yet they not only have made it, but today the F-22 is not just stealthy but extraordinarily maneuverable, because we have learned how to use both Maxwell's equations and Navier-Stokes to build aircraft. So DARPA really led the way on stealth in the early days, but the Air Force took over in the mid-1980s and has carried it since then.'[2]

Although Have Blue was not among them, from the beginning DARPA pushed the boundaries of aeronautics through a series of X-planes – flying testbeds used to transform theory to fact for high-risk, high-payoff technologies. While not intended to be prototypes for any specific platform, they provide necessary breakthroughs in aviation design that would be difficult, if not impossible, for any other organization to accomplish. Since its inception, DARPA has been the driving force behind more than half of the nation's X-plane programs – typically, the most Far Side technologies with the highest risks, but potentially the highest payoffs, as well.

'Throughout its . . . history, DARPA has been involved in aviation development. In fact, if you wanted a glimpse into the proverbial crystal ball of aviation, you only had to take a look at what DARPA was doing,' Van Olinger, a DARPA program manager at the time, told DARPATech 2004. 'DARPA has been involved with both manned and unmanned X-planes. These revolutionary aircraft demonstrated significant advances in aerodynamics, propulsion, structures and flight controls.'[4]

11.5   DARPA's aviation-related programs

On a decade-by-decade basis, DARPA's aviation-related programs are described below.

11.5.1 1990s

- Joint Unmanned Combat Air Systems – Air Force X-45, Navy X-47,
- A-160 Hummingbird rotary UAV,
- Affordable Short Takeoff/Vertical Landing (ASTOVL), later renamed the Common Affordable Lightweight Fighter technology demonstration program, precursor to the F-35 Joint Strike Fighter,
- Schottky IR Imager for the B-52 (replacement for the AAQ-6),
- Materials Technology for the F-22,
- X-50 Dragonfly canard-rotor wing, with rotor stopping in flight to create a fixed wing capable of high-speed flight,
- Technology for Transport Aircraft,
- X-31 Experimental Evaluation of Major Innovative Technologies (EEMIT) aircraft, featuring integration of conventional aerodynamic control with multi-axis thrust vectoring, post-stall flight, tailless flight at supersonic speed, advanced helmet-mounted display to improve pilot situational awareness, digital fly-by-wire flight controls and significant advances in high-angle-of-attack,
- Programmable matter – smart material that adapts to changing conditions to maintain and optimize new adaptable functionality, such as an airplane wing that adjusts surface properties in reaction to environmental variables.

11.5.2 1980s

- Stealth Fighter,
- Tacit Blue (Whale) Stealth Bomber testbed, new stealth approach leading to B-2 Bomber,
- Joint STARS,
- X-29 Forward Swept Wing Aircraft Technology,
- Pilot's Associate,
- Materials Technologies for the F-15 and F-16,
- Low Probability of Intercept Airborne Radar,
- Teal Rain high-altitude, long-endurance, extended-range ISR/Target Acquisition UAV,

- Condor long-range UAV, all-composite honeycomb structure, autonomous controls, high-altitude aerodynamics, fuel-efficient propulsion system – considered Global Hawk conceptual prototype,
- X-wing Rotor Systems Research Aircraft,
- No Tail Rotor (NOTAR) helicopter.

11.5.3 1970s

- Materials technologies for the SR-71,
- Mini-RPV.

11.5.4 1960s

- Navy Drone AntiSubmarine Helicopter (DASH).

Some of those programs transitioned to the military for development as real platforms and systems, others set the stage for later DARPA and other research efforts, and some were shelved for one reason or another, because the enabling technologies were not yet available, military requirements had changed or the theory in question failed to prove out.

11.6 Conclusions

'It is the freedom to fail that gives [DARPA program managers] the boldness to go for the big payoffs. And fail we do!' Tether told DARPATech 2007. 'But that's OK – failure sometimes happens when you're bringing new capabilities into reality; you only really fail if you don't learn what happened and stop trying to succeed. You have to try again – and again – and again.'[5]

Despite tightening federal and corporate budgets, especially for extreme cutting-edge research and development, DARPA has entered its second half-century continuing to resist pressures for more conventional, short-term research. For DARPA, the Far Side of miniaturization, multi-year flight endurance, hypersonic speed, in-flight shape morphing and revolutionary aircraft designs remain primary, pushing both military and civilian aerospace stakeholders to envision futures still considered the realm of science fiction.

'Without DARPA, a lot of things might not have come about as fast. The whole idea of UAVs that could fly very high and stay up a long period of time, for example. And stealth was something the Air Force didn't believe could be done. A lot of engine technology, such as the concept for the JSF (F-35 Joint Strike Fighter) engine, was a DARPA program,'

Tether concluded.

'Most of what DARPA does gets lost, in terms of DARPA's role, often because those who did it aren't around DARPA when it becomes a success.'

11.7   References

1. Warwick G. and Norris G. (2008). 'DARPA at 50: Blue sky thinking'. *Aviation Week*, 18/25 August 2008.
2. Wilson J.R. (2008). 'Fifty years of inventing the future'. *Aerospace America*, DARPA 50th Anniversary Supplement, February 2008.
3. McCormick R. (2005). Closing remarks. In *Proceedings of DARPATech 2005 Symposium*, 7–11 August 2005, Anaheim, CA. Available at http://archive.darpa.mil/DARPATech2005/proceed.htm.
4. Olinger V. (2004). The future of aviation'. In *Proceedings of DARPATech 2004 Symposium*, 9–11 March 2004, Anaheim, CA. Available at http://archive.darpa.mil/DARPATech2004/proceedings.html.
5. Tether T. (2007). Opening remarks. In *Proceedings of DARPATech 2007 Symposium*, 6–9 August 2007, Anaheim, CA. Available at http://archive.darpa.mil/DARPATech2007/proceedings/dt07-dir-tether-welcome.pdf.

12

Revolutionary ideas about the future of air transport

M. HIRST, Independent consultant, UK

Abstract: This chapter presents a selection of revolutionary ideas for air transport in the latter part of the twenty-first century. The concepts look beyond the evolutionary developments of the current air transport system, by responding to defined forces for change that are evident today – such as climate change, hydrocarbon fuel shortages, communication systems in society, and large-scale demographic shifts. Drawing on scenarios for future air transportation demands and new technologies that may be available, innovative concepts that respond to change are outlined.

Key words: systemic process, technological change, aircraft concepts, fuel sources, future influences, information technology, transaction integration across transport infrastructure.

12.1 The mind set to find revolutionary solutions

Now is the time for civil aviation to face revolutionary changes. The same questions are on the lips of everyone involved in aviation. This includes a wide swath of non-aviation businesses that are intractably linked to aviation: from manufacturing, through commercial to service, and ranging from multi-billion goliaths to corner-shop travel firms, or even virtual businesses, whose operations are hidden behind website pages. In fact, considering the scope of the challenge in this manner and reflecting how differently this would have been perceived merely 20 years or so ago, whilst also looking even further ahead, encourages activation of some very broad thought-processes. Finding a route for survival can be a process of continual refinement, always placing emphasis where it is anticipated that progress will take the traveller in the right direction for their perceived, but invisible, destination. It is like embarking on a journey with the attributes of one of the oldest of disciplines, education, of which it is said – it begins the day we are born, and it never ends.

To consider revolutionary ideas in air transport is to attach to a very similar tale. Revolutionary ideas have been the key to a host of major developments already, and it is hard to find anyone who can be trusted to have predicted the future with more than just a serendipitous degree of success. Twenty years ago, climate change was on the edge of the agenda. Now it is, virtually, centre-stage. It has displaced little, however; it has simply added to the facets of a business that its leaders need to address, and its practitioners are expected to solve. Head forward 40 years and the environmental questions may well be the greatest facet of all,

233

with technology either demoted or recruited to meeting environmental needs. Technological capability cannot be channelled simply into responding to forecasts that demand will be driven upwards. This is akin to cosseting solutions narrowly, using justifiable but weakly expressed trade-off parameters, and it is already unthinkable to most technologists. Much more so than most commercially orientated decision-makers, the majority of engineers today wish to present credentials that are not just 'green' in respect of climate change and carbon emission policy, but are also based on holistic objectives that give room to embody views to accommodate inspiration as well as incremental discovery.

This is a tremendous turnaround. Forty years ago, air transport's future was entirely predicted on extrapolation of attainments, especially in aircraft speed. As the average cruising speed of the world's leading airline fleets rose from 180 mph or so post-war to almost 400 mph in barely a decade and a half, it was easy to imagine that there would be an equivalent two-fold or more improvement in the same parameter within another 15 years. In fact, the prediction was almost spot on – barely 30 years after the end of World War Two, Concorde entered service, providing a 1400 mph trans-Atlantic link between what were the world's leading communities. But it came, and it went, and to all intents it is now history. The fastest airliners still in service now wend their way around the world at subsonic speeds.

Even so, much that can be held as technologically revolutionary in its time has still occurred. The high by-pass ratio turbofan has become the accepted norm for propulsion, because it is efficient (lesser fuel burn per unit of delivered thrust) and effective (reduced noise, easier to maintain, etc.). Airliners' aerodynamic surfaces have matured too; starting with names like 'top-loaders' and 'supercritical', they have silently morphed into almost invisibly adroit high lift-to-drag ratio technological works of art. Their efficiency has been enhanced too by newer materials; aluminium–lithium, and more recently carbon-fibre, have let aerodynamicists win significant advances in the war of combining a good cross-section with a higher aspect-ratio planform. Cabins have developed in girth, with the wide-body airliner establishing itself across a broader range of markets as time progresses. These are ongoing revolutions, sure to continue for some time to come, but they are largely incremental growth in known areas. That is not to belittle the effort needed to establish these gains. They come at a price; they are tried and tested before they are adopted. They vie with other ideas that have emerged and withered on drawing boards and in laboratories.

What is there that is revolutionary out there? Are there developments that will do for civil aviation what DNA profiling has done for forensic science? It is worth noting that the example quoted, whilst it has revolutionised the way that criminal investigators handle evidence, was a product that emerged from research pursued within medical laboratories, and without an eye on the attainment of criminal convictions.

The aeronautical thinkers today have to open their thinking processes equivalently. They can think innovatively in a technical sense – as they are encouraged to within conventional research circles – and also think innovatively in a more holistic sense, by considering what role aviation plays, or will need or be expected to play in the future, and what it will be necessary to address to meet the needs of future societies. This is out-of-the-box thinking, and it is unavoidable. It will become the centre stage topic in this analysis of revolutionary change. The scope adopted in these notes is that in the short term there will be evolutions from current technologies, almost without external influence, and that technology will gradually transform the user requirements within society, precipitating developments across a wider area of influence.

12.2   Technological change

The businesses that create, supply and support, and utilise aviation products in use today have to view all scientific endeavours. They need to address the core technologies of the existing businesses, and the more fundamental technologies, ranging from pure (physics and chemistry), to applied (information technology – to quote such a widely used euphemism that its boundaries are difficult to define). They may need to mine in fields that are, in catalogue terms, a world apart, and where they look to find such inspiration or products is not easy to describe.

The difficulty that perplexes the blue-skies researcher straight away is that there is no framework against which such thinking can be benchmarked. It leaves connection between fundamental researchers and free-thinking in a serendipitous state. This is hardly the way to convince large engineering corporations, who have become used to investing time and effort in ever increasing proportions to get diminishing returns in terms of leading vehicle attributes. The vehicle is the centre-stage object, however, its good and poor attributes dwelled on at length by almost everyone with a vested interest in the future of civil aviation. It is also the artefact within which the most fundamental expertise has been devoted, so it is the logical starting point.

A framework upheld throughout the history of aeronautics is that the vehicle, whatever its form, is a synthesis of propulsion, aerodynamic and structural endeavours. In some cases the propulsion system is entirely passive. For instance, a sailplane or a balloon will extract the energy it needs from the atmosphere. The methods vary, with the balloon merely taking a ride on currents, and accumulating potential energy wherever it can, and the sailplane is more active, employing aerodynamic processes to convert kinetic and potential energy, and thus exercising more control over its ability to determine destination. These are vehicles that are so intrinsically at the whims of natural elements that they do not interest those who desire to contribute to a passenger-carrying business. But they are examples of the limits within which an open-minded thinker, perhaps, has to work.

If civil aviation does decide to move away from the high sub-sonic flight regime towards lower speed, and more efficient vehicles, the limitations of these existing, and almost wholly recreational, forms of aerial transport will begin to be felt in operations. They can set limits on, or contribute solutions to, the properties of the vehicles that emerge in the future. In fact, slower aircraft are not so much of a drawback on shorter routes as the relative cruising speed differences of current and proposed vehicles might suggest. However, a slower vehicle invites poor comparison in long-range operations. That is in part because of the extra time required, but also because the impact of meteorological effects, wind and turbulence, can be much more significant. Speed might be traded for a more sedate travel experience in more environmentally refined times, however. Nothing should be traded in, or out, until a reasonable assessment framework is in place.

For the moment almost all interest is focused on air vehicles that operate at speeds similar to current-day vehicles. To provide a benchmark, the possibilities of developments in propulsion, aerodynamic and structural areas are briefly reviewed, simply to consider if there are any evident hidden avenues through which progress, potentially, will emerge.

12.2.1 Propulsion

There is little doubt that, barring a landmark volte-face by industry, the gas-turbine engine will remain the dominant commercial aviation propulsion source for several decades. It is an energy converter that is suited to the stratospheric regions of the atmosphere. In those higher reaches of the atmosphere the air is smoother than in the lower troposphere, and thus it is where societal pressure still forces aircraft designers to maintain their target cruise regime.

Throughout the twentieth century the aero gas turbine was considerably enhanced, largely through aerodynamic and material innovations used to enhance the efficiency of the thermodynamic cycle. The gas path has to remain a four-part process – compression, combustion, expansion and exhaust – but for a given thrust unit the mass flow has been considerably increased, by using higher internal gas temperatures and increased internal flow velocities, yet moderating the external flow velocity of the exhaust, thus increasing propulsive efficiency. Most of that change has been associated with compressor fans, and increasing the bypass flow ratio (the proportion of flow not expelled through the engine's core). This brought what was, at first, a welcome but not primary benefit, in the shape of reduced noise, and it was in the fullness of time that the societal value of this improvement was really appreciated. Even further noise attenuation has been achieved through innovative material applications, such as honeycomb-chambered cowling panels. Meanwhile, it has been the ability to develop engine cores that were able to operate at increasingly greater temperatures, thus liberating almost all the conceivable energy held in the fuels that are used, and within the temperature limits of the materials available, that has allowed this engine configuration to emerge.

The gas path in propulsion systems may become more complex – perhaps with heat exchangers, for example – but the principles will remain unaltered. A widely anticipated development is the aft fan (internal aft fans were used for a while in the 1960s and 1970s), and an external variant, dubbed the unducted-fan (UDF), was amply demonstrated as long ago as 1982. In both forms, the system offered fuel efficiency improvements. In many respects the UDF is causing jet and prop engine technologies to converge, with the turboprop tending to be preferred already as the lower cruise speed propulsion unit. Long-term interest is assumed to lie with high subsonic cruise, to which geared-fan development points as an increasingly attractive solution, with the mechanical efficiency and reliability of the gearbox design absorbing the development budgets in those companies where this development is being pursued. These are sizeable and admirable development programmes, but they are still incremental developments of a concept that was revolutionary in the 1930s.

Given that there is a finite limit to the amount of energy that can be liberated from a given quantity of kerosene, the use of this petroleum product as the sole fuel for aviation use should itself come under scrutiny. The best possible alternative solution meeting the current-day environmental framework is bio-fuel, generated from cultivated areas that are classified as renewable sources. Provided it has the same, or better, energy content per unit volume as kerosene, this is acceptable because the aircraft's kerosene-based payload-range capability should still be attainable, and the carbon liberated during flight is accountable against the carbon consumed by the bio-fuel source during its growth period. However, if this solution is used to justify the continued used of internal combustion and gas turbine engines across a wide range of transportation modes alone (it could be extended to include surface-based power generation too), the cultivation area requirement is expected to compete with cultivation requirements to meet human food consumption requirements. In that case, the end does not secure the means. Bio-fuel might be one of the future's only acceptable fuel concepts, but its limited supply capacity could be one of the most constraining factors limiting its use in transport and elsewhere. The fact that kerosene will run out, so to speak, but bio-fuel can still be produced in the future is the factor that makes it a potential long-term fuel solution.

More radical are the prospects offered by alternative technologies: fuel cells, solar energy, and energy storage devices. These abandon the gas turbine, and are revolutionary. At the present time it seems inconceivable that these can be used for anything other than low-speed vehicles, and as such they will bring the air transport products they power back into the clutches of the atmospheric maladies that affect existing low-speed air vehicles.

There are dichotomies that researchers in these fields need to reconcile with aircraft designers. The best that the latter can do, probably, is to examine any proposals in terms of what will represent an acceptable proportion of air-vehicle mass devoted to propulsion. In essence, a future 'fuel-less' system of equivalent propulsive capability can be a much heavier engine – with the current normal proportion of fuel and engine mass in mass audits setting a limit to respect.

For most designers in this revolutionary design field, the debate always seems to roll back to energy storage, which is almost inevitably a high-mass endeavour. The potential for a revolutionary solution emerging from knowledge of the mass–energy relationship is easily recognisable. The fact that nuclear physicists can liberate the energy of a nuclear explosion from a relatively small warhead is the prime example to quote as the ultimate goal. The quest for fusion rather than fission processes for ground energy generation is still being pursued, but for some researchers the 50-year or so dream that has been invested in this endeavour is at an end. There are some who persist, however, and believe that with super-conducting materials in prospect there will be mass and cost benefits in the future that will make the fusion reactor a reality. Some commentators have ventured so far as to express the belief that, without this breakthrough, society as we envisage it in the future (and that is based on still expanding population data) will be doomed. When one finds such thoughts in vogue, the question of whether air transport will even survive can become a realistic point to debate.

If energy cannot be carried and liberated during a flight, the only solution is surely to draw energy from the environment. A fine example is the solar cell, but the power that can be collected per unit area of cells tends to require vast arrays in any aeronautical application, and the system needs sunshine too. Since the mid-1970s some researchers have championed the idea of having massive solar collectors in high-Earth orbit, and of sending the energy down along microwave beams whose energy density is too dissipated to harm organic life, thus allowing a large array (several kilometres square) to yield many megawatts of energy, and to fulfil the role of a modern power station. Given the power requirements that must be anticipated for an aircraft application, whilst an orbital collector of adequate magnitude is conceivable, the energy beam from space to an atmospheric vehicle with a management receiving antenna will be of potentially hazardous intensity. The failure cases, and the potential for power interruption, mean that, at best, the concept can be down-graded from possible to 'only if necessary.'

Do we look therefore to fundamental physics, and question whether we will learn not just to live with gravity, but to beat it? This is a very pertinent question, and there are authors who believe we have scientific communities, practising in secret, who know how to do this already. They call it anti-gravity, without explaining whether it is as simple as it sounds, or not. There is no doubt that the fundamental anti-gravity particles that are theoretically unavoidable, if they can be harnessed and supplied to potential users in the manner of a saleable product, will re-incarnate the aeroplane in such a fundamentally different guise that the aeroplane as we know it might well be dead. This is revolutionary thinking – but it will be like the discovery and description of fundamental atomic particles themselves some 100 years ago. Indeed, the description of these still theoretical particles remains subject to debate, and their apparent short lifespan makes them unquestionably a more demanding set of particles to harness and control than even existing gravitic, as distinct from anti-gravity, atomic particles.

The ultimate anti-gravity capability, at a personal level, could be the first step on the way to science-fiction 'beam-transports', and take air transport off the human industry scene. It perhaps pays for innovators to keep an eye on the fundamental sciences to read across the relevance of such developments.

12.2.2 Aerodynamics

The over-riding concern in the design of an airliner is to achieve a high lift-to-drag ratio (L/D) in cruise. Conversely, if a ludicrously low-cost energy source is found, such that the vehicle drag can be overcome by thrust obtained at virtually no cost, the logicality of this requirement is overturned. Expecting that sort of development to occur, even within a few decades, as the previous section has shown, is like backing an improbably low-odds bet. It does exemplify how something like anti-gravity, or anti-matter, could be the development that makes the air vehicle concepts utterly different from what they are today.

Meanwhile, conventional physics forces the designer to assume that the best possible aerodynamic efficiency will be achieved with laminar flow over a large proportion of the wetted area, high wing-loading, and minimised vortex-drag. Today's airliners achieve an L/D over 20 at Mach 0.85 or so. The product (L/D) × M is 17+, and that is the numerical target to beat. Note that Concorde achieved $M = 2$ with a cruise L/D of approximately 8, so with (L/D) × $M = 16$ it was every bit as clean in technical terms as it looked to the untrained eye. Sailplanes have achieved $L/D = 50$, but they need to carry ballast to achieve it (forcing up wing loading), and to penetrate air relatively fast. Their (L/D) × M product at their best penetration speed is much lower than most airliners.

This has not stopped researchers from trying to find ways of reading across from sailplane to airliner technology. Blemish-free surfaces, and configurations so accurately constructed that configurations (such as aircraft span) are measurable to microns, are some of the palliatives in place. The latter capability has evolved in the military stealth industry, and requires laser measurement of major sub-assemblies during assembly. The blemish-free surface has been established for many years, and requires, again, meeting tough production standards, but also a material that will form a smooth surface. It is not impossible with aluminium alloy, but it is more repeatable with reinforced plastic structures. The civil airliner industry is already moving into the latter area, and higher-accuracy production technologies are also being developed. It is a step ahead, but it will not provide anywhere near enough to lift (L/D) × M significantly.

Revolution can sometimes be retrospective too. Control surfaces that disturb the flow over the aft section of an aerofoil are widely accepted. This can be alleviated by using wing-warping – although this does have some disturbing aero-elastic consequences (exactly why this kind of surface was abandoned). Modern design tools can be used to tailor design, so that such limitations are no longer severe. More radical has been the idea of affecting flow, and thus pressure field, by applying electromagnetic forces. It is not necessarily energy efficient, and, whilst it might prove to be essential in some parts of the military sector, the better solution in civil applications, where commercial influences are more prized, might be to minimise the need for control surface deflections, such as simply by planning aircraft to fly direct routes.

Just as military stealth design has concentrated on the minimisation of radar cross-section (and that does not read across to civil applications), there are new aircraft configurations (overall shapes) that can be considered, or re-investigated for airliners. The blended-wing body (BWB) is a pragmatic re-invention of the all-wing concept, which promises best possible cruise L/D. If high subsonic Mach cruise is attainable the $(L/D) \times M$ product should be unassailable. A compromise is the tandem-wing, where the stabilising surface is at the front, and is arranged to combine with the main plane to provide vertical lift across the full flight regime. A conventional fuselage is retained, and there are modest advantages, such as the opportunity to accommodate large CG changes with only a small impact on cruise L/D. Even so, this long-since utilised layout, and its related three-surface design (as used in the Piaggio Avanti business turboprop), have been slow to influence designers. The most pragmatic designers with an eye to the future are showing straight-wing aft-engine configurations, with V- or H-tail surfaces. Overall, compared with current airliners they tend to trade a better cruise L/D for a lower cruise Mach number, and thus achieve better fuel efficiency, but they do not really break the aerodynamic mould.

If the world does start to use slower aircraft, there will be an impact on the market for air travel. Subsonic travel, at Mach 0.8 approx (around 470 knots) in the stratosphere, has become widely expected for medium- and long-haul travel, and it has to be expected that there will be tremendous consumer pressure to maintain similar speeds.

If engine efficiency is achieved with higher by-pass ratio, geared-fan or even propeller technology, there is little doubt that cruising speed reductions will be beneficial. There could be more environmental benefit from this than from any other potential development. The degree of speed reduction will depend on which propulsion technology begins to dominate. Environmental pressure may desire a large reduction, to reduce power needs and thus fuel usage, and because it could be used to trigger other benefits, such as lower altitude cruise, with sub-stratospheric cruising reducing, and perhaps even eliminating, contrails. The passenger will not want this, because it does bring the cruise regime into more turbulent regions of the atmosphere. This trade-off highlights how the current Mach 0.8/35 000–45 000 ft cruise regime reconciles many objectives, and any change leaves the users, designers and environmentalists almost certainly needing to exercise some compromise.

12.2.3 Structures

All else being equal, the lighter an airframe, the better it is. Some of the propulsion and aerodynamic requirements discussed will militate against success in this regard, whilst some will aid such an attainment. Overall, the design synthesis of an aircraft has to balance these needs. Meanwhile, through using CFRP, polymers and advanced metals, considerable progress has been maintained over many decades. The rate of advancement is steady but slow, and a revolutionary push is always eagerly considered. As aircraft were expected to cruise faster there was an expectation that ceramic materials would begin to influence aircraft design, but this has never materialised. There have been some developments of this nature that have improved engine operating temperature capabilities, however, and such forecasts, whilst poorly targeted, were nevertheless looking in the right direction. Such are the difficulties of long-term forecasting.

The wider acceptable use of polymer-based materials, and the opportunities this offers for the development of intelligent materials to meet needs, has become a research opportunity. The most widely sought-after possibility is a self-healing material; one that will affect its own repair in the event of being damaged. Chemical molecules derived from shellfish have been adapted to perform this function in paints, so that scratches are rendered blended out for cosmetic purposes. These will clearly benefit an aircraft that requires a blemish-free skin, but whether they will work effectively with regular large ranges of temperature variation could be one of several misgivings. This could be a long road, along which the first steps have just been taken. It could also be a technological cul-de-sac, proving to be too expensive to implement. This is one clear example of the sort of technological development that will, even if unsuccessful in its current guise, leave a legacy that will guide later developers, or might become feasible through unrelated chemical developments in an unexpected scientific area, such as in medical development.

Smart material applications that have been taken seriously by the aerospace industry include skins with embedded glass-fibres, which if disturbed will indicate the location and extent of damage, and thus provide data to a computer-based system, such as the aircraft flight controls, alerting the need to change aircraft control trim settings, or to impose manoeuvre limitations, etc. These are a further example of where the cost–benefit will stack up more quickly in favour of the development in a military application, and where in due course a commercial application can be expected, but is risky to predict.

Overall, the materials and structure area is difficult to predict. There is one development that does indicate a way forward, potentially, to bring vast benefits; it was heralded in the 1990s, when almost every engineer's best desk accessory was a complex, albeit fragile, rendition of a complex object – this was laser lithography development. The object is grown from a liquid that solidifies as a laser beam creates the desired three-dimensional shape. The basic material is a liquid-suspended resin. This is engineering, as yet with few applications beyond the fast-prototyping of complex shapes from computer-aided design (CAD) files, that allows rapid evaluation of the physical properties of a difficult-to-visualise design concept. Even so, in laboratories that might be only a stone's throw away on many an academic campus – classified as medical science rather than engineering – equivalent processes are being evolved that produce examples of complex human bones (such as those in the ear). These are often developed from cartilage material taken from a human, and can be transplanted without fear of rejection. Do these developments indicate how aircraft will be built in the future? If the structural design and construction profession has a surprise at all in the twenty-first century, it could be, with material scientists alongside them, the ability to grow whatever structure an engineer wants, *in situ*, and at acceptable cost. It could revolutionise the way that bridges, buildings and almost any kind of vehicle are converted from drawing to reality.

12.2.4 Aircraft systems

All the above developments will depend on the electronic sector (avionics). Its influence is largely in the operational sector, rather than in technical development as such, although fly-by-wire (FBW) airliners have expanded the horizons of designers more than is easily appreciated. This will not be dwelled upon, as the incremental benefits of reducing stability margins to improve cruise L/D and to engineer acceptable handling qualities have been driven close to the design limits allowed by long-standing safety requirements, which are likely to remain in place. There will be systems benefits in the daily routine of aircraft operations – obtaining knowledge of the most fuel-efficient path from A to B, sharing that routeing with other airspace users, and so on, being indicative of what can be considered. There are other sections of this book that look at such issues; the remainder of this chapter sets out to define how research will be categorised and managed according to a framework that will facilitate the coalescence of aerodynamics, propulsion, structural and systems forward thinking within what could be revolutionary projects.

12.3   A framework for assessing revolutionary ideas

Revolution will not necessarily be shaped by technical innovation, and the routes explored here are those that impinge on the businesses, emanating from three areas – societal, economic and politic change.

These are topics that are greeted in many a traditional design office in a manner that is akin to an organism acting to repel a viral attack. Designers who adopt such a traditionalist viewpoint are often accused of protectionism, or a fear of change. The latter view is the most illuminating to debate, as in every design organisation, and indeed in many of the operational professions that support their product in service, there is a vast amount of interaction already with societal, economic and political will, and change is what, more often than not, they will combine to face. They will be familiar already with approvals and licensing requirements affecting the organisation and personnel, not to mention noise regulations, statutory consultations and other modes of public accountability affecting operations. It is these areas in which the technically adroit feel most vulnerable, have accepted (but rarely applauded) change in the past, and are ready to object to creeping changes, seeing the cumulative impact of changes as amounting to a persistent series of attacks.

They have a duty, where such change is justified by public demand, to learn to regard them more objectively than with a withering or scornful manner. They have to develop evaluation processes that will allow them to react with appropriate and justifiable responses. There will be some proposals that are well expressed, many that may be poorly expressed, and in some cases proposals can be shown to be simple impositions. In this way, designers can seek dialogue with, and even hope to recruit to the cause of finding compromise, the very sources they otherwise will be obliged to treat as enemies. Recruited to assist, either to repel impositions, to polish the poorly expressed, or to integrate the best thought-out calls for change, legislators and implementers can become travellers with a common set of causes, and necessary change will be achieved without scorn or rancour.

The most pressing political debate outside design offices is about global warming. How will aviation respond to this debate – will it see this as a threat or a challenge? There is no doubt that, so far, it has seen it as the latter. However, as aviation is a business that serves society, if it fails to accept this challenge, effectively it will fail the people it serves.

This is an area well beyond the pre-existing technological framework, but there is no need to ditch consideration of the synthesis of propulsion, aerodynamics, structures and system design capabilities, in aiming to create the most efficient vehicle. The newer thinking requires the insertion of this framework into a more rigorous set of design rules that will allay the fears of those who depend on preserving the appropriate niche for aircraft as future travel vehicles. This includes the public at large, the financial institutions that will underwrite the manufacturers' development plans, the operating plans of the airlines that will use the vehicles to provide essential services, and the airports and airspace service providers that will tailor their facilities to accommodate demand. Whereas an aircraft specification has been long held to be the result of manufacturer/operator debate, the extension of this philosophy has been only piecemeal. Airports were consulted as very large aircraft (VLA) design work commenced, with the 80 m box adopted to define maximum span and length criteria that would constrain designers, but ensure that the final design (so far the Airbus A380 alone) incremented, and did not destroy, existing airport operations frameworks. In the future all of the requirements of the parts of the air transport system – operators, providers and manufacturers – have to be associated with the design process and, not least, the public perceptions of the impact of these operations on the world overall must also be taken into account.

It is not just a 'green' debate. Taking such a one-facetted approach is like believing that the froth of a beer or a coffee is the taste itself. It is an essence, and it is a part of the whole. In the new order of design priorities, the contribution of the whole has to be synthesised. There has to be a way to expand from the current, largely technological property-led viewpoints, to a wider set of perspectives, and the most expectant gazes have fallen on systems thinking.

This is not as radical a regime change as many people seem to fear. It is a common error of many, when adopting what they purport to be a systems approach, to fail to accommodate the breadth they need to address so that they recognise the most pertinent properties that must be incorporated in their new design processes. A framework capable of harmonising the perspectives that all components of the air transport system will address and incorporating vital public debate input can be expressed by addressing four perspectives:

1   Promotion of developments that are financially viable,
2   Attainment of statutory requirements,
3   Definition of an efficient process, facility or vehicle,
4   Coalescing these needs with user service expectations.

The traditional designer – they can be an air-vehicle designer, an airline's or airport's strategic planner, or an airspace planner – will deal the cards they have to play with all these aspects in mind. For example, the airspace planner, in creating a design for a new airport, is constrained by:

• Geography,
• The needs of other airspace users in the vicinity,
• The cost of the service provision,
• Legislative edicts that express noise and other environmental constraints,
• Agreements and design rules imposed by political and technical overseeing bodies,
• The results of public consultation.

They design the routings to serve the airport(s) they consider as efficiently as possible – perhaps aiming to determine arrival and departure routings that are as short as possible (to assist users to meet efficiency targets), whilst judging how they can influence the attainment of best aircraft performance practice, and assessing the need to interlace arrivals and departures themselves. They also aim to provide a framework that will allow the final operation to be sufficiently flexible to cope with a host of recognisable failure cases – within vehicles, in their own service equipment, and even failures that affect the system being designed, and that may not occur within the patch over which they exert an influence, but through disturbances transmitted from some external influencing system. The butterfly flapping its wings in the Amazon basin being the seed that creates the air currents that batter European cities with storms a long time later is an analogy that does hold water, although its direct relevance deserves to be modified by giving credit to the fact that the butterfly will have been one influence – not the only one.

The four-aspect framework suggested previously provides the basis of a paradigm that will allow the many businesses, whether they be societal (largely environmental), political or economic in nature, to identify system properties that can be extended into the fields where what are external influences to them can be recognised and stitched into the thinking processes of others. This is a route map that will allow current designers to express their confidence in the capabilities of what they will bring to the future with the ability to say, 'this is representative of the way all needs can be coalesced' (Fig. 12.1). The green debate will be as much a part of the arguments that determine capability as the political agreements of service sharing, on safety management edicts applied globally, and regionally, and much more. Amongst these will be the technological capabilities that pre-exist in each discipline.

There is no reason why designers cannot develop plans that address population change, that pre-empt demand variations, that forestall arguments about source of pollution, attribute responsibility rather than blame (absolutely crucial!), and address the costs, in both financial and societal terms.

The consequences involved in developing what is not so much a revolutionary approach as a process that provides reasoned, accessible and justified answers are staggering, but difficult to describe. The consequences of failing to do so are the more easily expressed outcomes, but this is negative, and not acceptable.

The next generation in the business is likely to live through, and should be ready to take part in, the industry's most radical revolution to date. If they fail to do so, the loudest voices on the stages being considered will reserve their undeniable right to preach what protects their desires. As a consequence, everyone else will need to accept their constraints. As an industry that has, throughout history, been able to provide so much, so fairly, and so widely, without prejudice, the air transport business has to assemble a stable, not a stockade. It must facilitate and improve debate in order that it will furnish solutions that are reasonable, justifiable and equitable with other needs of society. The ultimate alternative is to stop flying.

Stepping into such debate exposes any contributor to many regions of terror. This usually stems from ignorance, and not from a lack of knowledge. For example, the engineer does not speak in the same way as the environmentalist; the politician does not have the same perspective as the business leader; and the accountant does not see eye-to-eye with the technologist. There are many other aspects, but the latter is one area where there is evidence of what needs to be done.

Accountants and technologists have always been an inflammable combination. The one sums their professional interest in monetary terms, the other in service capabilities, societal expansion and other attributes. Those technologists who are new to this field often complain that they are wasting time and effort because the

12.1 The timescale in which change needs to be assessed is well beyond the technology horizon of aviation vehicle designers.

accountant's interest is 'value added', and nothing else. Undoubtedly, in some places, this is true.

In any business where there is a high dependence on technology, its relationship with fiscal, and other, properties has to be balanced. In the technology and fiscal trade-off arena there are, especially, questions of competitive threat, and of risk. In general, solutions that have been governed by thrift, and yet guided by prudence, have always been the most successfully balanced fiscal and operational solutions. Entrepreneurs and industrialists who have integrated their aims tend to believe that in the early stages of their projects they created a framework within which their individual perspectives were not just balanced, but their viewpoints were also appropriately expressed to allow trade-off across all the disciplines involved. The search for what this experience has constituted has led to a systems engineering expression for this early project task that is universally called 'requirement capture'.

Requirement capture should commence early in a programme, such as when research is about to be turned into actions, and, ideally, it is completed before the product – be it a vehicle, a computer system or whatever – is designed. Where a product is complex, the design process can be sub-divided into phases too, and the depth of the requirements captured for design refined at the various stages. In an aviation context, whether for an air vehicle, an airport or an air traffic system, the design phases will comprise conceptual, preliminary and detailed design. At each stage, the process is terminated at a progress point (sometimes qualified by a formal review process, and on a pre-ordained date to which very serious intent is declared). Note that all participants are, to some extent, fallible, and where requirement capture is incomplete, through lack of knowledge rather than ignorance, there can be a need to reconsider the expressions of needs and thus to offer, at later stages, guidance on project evolution: leading to mark 2, model B, etc.

An example of the kind of techniques that can guide those attempting the requirement capture of a concept or implementation that is well beyond current practical limits is one that embraces what its adherents call the three Cs. This refers to consistency, completeness and clarity.

12.3.1 Consistency

Consistency is easy to describe for a simple item, but even then there are many descriptions that could be used. For example: describe a door handle.

Just its function can be described, e.g., it allows a person to open or close a hinged aperture in an obstruction or wall. Or is much more detail necessary? There may be merit in adding that it must be at a distance from the floor that is related to the distribution of height in the population of users, that it is suitable for being grasped by the hand (or not, because not all handles are), and so on.

The message to get from this approach is that there is a first-level description – the function of the device/object – and then there are increasingly detailed levels of description, ending up with a very detailed design description, which will have diagrammatic illustrations, notes about materials' specifications and manufacturing processes that need to be used, cost estimates and so on.

This is fine for a physical, manufactured, item. What if it is a piece of software? Again, the function might be readily described, e.g., this software will reside in processor XYZ and it will respond to information from source A such that it opens or closes a door by sending an electrical signal to a device at location B.

From this first-level description program flow-charts with decision branches might evolve, and in stages; eventually, the program code that will perform the function will emerge. This is not enormously different, philosophically, from the hardware description process.

In each case considered there will be a point where there is a need to examine properties that are emergent from the design. These can have common complexions; for example, with hardware and software products there is often the need for a failure mode and effect criticality analysis (FMECA).

Accountants and technologists have learned this process, and their route maps (often varying significantly in different organisations) offer the essential guidance for technologists to take to the table in their debate with the newer influencers. The designer has to respect their desire for certain outcomes, and they have to accept the necessity to describe the consequences of their own contribution within a scope that also encompasses the traditional discipline needs.

12.3.2 Completeness

Completeness is a thorny question for most people who accept responsibilities for creating a product. The previous notes on consistency will have helped to scope debate, but knowing at which point to stop pouring detail into a description is often perplexing. There is the tendency to fall into a traditional route, following 'this is how we did it before', which can be very misleading and is useless in a revolutionary design project.

Those capturing requirements have to take the capture role to the specialists who are most likely to have the appropriate knowledge of the necessary emergent properties, to seek definitions of needs that are sufficiently complete for there to be no likelihood of the design process overlooking a vital attribute, or relationship between properties. This, again, is very difficult because determining who is an appropriate specialist can be a thorny decision point, perhaps hindered by barriers imposed internally by rank and seniority, or externally, within a collaborative project, by inter-company relationships, where intellectual property rights and commercial confidentiality may impinge on the degree to which relevant information can be exchanged.

To assist, several methods of syntax-based analysis programmes have been developed, and are embedded in formal, software-based, requirement capture programs. These will assist by forcing consideration of sub-headings –

performance, maintainability and many other 'ilities' being typical – and they will rank and identify (usually number) requirements. The progressive development of the project through review stages is greatly assisted where such tools are applied, as they will force the reconciliation of all requirements, with termination not disallowed, but with the interdependence of requirements tracked too, thereby creating the capability of a traceable audit.

No matter how capable the designer, and no matter how much they bolster themselves with such assistance, it is undeniable that, as they seek data, they will still be made to feel like the whinging lesser half as they search for a degree of completeness that is expressed, by pundits and laymen alike, with more passion than precision. This is a crucial and difficult reconciliation process.

12.3.3 Clarity

Clarity is the final issue at stake. It is essential to eliminate ambiguities, and yet it is well known that, if requirements are expressed in plain text, language will often introduce ambiguities. A condition on a performance requirement is least ambiguous if there is a precise definition of the conditions. These will usually be found in standards, so a qualification that might look innocuous, e.g., this shall accord with ICAO Annex xx, para 12.10, can be the best way of all to express a need with clarity.

These guidelines explain why it is that getting interested parties to become involved, and holding their attention for sufficient time to get them to contribute fully, is difficult. It explains, too, why they will often complain that the incumbents do not really want to listen to them, but force them to do what they want. The skilful negotiator will ensure that they feel there is progress all the way, and will have some way of providing a route map that presents the scope and the depth to which they will need to offer guidance, and with an indication of the necessary level of precision.

12.4   Carrying forward requirements into design

With the principal elements of a process defined, the questions remaining are those that address what needs to be controlled. There are differing opinions here, with some political and economic-orientated adherents believing that change has to be imposed. Rather than look for a way of encouraging appropriate change, they will insist on using a governmental initiative, such as taxation. This begs the question whether there is an understanding of what problems are being addressed. Most commonly, the answer to any question attributes the reasons for forcing a change to an environmental cause. The technology innovators who see through this are, as often as not, following a line of thought that the majority of environmentalists will follow too, as they know that direct action is essential, rather than casual fiscal tinkering.

Taxation not only fails to address the emergent properties; it simply imposes a penalty. Credible legislative development should include the possibility of reward as well as penalty, arranged ideally so that, if a genuinely pollution-reduced solution to an air transport issue is found, there will be no pre-disposed tax held against operations. Indeed, there should be incentives to innovators to find solutions that address emergent properties, so that they can be encouraged to find those who are willing to invest, in terms of time and effort, and be rewarded by results. If this does not sound like revolutionary design, or equivalent thinking, imagine the impediment that any simple blockage, based on historical rather than forward-thinking policy, can have on a design team. Revolutionary progress can often require some radical overhaul of what is too often regarded as unchallengeable.

At the present time, a popular theme is bio-fuel. The reasoning is that the fuel used is derived from a source with a sufficiently positive impact on carbon emissions for the emission to be negated. This can be farcical, as the usage of a bio-fuel will be encouraged to the point where the carbon emission is not addressed, so much as its neutralisation is assured. Air transport does not get capped, so the adherents say this is fair, as it lets the industry expand as demand requires. They do not address the fact that they now need to allocate vast areas of the Earth's surface to producing the bio-mass that will become the source of the fuel used, and that in turn the solution is thus addressing aviation without addressing the ecological impact. This is an incomplete solution. It lacks clarity and its drawback is its inconsistency. It is indicative of debates between technological problem-solvers and groups that seek to impose, rather than define, change. Many economic observers note already that bio-fuel demand is encouraging de-forestation, and the crops that are grown are for these new initiatives, as they have more value than food for feeding the hungry. Even worse is to be informed that the land often falls into neglect after a short period too, its nutrients being rapidly depleted. It is a clear example of a non-systemic approach to problem-solving. The scope of the programme has been too focused on one aim – in this case, the production of bio-fuel – which has been justified through a requirement definition that is not complete, consistent or clear.

Consider the scale of aviation operations overall. If the annual air trips per head of population in the USA are 20 times greater than in China, why should Chinese development be capped below the usage level in the USA? The USA, and indeed most Western European and other equivalently developed economies, appear determined to maintain current usage at least, so the desire of other nations to grow their demand to equivalent levels is perfectly reasonable. It would be more plausible if the developed nations considered capping, and re-distributing, demand. The process should see the pleasure that developed societies already get from aviation being diminished, and the pleasure that other nations get increasing, as they take a larger, more proportionate, portion of a limited capacity set through mutual consent.

If that capacity limit were then allowed to vary, coming down as carbon neutrality was diminished, and rising as the converse was true, then a change paradigm would be imposed that would share the pleasures and the pains of justifiable aviation usage. This would be a more complete, and consistent, global solution to a global problem. At the present time, there is no forum or global organisation that has claimed or sought to have the teeth to bite such an initiative. If this remains the case then air transport operation demand will continue to grow, and, when judged via every perspective given credence in current debate, the public consensus is likely to be that the situation in out of control, thus inviting, at some time, the imposition of savage regulations. This may sound draconian and inconceivable, but it is a very close parallel to what has happened to global economic systems.

12.5   Telecommunications and IT in society

The external influences on civil aviation are manifold and not insignificant. Information technology (IT) has been one, partly because it already has had a profound effect on society. It was in its infancy as aeroplanes first flew, but merely as a telegraph and occasionally a telephone system. It has developed from wire to wireless, and if communication and media developments that have used electronic formats and transmission methods are also considered, whilst facsimile, inter-continental telephone and television were all regarded as profoundly influential, IT took an unprecedented transformation when silicon-based integrated circuit technology was introduced. Since the mid-1970s the complexity of these electronic devices, initially termed microprocessors, has increased at an almost exponential rate, and apparently without boundaries. The desktop PC, the laptop, and now the notepad have become personal computing centres. Mobile telephones have, in many respects, usurped the telephone, have outshone the two-way handheld personal radio or walkie-talkie, and have begun to integrate into PC culture, with pocket-size devices now available on which the internet is readily accessible. The ability to communicate, even navigate, and to be covertly under surveillance, almost globally, has transformed the way in which future generations will organise and conduct their lives. The adage 'out with the old and in with the new' has never been more appropriate.

Already, these developments have had a great influence on almost every element of the air transport system, and where this will lead is difficult to chart; but it has to be a defining aspect of what will happen to air transport in the future. The way service provider (airline) demand is solicited, processed and nurtured has taken on a new complexion in the first decade of the twenty-first century. The once ubiquitous IATA flight coupon was the recognised ticket for air travel, and yet it disappeared in 2008. The idea that one has to visit an airline office was buried when computer-based reservation systems (CRS) made their debut in the 1960s, and transferred into travel agency offices in the late 1970s and throughout the 1980s. The airlines still advertised their service, not just with billboards and in newspapers and magazines, but on television, on radio and with promotional activities too. Not so much coincidentally, but with demand burgeoning and a world-wide desire to loosen the grip of government control on such commercial assets as airlines throughout the largest nations, the international laws that prescribed rights of carriage and points of service were often considerably amended, or even removed.

Facing greater competitive pressure, the incumbent airlines, with their vast CRS databases, pioneered ways of trading financial incentives with efficiency and service-quality attributes. The key development area was in yield management, where yield was defined as the price obtained per seat on a service (which clearly should exceed the cost of provision to ensure profitable operations), and that divided passengers according to their travel needs. The once-established first and economy cabin inherited a mid-cabin class, called Business class; and it was Business and First class that were regarded as deserving of loyalty benefits – frequent flier programme (FFP) initiatives – with the profit they generated potentially diluted by the lower-price economy passengers, who got a price incentive for booking early. The latter was a time-based distinction, with later -booking passengers helping to boost seat occupancy – thus aircraft passenger load factor – and contribute equivalently to the business and first-class passenger profit. This is referred to as yield-related market fragmentation, and it was instigated in part by having larger aircraft with additional seats, but also by the desire to achieve equitable service qualities, such as service frequency.

In recent years the airline sales model has changed from the fluidic form that appeared following the rigid systems of the twentieth century through a maelstrom of developments. The airlines will track potential customers, not just for routes and desires, but through the internet, and access individuals and groups of potential passengers, unsolicited, through their personal IT ports – the computer and the mobile phone. Getting a text (SMS) message was startling, but within a short period the email was there, and the pitch within the message was sharper than ever. The airlines are now independent of advertising streams, and have the ability to initiate and construct unique advertising campaigns. Without some form of control this can become part of an IT epidemic. Many older passengers are already feeling vulnerable to the rate at which they are solicited, because it is so intense compared with what they saw in the past. The younger generations see it as less threatening, and less invasive. This is change. It is a societal change that is seeing the boundaries of personal freedom expanded, but with new conditions that depend largely on individuals exercising choice in ways that were unnecessary in the lower-IT era.

The impact on air transport will be even more profound as time, technology and business techniques evolve. The extent of these changes will explode, emanating from the airlines to affect all the other components of the air transport system. Airports are seeing this already, and the customer too, as remote check-in (using a plethora of IT ports, from mobile phones to internet cafés) replaces the airline desk at the airport.

Consider that the airline might not be the sole custodian of service provision. The technology is within the industry's grasp to lead to the demise of a published timetable. Where will the airlines stand in that situation? They become little more than aircraft operators. They might even be just transport service providers, as the timetable available to potential travellers might not be solely airline-based, but point-to-point-based. A request for a service might introduce transport options – coach, rail, ship, aircraft, combinations, etc. The request for a service might be tailored by knowledge of whether the user's travel purpose is business or leisure, and could even reflect whether the journey is a means to an end or an experience that the user will value. The boundary between airlines and airports, and the interaction with airspace, will become less well defined. The airspace service provider might even become a transport service control authority whose mandates include collaborative decision-making within a global transport context, which will take account of how competing travel modes will be prioritised. If surface transport can become more competitive over shorter ranges, and is more acceptable in societal terms, when such matters as economic and environmental effects are considered, even if time is of an essence and valued, the airspace service provider might be duty bound to close its capacity to such users, and award priority to longer-range operations. Of course, price will play a part too.

Stemming from the above, it is clear that in the near term, and throughout decades to come, there will be radical changes in the way that demand is addressed, and service response is promulgated. As has been intimated, it might not end at airspace (as now – with demand threatening capacity). Instead, capacity-led design of service provision could lead to greater airline, airport and airspace symbiosis, and even broaden out to consider the surrounding competitor transport modes. These will be the new environments, no doubt scaled and managed according to the natural environmental issues that comprise the majority of this debate at the present time.

12.6   The revolution – far beyond the air vehicle

The way to analyse the recognisable future influence will be through attribute-led systemic debate. It must address financial viability, statutory compliance, service efficiency and effectiveness (service quality) requirements that will integrate the viewpoints of stakeholders. In academia, and forward-looking industrial empires, there is an urgent need to identify change patterns, where they will originate, and how they will propagate throughout – even revolutionise! – the transportation system, of which aviation will be just a part.

The breadth of the new thinking has to embrace the full extent of air transport influence. It will be user-centred (not business-centred). The carbon emission issue will not disappear, but neither will it remain centre-stage. There are finite limits on what can be achieved using technology alone. What must happen is that the air transport businesses have to recognise and respond to demand that is natural, and that can be served responsibly. It is what many participants of eco-debates call sustainable development, but in many respects it is better to regard this as societal will.

The most promising path for the air transport specialist is almost certainly one that provides a view of the prospect for transportation overall in the coming decades. Civil aviation will face stiff competition. It will come principally from surface transportation, with the best of the environmentally friendly designs that emerge from developments setting benchmark performance in the statutory arena (where the constraints will still be carbon emission and noise emission-conscious). The aircraft will be hard pressed to beat the road/rail offerings of the future, in terms of environmental impact, and time (door-to-door) as the surface systems improve their performance. This is almost certain, because surface transport vehicles face fewer design constraints than aircraft. The competitors' successes, where they can be incorporated beneficially into civil aviation strategies, deserve to be transplanted. This is an acknowledgement that the aircraft is no longer, necessarily, the prime target for innovative research, at least on a comparable scale to that expected to be devoted elsewhere (Fig. 12.2).

A general effect will almost certainly be that the aeroplane gets beaten back into longer-range operations, but it would be complacent to imagine it will be unchallenged over longer distances – especially over water. Water transportation is still a diverse and technologically receptive area of endeavour. Large passenger-carrying ships have been predictable, in that there are many even larger cargo/freight ships in service already. Shipping businesses must expect, based on environmentally based predictions urging the local production of food wherever possible, to see the current very large bulk carrier vessels that transport energy from point to point on the globe disappearing from their fleets. The desire to source energy locally (as well as food and other resources) will be one driving factor. The fact that one can generate energy using natural resources at or close to the point they are mined and then transfer it to anywhere in the world using radiation, etc., is indicative of the kind of sea change we have to expect in society. Such change will leave shipping magnates keen to exploit a new form of environmentally friendly sea-going vessel – one that can be so large that it offers superb comfort and can collect the energy it needs from orbital distribution centres, or from on-board solar power source systems, and thus be lighter, more efficient and pollution-friendly than the gas-guzzling ships in use today. It is difficult to be sure that such vessels will be fast enough to meet all requirements, but the scale and orientation of taxation and other socio-economic influences could see the long-range travel migration of the 1950s/1960s, when aircraft almost eliminated the passenger-carrying ocean liner, reversed – through technology in part, but mostly owing to the mood and requirements of the times.

A pragmatic, and not too alarmist, judgement would be to assume that the financial incentives that will seek to allay public fears will mean that air travel becomes more and more dominated by the traveller for whom time is of the

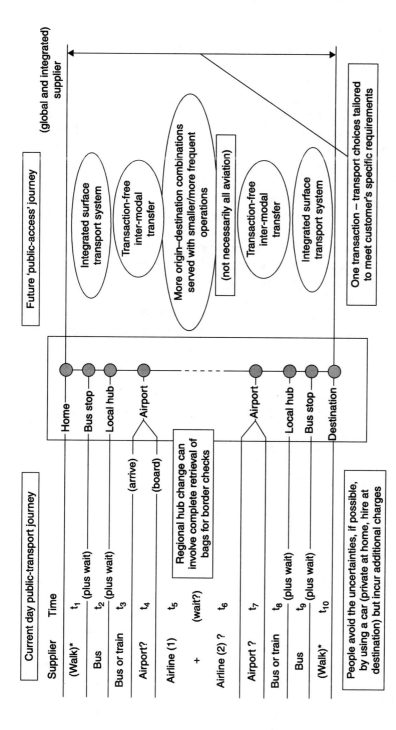

12.2 It is unlikely that air transport will retain a unique identity. The integration of products through information technology advances could be the most decisive factor in determining what share of the travel market air transport is selected to serve in the future.

essence, and the price is, accordingly, of lesser concern. It could be that this kind of market fragmentation is already beginning to occur. The fractional-ownership business jet and high-end business air-taxi market is hardly recession-proofed, but constitutes the less affected end of the existing market. A present-day judgement might be that this just is not the case, with many business chiefs being hounded by public and press-amplified pressure to forego their status symbol business jets. This is only because they are being forced to act retributively. There will be plenty of opportunity to present a converse case once the public-use air transport market begins to collapse under the pressure of environmental responsibility. Businesses will be able to justify a high price tag to get influencers and decision-makers into commercial hot spots rapidly anywhere on the globe, and there are undeniable equivalent arguments for diplomats and statesmen to have access to solve stress points within politics.

One interpretation is that a significant proportion of future traffic will be most easily met by faster and smaller aircraft. Smaller can mean as small as a modern business jet. Faster is more debatable. Leaving aside the carbon emission issue – on the grounds that this is a demand that is being justified by commercial imperatives – the entrepreneurially motivated user will want to be at their destination as promptly as possible. Their sights are set on door-to-door. They want to leave their office, attend a business meeting, do the business and return (or proceed elsewhere), without having to regard a second of time as being unnecessarily wasted.

Airports should therefore be extending their ground uptake requirements into off-airport (actually adjacent to airport) developments that will accommodate global company headquarters or continental/regional area offices. The business leaders can leave their office and be on their flight with no time wasted. The flight is unlikely to be scheduled, but it might be provided by a public-use operator whose clientele will be sufficient to ensure that the aircraft being used is attaining a high level of utilisation, and thus is not prohibitively expensive to operate. It might be flying to a destination that is specified by the client, and that destination might be a smaller airport in the region being visited; convenient for the user in that it is again near to the essential business meeting point. The aircraft should get there quickly, and Mach 0.85 is the benchmark. Why no more quickly? Any supersonic aircraft that can achieve a cruise $(L/D) \times M$ value of about 20 should be acceptable, but beyond Mach 2 the aircraft, assuming it is air-breathing and within the Earth's atmosphere, will need to be constructed of heat-resistant materials, and the system on the aircraft would need to protect those within it accordingly. Mach 2 is a break point, beyond which the attainment of additional requirements will need to be addressed. The revolutionary must look ahead to when these issues can be addressed, and perhaps seek solutions that balance the attributes for and against the solutions to various issues. Achieving this balance across appropriate financial, statutory, efficiency and service-quality attributes is the only way to suggest that a guidable solution can be addressed.

This viewpoint suggests a long-range capable aircraft, with a small passenger capacity, yet suitable accommodation for a high-yield passenger to conduct work whilst in flight, capable of operating on/off relatively small runways, assumed to be almost weather-independent, embedded in a timetable-less aircraft operator's fleet, and accessing a globally managed airspace system. The airspace system will be configured/managed to minimise disruption of operations that are planned with only minimal pre-knowledge of their demand on the airspace, and services will be to/from airports that are not so wedded to timetables, and tailoring hourly slot capacity, as they are at the current time.

It is a vision that envisages civil aviation being, in passenger–kilometre terms, a fraction of the size it is now. Not what airlines – or even airport, aircraft and airspace service operators – want to hear. So what if this is the vision that is targeted and the prognosis is wrong?

Researchers just do not know. But what they do know is whether the goal they pursue is one that challenges their convictions. Given the opportunity, they are able to express how well they understand the requirements they are addressing against the three Cs. If they employ a requirement checklist that forces the consideration and rating of the three Cs, and the proposed attribute-based checklist with financial, statutory, efficiency and customer service benchmark properties, they can soon be spotted if they have strayed from the feasible to the unfeasible, or even risky, zones in which research has to be classified.

Revolution is about risk: commercial risk and technical risk, and not just getting the balance assured, but getting a leap in understanding implemented within the leeway that is available between the boundaries set by the risk criteria.

There are well-proven ways of managing risk. One that has been popular, and used successfully, in aviation is the use of proof-of-concept (POC) demonstrators. Nowadays the POC vehicle can be not just well instrumented, but unmanned, and smart. It can be flown from a control room, using technology well proven and borrowed from military covert operations, so the risk, timescale and cost issues in the demonstrator approach are not necessarily onerous.

Keeping the thinking cap on requires some determination to question the limits that statutory requirements can place on an operation. If the payload is smaller, the crew as a proportion will be greater, so, if the POC vehicle is unmanned, there is surely nothing to stop the future air transport vehicle being unmanned? Steps were taken, in the late 1980s, to show that flight envelope protection can be handled in a fly-by-wire (FBW) flight control system, effectively taking safety decisions on behalf of a crew on board the aircraft, so extension into full take-off roll to landing control – perhaps even ground-movement phases at the beginning and end of the flight as well – do not seem unreasonable long-term objectives. The security benefit of having an aircraft that cannot be hi-jacked is accrued, albeit the fear will be there that the aircraft will have an active uplink/downlink communication system that will be vulnerable. The ways to combat attempts to disturb this link and to transfer control to within the vehicle are well understood. As a development option, the unmanned airliner could be preceded by a one-place crew workstation airliner. This might be the moral and spiritual norm that certificators will respect anyway. On a long-range operation, the cruise phase could be handled fully automatically, and the on-board crew member would handle the taxi-out, take-off, climb, and then descent, approach and landing, and taxi-in phases of the flight. This will minimise the investment needed to establish autonomous ground-movement systems, which would be very expensive at all airports, and could be an example of where technical capability is curtailed by commercial constraints.

Clearly, the air transport business is complex, and the idea of moving towards a revolutionary new operational paradigm is an undertaking that will require courage, time, and commitment, and that is backed by a well-resourced and impeccably conducted series of research initiatives.

The run of any thinking is usually put under greatest stress when it comes face to face with operational realities, and in the route map and the latter example presented here there has been a desire to show that technical innovation will be an enabler, but that operational reality will be the testing ground of the feasibility of any product.

In the past there have been designs that have been bigger and better and that have had their commercial good quelled by the emergence of technology that has revealed the narrowness of the instigator's approach – the large passenger-carrying flying boat is perhaps a prime example.

There have been concepts that have been technologically superb, and that have equivalently fallen bloodily on the sword of commercial common sense. Concorde is the most extravagant of the aircraft in this category.

There have been designs that have promised more than they could deliver too. The Dassault Mercure was lighter and cheaper than the Boeing 737-200, to which it bore a resemblance that was stunning, until one looked at the payload–range diagrams of the two types. The French-designed airliner was well matched to French regional operations, and this revealed that its lightness was a function of smaller fuel tanks. Without the range versatility of the American competitor, it was unable to sell outside its home country. The Trident was Britain's mistake with equivalent consequences, scaled down to a local airline's needs, and its technical Achilles heel was the over-specification of what was a stunning wing (the world's first super-critical wing), that gave it a cruising speed it could not use amongst the rest of the population in the sky, and a field performance that limited its ability to use other than very large airports. But there is a sting in the tail for those who would willingly write off these two extravaganzas, as it was the confidence, and lessons learned, from such projects that was embedded in the psyche of the team that created Airbus Industrie. With UK-designed wings, French pragmatism and US-developed engines, the European consortia slowly, but resolutely, established a foothold, and climbed from obscurity to pole position in the aircraft market.

The legacy for all aircraft manufacturers has to be that, whilst the product is an item worthy of great respect, it is vital that it also respects the needs of others, from passengers, to airlines, to airports and the air service provider community. It has to be as operationally mature as it is technologically mature, in order to fulfil its revolutionary dreams. So – balance the books, across financial, statutory, efficiency and customer-service requirements – and take on board processes that will establish a route map in research that will allow goals to be expressed clearly, consistently and completely.

In determining a route as an example for the text it is has been a lesson for the author to uphold these principles, and it is at this stage that reflection can allow observation of the fact that these represent, without a doubt, an 'easier said than done' approach to the topic of revolutionary development. Nevertheless, this is perhaps the only way to let out-of-the-box thinking survive, and be managed, rather than scorned.

The starting point, therefore, is almost certainly to get the managements in all the operational elements of the air transport industry to satisfy themselves that they can fulfil their own role in their support (in these cases not leadership) of the development of new research paradigms. The leadership comes in the implementation stages, and by then the researchers should be looking at the next revolutionary developments. By 2100 the world could have morphed all transportation endeavours technologically, commercially and operationally. It will be a vast undertaking.

12.7   Further reading

Hanlon, G. (1999). *Global Aviation*. Butterworth-Heinemann, Oxford.
Hirst, M. (2008). *The Air Transport System*. Woodhead Publishing, Cambridge.
Owen, C.A. (1982). *Flight Operations*. Granada, London.
Shaw, S. (2007). *Airline Marketing and Management*. Ashgate, Aldershot.
Williams, J.E.D. (1964). *The Operation of Airliners*. Hutchinson Scientific and Technical, London.

Part III
Challenges

13

Intellectual property, patents and innovation in aeronautics

D. A. MCCARVILLE, Oregon Institute of
Technology, and The Boeing Company, USA

Abstract: This chapter begins with a general discussion of intellectual property, patents and creativity/innovation. A historic case study of automated material placement (AMP) equipment is presented. As the large commercial aircraft (LCA) industry attempts to shift from mostly metallic to composite airplanes, the ability of airframers (prime contractors or system integrators) and equipment vendors to innovate and transactualize technological advances (i.e. improved machine functionality, speed and reliability) will no doubt dictate the long-term health of the industry. It is estimated that, by the year 2020, 80% of all the composite material used in the LCA industry will be laid by machine, versus approximately 30% today. This chapter should be of interest to students and professional practitioners involved in the study, design and manufacture of composite components.

Key words: composite equipment, composite patent, tape laying, fiber placement, filament winding, robotics.

13.1 Introduction

This chapter begins with a discussion of human creativity/innovation, intellectual property rights and patents. Initially, the question of what constitutes an innovation is discussed in terms of how such breakthroughs propel cultural evolution and add to our overall level of social knowledge. Next, the mechanisms behind individual and systematic industrial creativity are reviewed. Subsequently, the patent process, as a means of capturing equipment innovation and providing competitive advantage, is discussed.

Typically each class of invention has a unique life cycle driven by technological acceptance and industry persistence. In industry, the ability to innovate and capture competitive advantage by establishing intellectual property rights often drives the pace of technological adoption. As a case study, this chapter discusses evolutionary steps in automated material placement (AMP) equipment that coincided with or preceded significant changes in the large commercial aircraft (LCA) industry.

In order to improve fuel economy and reduce maintenance costs, new-generation LCA (e.g. Airbus A350 and Boeing B787) are being built with less aluminum and more composites (up to 50% by weight). This shift in composition is requiring a fundamental change in the methods and equipment used to manufacture airplanes. Consequently, equipment vendors are being tasked with improving the reliability

263

and throughput of AMP machines. This chapter identifies technological changes in AMP equipment that have facilitated or will facilitate a shift from aluminum to composites during the manufacture of LCA. The chapter covers a 50-year period from approximately 1960 to the present. Data was gathered and triangulated from the industry's creative design spaces:

- Patent databases,
- Trade literature,
- Scholarly texts,
- Vendor web sites.

The nature of ongoing equipment evolution is analyzed in light of the AMP industry's attempts to meet steeply escalating airline industry demand.

13.2   Commentary on likely future trends

In an effort to improve fuel efficiency and reduce overall operating costs, the LCA industry's two major airframers (Airbus and Boeing) are launching airplanes that contain a large percentage of composites by weight (A350 and B787). This shift from aluminum to composites will require new facilities and changes in manufacturing methods. Most importantly, new-generation AMP equipment will be required to build large wing and fuselage skins and associated substructure (e.g. spars, ribs, stringers and frames). AMP equipment demand is being driven by:

- A shift from aluminum to composite in the design of new airplanes,
- An increase in the demand for airplanes (the global fleet size is expected to roughly double in the next 20 years), and
- An increase in the percentage of composites placed by machine (expected to grow from roughly 30% to 80% by 2030).

Composites were first used on LCA in the late 1950s. Shortly thereafter, in the 1970s, military airframers began dabbling with AMP equipment. For the next few decades, the AMP industry was dominated by a few key manufacturers and equipment vendors. During the last ten years exploding machine demand has increased the level of research activity, as reflected by the number of US and foreign patents, and several new companies have entered the marketplace. As discussed in this chapter, companies typically use research organizations to harness human creativity/innovation and drive the inventive process. In so doing, they attempt to address specific problems and create competitive advantage.

Typically, intellectual property generated during research activities is either held as a trade secret or patented. A patent provides a company with monopoly rights for a finite period of time (approximately 20 years). But, since patent rights only apply in the countries where they are submitted, as the global economy becomes more intertwined they are becoming more difficult to enforce.

Airbus and Boeing are shifting away from in-house part fabrication and towards large-scale system integration. That is, they are focusing on assembling high-value subassemblies made by partners and sub-tier suppliers. Simultaneously, they are utilizing more globally diverse supply chains than they did on legacy aircraft such as the A320 and B737. With more machine users, and hence equipment vendors, the AMP industry will most likely experience a higher level of technological growth in the next 20 years than it did during the prior 50 years of its existence.

13.3   Creativity and innovation as a mechanism for capturing intellectual property

As pointed out by Stevens (1999), society has become reliant on a steady stream of 'useful' inventions that focus on increasing our productivity. Inventors may initially seek monopoly protection in the form of a patent for a new product or process, but in a relatively short period of time (roughly 20 years) such knowledge becomes public property; a part of our social culture. Therefore, the individual inventor and/or assignee can be considered a change agent whose efforts add sequentially to the sum total of humankind's knowledge.

Some researchers have compared the push for innovation with the change processes involved in biological evolution. As far back as the early twentieth century, Gilfillan (1963) pointed out that, like spontaneous mutation, social innovation often occurs most rapidly during times of environmental stress. In nature, climatic stress may cause certain traits to become liabilities, preferentially causing new abilities to be bred into a population. Similarly, economic strife may spur new and novel innovations that allow civilization to overcome shortages or become more efficient at meeting basic needs. As such, our rise from primitive hunter-gatherers to a globally connected species can be attributed to the invention and ongoing refinement of physical devices.

Although the individual innovator sometimes holds the key to cultural progress, it is often impossible to conclusively trace major breakthroughs to any one specific invention. Rather, certain ideas merely plant a seed for development that may eventually lead to a new product or industry and even social change. In practice, technological progress may be as ruthless and unremitting as biological evolution. Instead of competing with teeth and claws, humans seek market and technological dominance using laws and business practices. As stated by Evans (2004), a true innovator is a creative individual or company that is willing to take an invention into the marketplace and risk the whims of consumer preference.

When considering industrial innovation, perseverance may be more important than originality. According to Stevens (1999), inventions do not always address marketplace demands; sometimes 'great inventions' create the marketplace. Since most innovations come from a sequential combination of previous inventions, it is not surprising that many new developments are aimed at making high-end goods and services (such as air travel) available to the general populace.

Creativity can be thought of as the ability of the human mind to generate new correlates (i.e. original and novel ideas). For a creative act to be useful to society, it must solve a problem and be adopted in such a manner that it accomplishes a useful goal. As proposed by Huber (1998), a creative idea must be 'remarkable' to the extent that its implementation significantly transforms a particular field of human endeavor. Generative innovations are creative inventions that provide inspiration for outstanding ideas within an era. The economic strength of a nation (e.g. output per capita) depends upon its people's ability to make innovative discoveries and transform creative ideas into useful products (Kock, 1978).

The psychological aspects of creativity are often studied as a potential mechanism for the stimulation of human innovation and subsequently the advancement of civilization. Individual creativity has helped our species advance from one possessed of random creative competency to one capable of deliberate inventiveness. This humanistic perspective leads to the optimistic belief that everyone possesses an innate creative potential that can be brought to light through self-actualization (Taylor, 1975a). Researchers such as Fauconnier and Turner (2002) contend that individual creativity is propelled by innate levels of imagination, intuition, insight and intelligence. As pointed out by Prince (1975), hard work (i.e. a person's outward actions) and the desire to fulfill certain needs (e.g. fix a given situation in a repeatable way so as to prove one's competence) may serve to mask creativity-based character traits. It is not always possible for a company to evaluate a person's work behavior as a means of determining which employees will be suited for development efforts that require creative innovations.

The corporate inventor is typically a technically trained engineer or scientist who is trying to improve a complicated system rather than achieve epoch-making breakthroughs. Such individuals must not only possess fundamental scientific knowledge in their field of study; to be efficient, they must often have years of practical experience solving similar dilemmas.

Inventions can come about through the exercise of individual or group creativity. Companies frequently recognize the extent to which inventors support business growth and have programs that foster and reward creative talent. For example, some companies assemble groups of creative individuals into research and development organizations as a means of

- Improving existing products and/or processes,
- Creating new products and/or processes, or
- Combating competitive innovations.

Ogle (2007) suggests that 'creative leaps arise from the imaginative and insightful transfer of powerful, externally embedded intelligence from one idea space to another'. These culturally embedded design spaces allow the extended mind to tap into such collective intelligence areas as research methodologies, accepted theories, laboratory techniques, historic practices, textbooks, journals and shared conversations.

13.3.1 The creative process

Inventions can be thought of as unique accretions or extensions of prior art. Therefore, it is sometimes difficult to attribute a specific advancement to an individual creative act. As pointed out by Garrett (1963), creative discovery can come about via any of the following mechanisms:

- A flash of genius,
- Trial and error,
- Planned research, or
- Lucky accident.

More typically, it appears that industrial innovations are less the consequence of a unique creative outburst than the persistent application of specialized knowledge in a systematic manner.

Huber (1998) found that creativity time patterns are typically random and not always within the control of the individual inventor or interested business entity. Similarly, Maini and Nordbeck (2003) contend that isolated incidents or 'critical moments' drive the creative process. Critical moments can be driven by interpersonal relationships, emotional changes, internal realizations, environmental constraints or research focus changes. As mentioned by Prince (1975), 'to turn an idea into a product or process demands enormous and continuing energy and commitment'.

Garrett (1963) pointed out that most creative activities contain the following elements:

- Problem definition,
- Frustration,
- Incubation,
- Illumination, and
- Verification.

More recently, Maini and Nordbeck (2003) further delineated these themes through the identification of the following phases:

- Initial phase – create study conditions, focus on broad concepts and gather data,
- Intensive phase – conduct structured evaluations and recognize hindrances,
- Pre-idea phase – reduce activities to a reasonable level, attempt to displace problems and stimulate or abandon the concept,
- Idea cognition phase – clarify new ideas and increase effort motivation,
- Post idea – evaluate concepts, add details and make necessary changes.

Taylor (1975b) contends that, although simple problems can be solved using direct solution processes, complex problems require transformational problem rendering methods such as:

- Flexible solutions,
- Lateral thinking,
- Contrasting viewpoints,
- Analogous study of prior work,
- Systematic association of divergent concepts, and
- Divergent thinking.

When using transformational methods, it is generally assumed that creative output is partially dependent on the degree of environmental openness. Taylor summarizes that the creative process involves the following phases:

- Initial exposure phase,
- Pre-divergent or input direction phase,
- Conversion or insight phase,
- Post-divergent or developmental phase, and
- Expression or final release phase.

13.3.2 Innovative products

As a basic criterion, innovative products, such as patents, must be novel (e.g. statistically infrequent, unusual or new). Technical transactualization involves repeated refinement of a product until it becomes a societally transforming resource. In other instances, products are deemed innovative because they fill a 'gap' or provide a solution to a troubling problem. Beyond these definitions, some authors have established grading systems for the degree of creativity/ innovativeness. For example, in the early 1970s Taylor and Sandler (1972) developed the creative product inventory (CPI) evaluation method that ranked products based on their 'generative power, transformation power, degree of originality, relevancy, hedonics, complexity and condensation'. Using the CPI, high rankings are assigned to products that generate subsequent research and future products both within and out of a given field of inquiry. Others, despite extensive search, have been unable to establish empirical correlation between innovativeness and product success in the marketplace (Chandy et al., 2006).

Innovative products or inventions may be physical, social or mental. Physical inventions (e.g. AMP equipment) involve the unique combination of materials or forces into previously unknown products and/or processes (Stim, 2007). Further, inventions can be subdivided into

- Basic or pioneering, and
- Improvement or intensive.

Basic inventions create products that fill new and useful applications, while intensive inventions provide improvements to existing products. Often, inventions are driven by a company's desire to create products or processes that are better, cheaper or faster than a competitor's.

Early researchers such as Rossman (1964) pointed out that most inventions

- are cumulative,
- complete a pattern that was previously recognized as being incomplete,
- assimilate preexisting elements into new configurations, and/or
- fill a particular need.

Or, as stated by Gilfillan (1963), 'A great invention is a conglomeration of detail inventions so vastly numerous that no one hero, nor country, nor century is enough to produce it, nor all the dictionaries enough to define it.' Quantitative studies of product creativity typically involve enumeration of items such as

- Peer reviews,
- Industry rankings,
- Turnover rates,
- Publication count, and
- Number of patents (Taylor, 1975b).

Although business people, scientists and engineers are constantly searching the horizon for new breakthrough products, it is often difficult to predict when and where the creative spark for such innovations will come from. Ogle (2007) stated that breakthroughs are typically discontinuous and driven by cultural dynamics and socially embedded intellectual systems. Specifically, he hypothesized two overarching principles:

- Network science can be used to understand the extended mind's ability to interact with self-organizing idea spaces, and
- Imaginative individuals, by engaging design spaces, can sometimes make discontinuously creative leaps of insight that lead to novel products.

Civilization depends on the synthesis (i.e. creation, capitalization and commercialization) of inventions as a mechanism for social advancement. According to White (1990), inventions may transform 'society, the general welfare of people, the economies of nations and the fate of individual industries'. Certain inventions are so creative and valuable that they shift the locus of economic and political power between countries (e.g. textile manufacturing, steam power, iron/steel manufacturing, railroads, electrical power, internal combustion engines, automobiles, chemical products, electronics, airplanes, pharmaceuticals, biotechnologies and nuclear power). As evidenced by rapid sales of the A 350 and B787, composite processing equipment, that is enabling a global re-vamp of the methods used to manufacture large airplanes, may fall within this category of transformational inventions.

As described by Gladwell (2000), tipping points can occur in a field of human endeavor when seemingly small things have great effects. One can imagine how industrial transformations could be driven by tipping points. For example, a rise in the price of aviation fuel by as little as 10% could reduce airline operating margins to such an extent that the demand for more efficient composite airplanes could skyrocket, and therein propel a need for new faster and more reliable AMP equipment.

White (1990) classified product innovation as technology-driven, product/production-driven or market-driven. Further, Crawford and Di Benedetto (2000) pointed out that new products can be focused within any of the following categories:

- New-to-the-world,
- New-to-the-firm,
- Replacement products,
- Additions to current products,
- Improvements/revisions to existing products, or
- Cost reductions.

Ideas for new products can come from any number of sources: customers, distributors, competitors, consultants, employees or other stakeholders (Lamb *et al.* 2000). The most useful ideas often solve existing product use problems or fulfill unmet customer needs. As stated by Crawford and Di Benedetto (2000), areas that may spawn ideas include: underdeveloped markets, new market segments, external customer requests and internal decisions to pursue a new product strategy. As pointed out by Christensen and Raynor (2003), disruptive innovations may involve the formulation of a new product from an unexpected source whose significance was never anticipated by a particular industry. For example, from its inception the AMP equipment industry has been dominated by a few key players. But, as new vendors apply knowledge from different design spaces (e.g. robotics), new machines with vastly different capabilities from those built in the past may reshape the industry.

As stated by Cooper (1999), successful product deployment requires a market-oriented business strategy and the resources to develop and launch the product. New products are typically identified and launched by using a product development process as shown in Fig. 13.1.

Strategic product goals may include such business-level aspects as

- Defending the current product base,
- Extending the current product base, or
- Pursuing opportunities that fall out from fundamental research.

Most companies stick with new products that fall within an area of core competency. Typical competency areas include:

- Technological,
- Product and/or field experience,
- Customer understanding, and
- End-use experience.

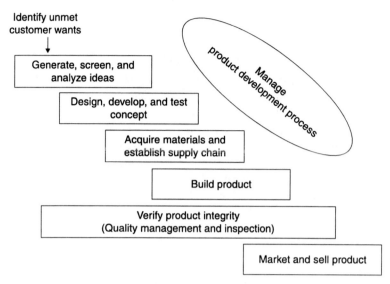

13.1 Product development process overview.

13.4   Intellectual property and patenting

Throughout the history of human development new problems arise. For society to be able to advance beyond such problems, it must identify the need for a solution and possess sufficient technical knowledge to solve the problem. Therein, civilization is dependent on inventions for continued health and wealth. 'Patents are regarded as an engine for technological progress and growth' (Llobet, 2000). Once adopted, inventions become part of the social culture and general knowledge fund. As such, technology is cumulative and the ability to create new and innovative combinations of prior art is continually increasing.

Inventions are not only a source of new products and processes, but also a foundation for economic success at many large corporations (White, 1990). As technologies advance, global markets become more open and product life cycles shorten. The rapid conversion of innovation into invention is one way for companies to remain competitive. As an early leader in the field, Gilfillan (1963) contended that the nature and pace of inventions may change over time, but certain social principles remain constant, for example:

- Inventions typically, but not always, occur through evolutionary processes (e.g. new combinations of prior art).
- Each invention has its own distinct life cycle dependent upon such features as:
    - Ability to fabricate,
    - Rate of commercial adoption (e.g. inventions ahead of their time may remain undeveloped and unused), and

  ○   The inevitability that many inventions will diminish their own worth by
      stimulating future innovations.

• A technology's rate of adoption tends to coincide with the volume of inventions
  being generated in the area, and high invention volume tends to stimulate
  further advances. Often, inventive preeminence allows certain companies to
  capture/centralize market segments.
• Since need often drives innovation, it is not uncommon for duplicate,
  equivalent or simultaneous inventions to arise in response to a given problem.
• Although disruptive or revolutionary inventions may be made by individuals
  outside a given field, the bulk of 'perfecting' inventions are made by specialists
  trained to work in a specialized area.

13.4.1 Reasons for patenting

A patent is a contract between the government and an inventor. In exchange for
build and sell exclusivity, the inventor publicly discloses the essence of his or her
invention. As stated by Romano (1987), monopoly rights help to stimulate
inventive activities. Inventions can arise as the result of varied impetus:

• An extension of ongoing scientific progress,
• Solution of a practical need,
• Serendipitous discovery, or
• As part of a focused research strategy.

Often it takes years before the value of an invention is appreciated and capitalized.
Sufficient market demand must exist to warrant production. As pointed out by
White (1990), inventions are a double-edged sword that can make a company or
industry profitable or obsolete. This fact is particularly true in today's global
economy (e.g. the LCA industry) where nations are aggressively attempting to
establish economic and technical distinction over other countries through the
application of technological innovations. Therein, some researchers suggest that
tax policies, not patent law, be used to promote innovations and inventions (e.g.
tax credits to offset research and development costs (Orkin, 1984)).

   As pointed out by Gallini and Scotchmer (2001), if the US did not have a patent
protection system, inventors would be at a disadvantage compared with non-
inventing competitors; perhaps to the point where they would choose not to invest
in new innovations. The typical investor understands that funds spent on
superfluous inventions/products will not be rewarded by the market and that
valuable intellectual property has the potential to garner high rates of return.
Essentially, the current patent process acts as an industry-wide research and
development investment screening mechanism.

   Since it is beyond the resources of a typical company to patent all innovations
that its employees may come up with, most large companies have an internal

invention evaluation system. The purpose of such a system is to determine how best to deal with innovations and invention disclosures, for example:

- File a patent application with the US or foreign governments,
- Protect the innovation as a trade secret, or
- Assemble a defensive publication.

The choice of whether or not to patent an idea or innovation will typically be based on the following factors (Stim, 2007):

- Are other companies interested in the invention?
- Are there other methods/technologies that could solve the problem addressed by the invention?
- If others used the invention could it be detected?
- How difficult will it be for other companies to solve the problem without the invention?
- What value does the invention have with regard to increased sales or licensing fees?

Whereas most industrialized nations have patent rights statutes (i.e. employer–employee negotiations, arbitration systems or invention review boards), US employees are typically required to sign intellectual property and confidentiality agreements as a condition of employment at technologically advanced companies (Orkin, 1984). Such agreements assign any inventions they may conceive over to the company. Although the corporate inventor does not typically reap windfall monetary benefits, such as a percentage of sales revenues or licensing fees, most companies reward employees through bonuses, incentive awards, promotions or increased pay (Kock, 1978)

Therefore, at least in the US, it can be assumed that most inventors do not invent for pure monetary reward. In an early survey of 710 inventors, Rossman (1930) found the following were significant motivational factors for the corporate inventor:

- Love of inventing,
- Desire to solve a problem,
- In search of monetary gains,
- Desire to achieve higher social status,
- As part of an assigned task,
- As a quest for peer notoriety, and
- To make the world a better place.

More recently, studies by LeBlanc (1994) and Henderson (2002) ranked problem solving and a desire to contribute to society as the primary reasons individuals invented new products and monetary compensation as a secondary influence. As pointed out by various studies spanning decades, rather than fostering invention, some companies inadvertently hinder innovation by not combating the following roadblocks (Rossman, 1964; Stim, 2007):

- Lack of upfront capital,
- Lack of technical training,
- Preconceived prejudices,
- Legal challenges,
- Lack of marketing resources,
- Lack of knowledge of similar competitive efforts,
- Time constraints,
- Lack of research facilities and
- Inappropriate obstacles.

'Making invention a systematic component of production and improving the probability of successful outcomes were major motivations for the creation of research and development operations in the modern corporate structure' (Suarez-Villa, 1990). Innovative/inventive engineers often become a company resource and are used in the capacity of internal consultants who help other programs and organizations reach viable solutions to troubling problems.

13.4.2 Intellectual property processes

Intellectual property (IP), also known as intellectual capital, can be thought of as innovative creations of sufficient value to warrant legal protection. As clarified by Stim (2007), IP can be classified into the following types:

- Patents – exclusivity rights (i.e. the right to sue if infringed upon) awarded by the government,
- Trade secrets – the process of keeping innovations confidential as a means of dissuading imitators,
- Copyrights – protection of artistic expression from being copied,
- Trademarks – exclusive right to a brand name as a means of market recognition.

As detailed by Huber (1998), to be patented, an invention must be

- New – the invention must be new to the world, domain or field to which it is to be applied,
- Useful – the invention must have economic potential, and
- Non-obvious – one conversant in the art would not readily anticipate the invention (i.e. the invention is heuristically surprising).

The patent and copyright processes provide a method for protecting innovations/creativity in the hope that inventors may receive monetary remuneration for their labors (Kock, 1978). In the US, a patent lasts approximately 20 years, dependent on laws at the time of filing and patent type. Further, once it expires a patent cannot be renewed. IP rights can be transferred by assigning (i.e. handing over IP as part of a business arrangement or in exchange for financial compensation) or licensing (i.e. allowing individuals or companies the right to use IP). As detailed by Gallini and

Schotchmer (2001), 'Patent scope governs the market price in the proprietary market and licensing prevents wasteful imitation'. Llobet (2000) contends that patents are a good tool by which a company can capture and protect research and development activities. Further, as stated by Gallini and Scotchmer (2001), patents effectively screen projects for which the value and cost are hidden from the investor.

Since a patent's value is essentially equal to its social value, a company typically estimates the cost to mature an innovation versus projected revenues and only seeks patent protection if it foresees a positive cash flow. Many companies maintain a 'portfolio' of patentable projects as a means of increasing the odds that at least a portion of their research and development expenditures will lead to new revenue streams (Suarez-Villa, 1990). Most industrialized nations have patent systems similar to the US patent system. Patents awarded in one nation do not typically extend to others. This makes it challenging and expensive for a company to protect innovative new equipment and processes in today's global marketplace.

Consequently, some companies may choose to keep items, such as new innovative processing equipment, as a trade secret rather than incur the cost of patenting. There is no governmental involvement in the trade secret process. Instead each company must set up its own internal trade secret policies and methodologies. Stim (2007) suggests that the following items be addressed when developing a trade secret system:

- Have employees and outside contacts sign confidentiality agreements,
- Restrict access to a 'need-to-know' basis,
- Keep sensitive material under lock and key,
- Establish a record-keeping system that documents new innovations,
- Reward employee creativity.

Protecting an innovation by maintaining it as a trade secret has the following advantages versus patenting:

- It may cost less and may be sufficient if the innovation cannot be readily reverse engineered,
- Since the innovation is not publicly disclosed, other companies cannot learn from it, and
- Unlike patents, which have a limited life, trade secrets can be maintained indefinitely. Conversely, attempting to maintain an innovation as a trade secret runs the risk that someone else will independently discover the innovation and patent it.

As incentive to disclose the claims of an invention, a patent holder receives exclusive rights to a divulged technology. Patents are broken down by the United States Patent and Trademark office (USPTO) into the following types:

- Utility – a technological innovation (e.g. AMP machines, composition of matter and manufacturing processes or art),

- Design – original shape, ornamental design or manufactured article, and
- Plant – a newly conceived biological plant (Stim, 2007).

Of interest to the current discussion are the following definitions:

- Machines – all mechanical powers and devices that perform some function or produce a certain effect or result (e.g. AMP equipment),
- Manufactures – articles or implements made by a human agency from raw or prepared materials (e.g. composite components: wings, fuselage sections and engine cowlings),
- Composition of matter – all compositions or intermixtures of two or more substances or ingredients (e.g. composite materials: epoxy/carbon, epoxy/fiberglass and polyimide/fiberglass).

In addition, some researchers categorize patents as either basic (i.e. new-to-the-world inventions) or improvement (i.e. inventions that improve upon existing inventions). As detailed by Stim (2007), a typical patent contains the following elements:

- Abstract – a brief overview of the invention,
- Field – a description of the technology area covered by the invention,
- Background – reasons and purposes for conceiving the invention,
- Claims – the legal scope of coverage, and
- Description – specifications of the preferred embodiment or the best way to implement the invention.

Numerous researchers have pointed out perceived deficiencies with the US patent process. These range from such items as high filing and litigation costs to negative effects on competition and society as a whole. A comprehensive list of cited deficiencies compiled from studies over the past two decades is provided in Table 13.1 (Fixler, 1983; Gallini and Scotchmer, 2001; Llobet, 2000; Markiewicz, 2003; Romano, 1987).

Some researchers suggest that other incentive systems might be more efficient than the current US patent process at stimulating and rewarding inventors in certain situations. For example, Gallini and Scotchmer (2001) provide details on the following alternative methodologies:

- Allow competitive market forces to provide incentives for inventing (e.g. increased revenues).
- Offer cash prizes for companies that address a challenging problem. For example, the X Prize Foundation offers checks for innovations that have the potential to benefit humanity in the fields of space, medicine and automotive.
- Have the government (i.e. taxpayer) sponsor research and development activities in certain fields through national labs, grants and/or sole source contracts.

Table 13.1 Perceived deficiencies with US patent system

Area of deficiency	Item
Costs	The patent system imposes invention costs on the inventor. In certain instances, other incentive systems (e.g. rewards or taxpayer support) might be more socially efficient (Gallini and Scotchmer, 2001)
	Even with a patent and associated monopoly rights, an inventor may not be able to generate sufficient sales to offset development and launch costs (Romano, 1987)
	High litigation costs for parties (i.e. patent holder and infringer) involved in legal disputes
Imperfect assignment of rights	Patents are often an imperfect assignment of rights. That is, it is sometimes hard to establish the range of innovations (i.e. breadth and scope) covered by an invention
	New inventions motivated by an existing patent may erode the viability or value of the original patent (i.e. creative destruction; Romano, 1987)
	Intellectual property rights may reward inventors beyond a level that is necessary to spur innovation
	In certain cases natural market forces would be sufficient to protect inventors. Therefore, a formal system is unnecessary (Gallini and Scotchmer, 2001)
	Knowing the specifics of a patent, a competitor can often easily circumvent a patent's claims
Cultural knowledge blocking	Patents sometimes hinder rather than facilitate the diffusion of knowledge into society (Llobet, 2000)
	Broad patents may discourage potential innovators from improving a product (Markiewicz, 2003)
	An innovation and associated patent may benefit private corporations at the expense of society's greater welfare. For example, a competitor may spend inordinate resources on research and development in order to develop a patent that will usurp a monopolistic market. Doing so may merely redistribute profits while being overall socially wasteful
	The patent system may prompt companies to limit informal knowledge transfer and pursue the bulk of their research in secret (Llobet, 2000)
	Instead of working on socially beneficial areas of basic research, companies may focus on projects that appear to be patentable (Fixler, 1983)

13.5   Converting patents into products

Within the last ten years, numerous changes have occurred within the US patent office that appear to have weakened its viability (*The Economist*, 2004). Most notably, the number of applications is rising at a rate of approximately 6% per year, or double what it was a decade ago. This growth is not limited to the US. The number of patents issued per year is increasing worldwide. For example, patent applications in China have increased fivefold since the mid-1990s. Not only has the sheer volume of patents increased, but the scope and complexity of technologies covered by patents has exploded.

Consequently, the effort required to review a typical application is roughly three times greater than it was ten years ago. Patent examiner workloads have increased and some examiners are not properly trained in all the areas they are reviewing. Typically, the time to review submissions has gone up by 50% (the current backlog is over 500 000). Not surprisingly, since it takes less work to accept a patent than to reject it, there may be a trend towards approving applications that may not have been approved in the past. All-in-all, one might contend that overall patent quality has, in general, declined; that is, claim accuracy and degree of novelty/non-obviousness has decreased (*The Economist*, 2004). Concurrently, due to high litigation costs, it has become more difficult to revoke a patent.

Chandy *et al.* (2006) found that companies typically have limited control over the conversion rate of application to awarded patent. This fact may have as much to do with the difficulty of assembling industry-wide conversion data as it has with the USPTO patent evaluation process. Additionally, a company's internal patent creation and review process may thwart data collection activities. The volume and value of records kept during project development efforts may vary significantly between companies. Projects that fail are often quickly purged from a company's collective memory and records on the reasons for demise may be destroyed. At other times, companies keep development projects a trade secret until they are patented. Subsequently, if a patent is not obtained, they may never share details of the work with the industry.

It is commonly accepted that speed to market helps foster marketing and distribution efficiencies (e.g. faster movement down learning curves, continued brand loyalty and lower supply chain and logistics costs). Such efficiencies may lead to a company developing competitive advantages and market niches. Chandy *et al.* (2006) identified four factors that impact the speed of turning innovations into intellectual property and eventually new products:

- Workload – quantity of tasks or programs that individuals attempt to simultaneously complete,
- Time pressure – the time frame in which critical tasks and/or projects must be completed,
- Expertise – the extent of area-specific technical knowledge that individuals working on a problem possess,
- Task importance – the value a company places on completing a given task or project.

Essentially, a company's new product conversion ability depends on its focus and deliberation in the above areas. In the current study, one might expect the number and variability of AMP equipment patents to wane at times when the technology is not directly supporting a new airplane launch activity.

The existence of available industry-wide knowledge often helps facilitate the idea generation process and decreases the time/effort required to produce sequential innovations (Markiewicz, 2003). Thus, one might assume that, as Boeing and Airbus launch new composite airplanes, the breadth and depth of AMP development efforts may increase, possibly to the extent that new equipment vendors are encouraged to enter the field. Such an infusion might be noticeable by an increase in the number of patents filed in a field and/or the volume of trade literature created about a specific technology.

Some recent US governmental policy changes have been aimed at extending or clarifying knowledge sharing and innovation conversion practices:

- 1980 Bayh-Dole – assures the right to patent innovations generated during government-funded research efforts,
- 1982 Federal Court Improvement Act – strengthened intellectual property protection laws and created a Federal Circuit Court of Appeals for trying infringement cases,
- 1984 National Cooperative Research Act – encourages companies to participate in joint research activities.

Such policy changes have been hailed as having helped:

- Increase university research funding levels,
- Create more industry–university collaborative research efforts and
- Cause many universities to start technology transfer offices (Markiewicz, 2003).

Other researchers contend that strengthened property rights may in fact impede collaborative knowledge sharing and/or the desire to enter into joint development activities (Nelson *et al.*, 2002). If 'non-patent' knowledge sharing policies are extensively adopted by companies and universities, one might expect to notice:

- Restrictions in external affiliation interactions,
- An inhibiting of certain types of networking and informal social interactions, and
- Implementation of policies aimed at keeping innovations as trade secrets.

Such trends could have an overall effect of restricting society's ability to build on new technological advances.

13.6   Establishing patent value

As stated by Orkin (1984), 'although patents are not an absolute indicator of innovation, their growth – or decline – provides one measure of industrial creativity'. Redelinghuys (2000) pointed out that inventions manifest themselves by adding, altering or eliminating system parameters associated with existing technologies. When studying such parameters, researchers often attempt to quantify an invention's total 'positive differential value' compared with previous work in a field.

Redelinghuys (2000) suggests that, when evaluating the potential benefit of an invention, an assessor should have an in-depth understanding of the current state of the art. Moreover, the assessor should understand the position of the interested entities (e.g. stockholder, manufacturer or customer). The 'dimensions' or scope of the 'invention metric space' should be defined prior to initiating the study, and patent categorization should be established by evaluating common parameters.

One of society's major objectives is to continuously increase aggregate innovative capacity as a means of addressing human welfare and population increase (Suarez-Villa, 1990). Consequently, social issues should be considered when conducting a macro-level study of inventions. For example, a researcher might try to ascertain whether an invention has led to a viable new product or service. If so, he or she might try to determine whether the new product has the potential to change socioeconomic trends or if the benefits appear to be merely short-term.

Some items are patented primarily for their licensing potential. Therein, it is possible that a significant portion of corporate innovations within a field are not patented if it is thought that they will lead to a licensing revenue stream. Alternatively, if a company already holds a monopoly market share in an area or can easily keep an innovation as a trade secret, it may not think the benefit of patenting worth the cost.

Markiewicz (2003) pointed out that a comprehensive knowledge base and efficient search tools make the evaluation of a patent or branch of technology more efficient and potentially successful. Such preexisting knowledge can be held in-house by the various companies that make up a field of inquiry or it can reside at centralized industry or governmental collection sites. For example, the USPTO and European Patent Office (EPO) maintain on-line patent databases searchable by patent number, name, assignee, date, classification, etc. In the current study of AMP equipment, patent databases were used to trace the movement of knowledge and the maturation/evolution of certain technologies across inventors, companies and locations.

Patent prior art citation scanning is an efficient means of ascertaining the age and scope of classes of innovation. As described by Markiewicz (2003), the median age of patents cited by other patents can be used to estimate the technology cycle time for a given class of inventions. A decreasing median citation age generally indicates an increase in the speed of adoption of a given innovation. When performing such calculations, one should account for citation inflation (i.e. the number of recent citations increases primarily because the number of patents is increasing) by weighting the data.

As pointed out by Suarez-Villa (1990), study of patent databases may yield insight with regard to a particular patent or class of innovations. Sometimes ideas that appear to be breakthrough discoveries fail to turn into viable products. Alternatively, 'pathbreaking' innovations can be applied across numerous fields of human endeavor. Some useful innovations only enjoy widespread implementation after they become public knowledge. Thus, even if a patent is not an economic success, it may play a part in reducing knowledge uncertainty in a given field of study.

In certain instances, the value of an invention can be ascertained directly by looking at industry statistics or indirectly by observing the number of patents that 'accessorize' or build off the technology. If an invention provides a company with an advantage over competitors, the firm may experience a growth spurt (i.e. a triggering or precipitating effect). Alternatively, some useful inventions experience long delays between issuance and practical application. This may happen if a patent is somewhat redundant or premature, or requires additional developmental work to realize maturation or transactualization. Sometimes, the implementation of an invention depends on the maturity of related patents. Consequently, the incubation period (i.e. the time from patent issuance to significant usage) or cultural lag can be a useful measure of a patent's social value.

13.7 Trends driving innovation within the commercial aerospace industry

In the duopolistic LCA marketplace, the success of each firm depends on its ability to conceive of and build products coveted by the airline customer, while anticipating the other's 'price and output decisions' (Carbaugh and Olienyk, 2004). Since the cost to develop and manufacture a new large airplane runs into billions of dollars, neither company can afford to bring a product to market with limited sales appeal and/or performance deficiencies. Large global corporations, such as Boeing and Airbus, are of importance to the populations of their home countries and supply chain partners' home countries because they:

- Promote high-technology developmental work,
- Contribute to economic growth,
- Help sustain employment levels, and
- Foster feelings of national prestige.

As stated by Kerton (2006), 'True free trade is a friend of competition permitting the range of qualities the consumers prefer.' The 2002 launch of the 787 Dreamliner signifies Boeing's attempt to reestablish itself at the forefront of the LCA middle-of-the-market (i.e. twin-aisle 200–300-passenger size airplane) by addressing customer demands for better fuel efficiency and lower operating costs, while maintaining current safety levels. New high-efficiency engines, advanced aerodynamic shaping and the extensive use of composite materials in the airframe contribute to a projected 20% improvement in fuel economy versus the 767. Fifty percent of the 787's airframe is comprised of composite materials (Wallace, 2007).

Airbus is fielding the A350 extra wide-body (XWB), thus with a slightly higher seating capacity. After lukewarm customer reception for a metallic–composite hybrid aircraft design, Airbus is now following the formula for composite and metallic ratio used in the Boeing 787, and the A350 will also use approximately 50% composites by weight (McConnell, 2005). The A350 is expected to enter service in the 2012 time-frame. Table 13.2 details how the use of composite materials is projected to increase dramatically in the LCA industry during the next 20 years (Red, 2007). As shown, the 787, A350 and A380 will use two or three times more composites as a percentage by weight than legacy aircraft.

13.8   The switch from aluminum to composites

According to Newhouse (2007), Boeing chose layered carbon fiber-reinforced plastic (CFRP) as the building block for the 787 because it has excellent strength-to-weight characteristics and resists corrosion. 'Reduced overall weight, improved fuel efficiency, longer lifetime relative to fatigue performance, corrosion resistance, reduced part count, even containment capacity for a loose fan blade rotating at 2600 rpm – composites clearly offer the airline customer significant performance benefits' (McConnell, 2005).

The customer expectation process is an overriding concern during any new product launch activity. According to Kerton (2006), maximizing quality is the best way to assure customer satisfaction. Similarly, continuous innovation is a proven tactic for fielding quality improvements. By focusing on issues that are critical to quality, cost, delivery and responsiveness, a company can transcend business processes and create an enterprise focused on product performance and service consistency (Anthony *et al.*, 2004).

Emphasis on the use of composites, even though it introduces significant cost and manufacturing risk, is an example of the total quality management (TQM) practice of striving for outputs that exceed customer expectations. According to Stonecipher (2004), 'The 787 is a true game-changer. It will be the first large airliner built with a composite fuselage and a composite wing and it will provide economical and comfortable nonstop service connecting scores of new city pairs around the world.' Boeing claims maintenance intervals for the 787 will be half as frequent as those needed for mostly aluminum airplanes such as the 767 (Red, 2007).

13.9   Conception of AMP equipment

When first introduced in the mid-1960s, pre-impregnated (prepreg) resin/fiber composite materials were used to reduce weight on relatively small secondary structural members (e.g. fairings, flaps and lightly loaded door panels). As the size and volume of composite structural components grew, engineers began conceiving and developing machines capable of placing prepreg faster and more accurately than by hand.

Table 13.2 Large commercial aircraft market overview

Category	Passengers	Boeing		Airbus		Comments
		Model	Composite (%)	Model	Composite (%)	
Narrow body	Up to 200	B737	2	A320	15	Both products are successful, but have low profit margins
Mid-market	200–300	B757	3	A300	4	Airbus effectively captured the middle of the market in the 1990s causing the near demise of the 757 and 767.
		B767	4	A310	8	Boeing is responding with the fuel-efficient 787
		B787	50	A330	15	
				A350	50	
Mini-jumbo	300–380	B777	10	A340	15	The 777 set new comfort and operating-cost standards
Jumbo	366–467	B747	2	N/A		Nearly bankrupted Boeing during launch, but eventually became the company's signature aircraft
Super-jumbo	555–853	N/A		A380	23	A380 manufacturing difficulties and launch delays hurt Airbus's stock and caused management shake-ups. Boeing studied the super-jumbo, but concluded that the market was too small to justify two manufacturers

Composite components are created through the ply-by-ply addition of precisely oriented thin prepreg layers. The ability to create non-isotropic structures with direction-preferential load-carrying capability, along with excellent specific strength properties, is what makes composites attractive when compared with metals. A main reason to automate lay-down methods is to enhance the accuracy and repeatability of adding and dropping fibers (i.e. ply tailoring). Current-generation AMP equipment can consistently maintain true positional tolerances two or three times better than can be achieved by hand.

In addition to improved placement accuracy, AMP machines have the potential to increase material lay-down rates. Numerous factors can affect this rate (e.g. part size, degree of ply tailoring, ply orientation, material handle-ability and part contour). When integrating such complexities in a precise and controlled manner, manual rates top out at about 0.45–2.27 kg/h (Red, 2007). Even though some automated placement jobs don't do much better, due to equipment reliability and lay-up complexity issues, values ten times higher are regularly achieved. For example, on large constant thickness components that can be made using the filament winding process, rates in the 45–180 kg/h range are not uncommon (Campbell, 2004). Red (2007) estimates that AMP machines are used to place 30% of all composite materials in the LCA industry. This percentage is expected to increase to 40–50% as the 787 and A350 enter full-up production and eventually hit 80% if and when the 737 and A320 are replaced with composite airplanes.

13.10 AMP equipment definitions

Machines capable of wrapping, winding and weaving fibrous materials have been around since the 1800s. Initially, the companies that developed and patented such equipment were involved in the manufacture of textile products, toys and rubber articles (e.g. clothing, hoses, balls and tires). Over the past approximately 50 years, aerospace companies and partnering vendors have evolved these early machine precursors into four composite prepreg, part-specific types of equipment:

- Filament winding (FW),
- Automated tape laying (ATL),
- Automated fiber placement (AFP), and
- Special purpose (SP).

(Campbell, 2004; Grant, 2006; Miracle and Donaldson 2001; Strong, 2008.)

13.10.1 Filament winding (FW) machine

An automated piece of equipment that sequentially wraps continuous reinforcing fibers (i.e. tow) onto a rotating tooling surface in a precise geometric pattern. The fibers are pulled as a group or band, under tension onto the mandrel and guided to orientation (hoop (circumferential, ~90°), helical (typically ±45°) or polar (longitudinal)) by a synchronized oscillating delivery head. FW machines are classified as either wet or prepreg winding machines. A wet machine infuses resin onto the fibers on the fly while a prepreg machine uses composite material that has already been combined into the desired resin–fiber mix. Since machine motion is relatively simple, similar to that of a wood lathe, high material output rates of hundreds of pounds per hour are achievable. FW machines rely on tension wrapping and therefore cannot accommodate concave surfaces or add/drop material. FW machines can be used to create small to large bodies of revolution.

13.10.2   Automated tape laying (ATL) machine

A machine that layer-by-layer places courses of unidirectional prepreg tape to desired orientations (typically 0°, +45°, 90° and −45°) on a flat or slightly contoured tooling surface. Similar in outward appearance to a milling machine, an ATL machine consists of a computer-controlled delivery head mounted to a multi-axis gantry positioning system (parallel rails, a cross-feed bar and a raise-lowers ram bar). ATL machines are classified as either flat, contoured or high-contour tape-laying machines (FTLM, CTLM and HCTLM). Depending on part contour, ATL machines can place 7.62, 15.24 or 30.48 cm-wide tape. ATL heads contain:

- A composite material spool,
- A tape guidance system,
- A compaction shoe/roller,
- A cutter, and
- A backing paper take-up reel.

In addition, most modern machines have

- Optical position and flaw detecting sensors,
- A laser boundary trace, and
- A tape heating system to improve material tack.

13.10.3 Automated fiber placement (AFP)

A hybrid between a FW and an ATL machine in which multiple parallel strips of composite prepreg tow or slit tape (typically 3–13 mm wide) are layer-by-layer placed/compacted onto a tooling mandrel. Like a FW machine, the tool can be rotated to create a body of revolution. Like an ATL machine, the delivery head uses pressure to deposit the material, even into concave surfaces. Because tows (typically 12 to 32 are collimated to create a 10–40 cm-wide band) are steerable and can be added and dropped individually, an AFP machine can cover surfaces containing a high degree of complexity/curvature. To keep guidance mechanisms from gumming up with resin, most machines have an on-board creel (tow spool refrigerator) and head-mounted heating system (for deposition tack). AFP machines typically have axes of motion (including cross-feed, carriage traverse, arm tilt, mandrel rotation and wrist yaw, pitch and roll).

13.10.4   Special purpose (SP) machine

A composite material lay-down system designed for a specific application. Also commonly referred to as:

- Right-sized machines,
- Part-purpose machines,
- Customized machines, and/or
- Jobber machines.

Such machines may address any of the following special manufacturing needs:

- A simplified machine that manufactures only a limited shape, range or size of parts. For example, a delivery head guided by a robotic arm that can only place tow on a stationary tool in a given envelope. In high-volume cases, such machines may be designed to make only one specific part.
- A machine that can only perform limited placement tasks. For example, one that can only place a single width of tape onto a long narrow flat charge or plank.
- A machine that addresses the placement of a unique family of composite materials. For example, a tape-laying machine that *in situ* places and auto-consolidates thermoplastic prepreg.
- A machine that does not place prepreg layer-by-layer, but rather takes pre-kitted charges from a different machine and places them onto a lay-up mandrel. For example, a tape cassette (TC) kitting machine.
- A machine that attempts to improve productivity by having multiple load/unload areas, multiple composite material placement heads and/or multitasking heads (e.g. heads capable of lay-down, cutting, trimming and/or drilling).

Manufacturers of SP machines often try to fill unique placement niches by offering such functionality features as:

- Families of off-the-shelf processing heads and/or motion systems (e.g. commercially available robotic arms, standardized gantries and canned software),
- Machines that are faster than traditional AFP and/or ATL machines,
- Machines that are simpler and more reliable than traditional AFP and/or ATL machines, and
- Machines that reduce maintenance and downtime (e.g. self cleaning, automated material loading and in-process inspection/repair).

13.11 Evolution of AMP equipment

As the aerospace industry shifts from aluminum to composite primary structures, factors surrounding the optimization and facilitation of AMP equipment are of overarching concern. Typically, the time from released purchase order to full installation of an operational machine can be anywhere from one to four years. The cost of an individual machine includes:

- Initial purchase price,
- Facilities modifications,
- Support equipment,
- Start-up trials, and
- Employee training.

These can range from hundreds of thousands to tens of millions of dollars (Red, 2007; Campbell, 2004).

As evidenced by slides in the A380 and 787 programs, the wholesale simultaneous shifting of airplane composition, processing methods, partnering strategies and supply-chain make-up can involve considerable launch risk. Such delays are often accompanied by management turmoil and drops in stock prices. For example, in early 2006 Airbus announced the first in a series of A380 delivery delays. The initial six-month slide was blamed on wiring system length miscalculations. Later delays were attributed to supply-chain start-up difficulties, poor management and cost overruns (Newhouse, 2007). Consequently, upper management shakeups occurred at the European Aeronautic Defense and Space (EADS) company (at the time, Airbus's parent company) and the company's stock dropped 26%. Similarly, delays in the 787 program resulted in the change-out of Boeing Commercial Airplane's chief executive officer (CEO) and roughly a 60% drop in stock price.

Systems integration is a business strategy, wherein a corporation outsources component and subassembly work in favor of maintaining some degree of design authority and final assembly work. The idea behind the integration process is for a company to spread product development and launch costs across a network of vendors and partners. In the LCA industry, launch costs (e.g. research and development, facilities, capital and equipment) are typically in the billions of dollars. Therefore, systems integration is sometimes used as a part of a cost sharing strategy.

In a large global enterprise, outsourcing allows purchasing activities to go from 'vertically integrated' to diverse and distributed. A global supply chain can include a broad variety of innovative best-in-class partners and suppliers. Companies may pursue global outsourcing for any of a variety of reasons:

- To leverage purchasing power,
- To achieve economies of scale,
- To extend supply chain scope,
- To reduce costs,

- To improve quality, and/or
- To strengthen relationships with existing or potential customer countries.

The trend towards higher technology and value outsourcing has the potential to cause global supply chain complexities beyond what is traditionally experienced by multi-national corporations (MNC), which are used to outsourcing to reduce labor costs. As companies share intellectual property and trade secrets with foreign suppliers/partners, the opportunity for pirating and/or using this knowledge to produce competitive products increases. In such instances, short-term profits may be realized by the outsourcing corporation. But, in the long term, the company, inventors and employees may pay dearly for the loss of intellectual capital. According to Palmisano (2006), companies should seek a balance between outsourcing and knowledge sharing that maintains competitive advantage.

As detailed by numerous authors, composites were initially used on aerospace components in the late 1950s and early 1960s to reduce weight on secondary structural members (e.g. edges, flaps and lightly loaded panels) (Biggs *et al.* 2001; Campbell, 2004; Grant, 2006; Miracle and Donaldson, 2001; Murdock, 2006). As aircraft-level performance requirements intensified, composites were seen as a tool to lower airframe weight and they began to be used on primary structural members (e.g. doors and wing skins). Initially, in the 1970s and early 1980s, the bulk of these applications were on advanced military products (e.g. F16, B1B and B2) (Sloan, 2008a). Some components proved too large to safely and repetitively place by hand. Therein, the need for automated machinery that could lay composites to shape on large, relatively flat, surfaces arose.

Patents covering machinery capable of wrapping or winding continuously reinforced materials date back to the 1800s (US Patent and Trademark Office, 2008; Espacenet European Patent Office, 2008). Early FW patents discussed ways in which hand- or mechanically gear-driven tools and feed systems could be used to produce textile products, hoses, balls and tires. US patents in this field show that the technology advanced from:

- Winding constant cross-section tubular structures (1873), to
- Winding spherical shapes (1952), to
- Winding parts using polymeric resin systems (1953), and finally to
- Winding large pressure vessels (1968).

Authors are inconclusive as to when FW of polymeric resin-based composites first began. Campbell (2004) claims it was as early as the mid-1940s, while Beckwith (2008) and Blair (2007) put the date closer to the mid-1950s.

Development work on AMP machines designed specifically for the aerospace industry began in the mid-1960s to early 1970s (Beckwith, 2008; Grant, 2006; Miracle and Donaldson, 2001; Murdock, 2006). According to Murdock (2006), the first ATLM was built by General Dynamics (GD) and the Conrac corporation to place F16 vertical tail skins (patent #3 574 040: Taping Apparatus and Method).

AFP equipment trailed ATL equipment by 15–20 years and, according to various sources, evolved from a merging of filament-winding and tape-laying technologies (Grant, 2006; Izco *et al.* 2006; Kisch, 2006).

An AFP machine combines the ply tailoring capability of the ATL machine with FW's ability to place bodies of revolution. Unlike tension-based FW machines, AFP machines utilize a pressure head that can lay tailored slit tape or tow bands into tool concavities. This functionality makes AFP machines useful in the manufacture of large weight-critical hollow structures (e.g. fuselage sections and inlet duct cowlings). Murdock (2006) credits Boeing and Alliant for the initial development and 'acceptance' of fiber placement equipment, but Cincinnati is also seen as having contributed to the early art (Grant, 2006; Kisch, 1996; Sloan, 2008). Figure 13.2 suggests a plausible early AMP equipment evolutionary timeline, derived from literature review and patent search data. The following key events and milestones are notable:

- Hand-driven winding equipment was initially patented in the mid- to late-1800s.
- FW machines systematically evolved to be able to:

  ○ Place non-constant shapes (e.g. pressure vessels),
  ○ Use mechanical drives, and
  ○ Lay polymer based composites,

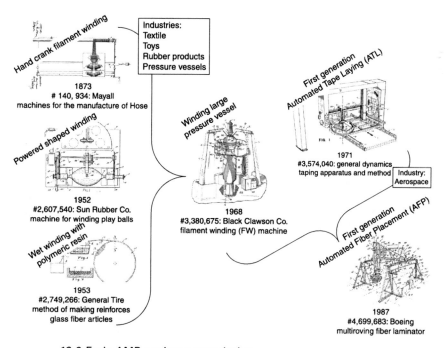

13.2 Early AMP equipment evolution.

- FW machines have seen limited application for aerospace components because of their inability to ply tailor,
- The first aerospace-focused AMP machines were ATLMs built for military applications in the early 1970s,
- Early-generation AMP machines exploited drive, control and guidance technologies taught by antecedent equipment patents,
- AFP machines resulted from a merging of FW and ATL equipment and were first conceived in the early 1980s.

Individual inventor or assignee companies often add incrementally to the existing body of knowledge to the point where patents tend to become 'social accretion' of prior art (Rossman, 1964). Taking this fact into account, the current study did not attempt to establish which airframer or equipment vendor first conceived of or brought to practice a particular piece of equipment or machine functionality. As shown in Fig. 13.3, in addition to General Dynamics, numerous airframers (Rohr, Northrop Grumman, Lockheed, Boeing and Vought) contributed to the early advancement of ATL technology (US Patent and Trademark, 2008). Other early patent-based studies have noted that, often, a common need leads to 'simultaneous innovation' by multiple companies (Gilfillan, 1963).

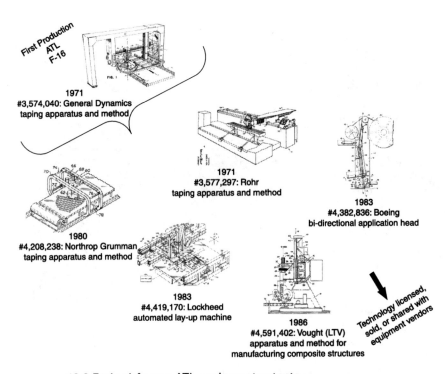

First Production ATL F-16

1971
#3,574,040: General Dynamics taping apparatus and method

1971
#3,577,297: Rohr taping apparatus and method

1983
#4,382,836: Boeing bi-directional application head

1980
#4,208,238: Northrop Grumman taping apparatus and method

1983
#4,419,170: Lockheed automated lay-up machine

1986
#4,591,402: Vought (LTV) apparatus and method for manufacturing composite structures

Technology licensed, sold, or shared with equipment vendors

13.3 Early airframer ATL equipment patents.

Airframers may have the need and the resources to develop a new piece of equipment and/or processing technology, but may lack the desire to build and maintain duplicate machines for ongoing production. Therein, they may license, sell or share their work with vendors who can take initial concepts and patents and improve upon them (Stim, 2007). Figure 13.4 details how numerous equipment vendors (i.e. Goldsworthy, Camsco, Cincinnati and Ingersoll) were involved with the launch of first-generation ATL machines in the 1970s and 1980s.

As pointed out by several authors, in the 1980s and early 1990s ATLMs were used primarily on low-volume military programs; see Table 13.3 (Grant, 2006; Miracle and Donaldson, 2001; Murdock, 2006; Sloan, 2006). Therefore, industry demand for new machines was so low that only Cincinnati continued building them. In the mid-1990s, in anticipation of the launch of new LCA composite aircraft, demand for large ATLMs capable of laying LCA skins escalated and Forest-Liné and MTorres entered the business.

AFP has a much shorter timeline than ATL. As shown in Fig. 13.5, the equipment was originally developed in the late 1980s to early 1990s.

Like ATL, most early AFP efforts were military-based. Table 13.4 provides a listing of development and production programs that used AFP equipment in the 1980s and 1990s (Kisch, 1996; Miracle and Donaldson, 2001). In the last ten years, with new LCA demand, the scope and pace of AFP patents have increased

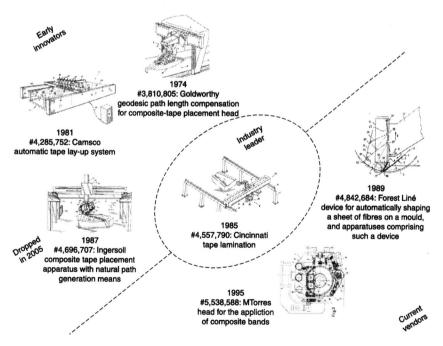

13.4 Large ATL equipment vendor patent evolution.

Table 13.3 Early ATL programs and components

Date	Company	Program	Components
1980s	General Dynamics	F16	Verticals
	Boeing	737	Elevator skins
	Rockwell	B1-B	Wing skins
	Dassault	A330 and A340	Skins
	Boeing	B-2	Wing skins
1990s	Boeing	777	Horizontals
	Lockheed	F117	Wing skins
	EADS	Eurofighter	Skins
	McDonnell Douglas	C-17	Skins
	Boeing	F-22	Wing skins

dramatically. Sloan (2008) attributes this boom to the fact that equipment vendors see 'a greater ascendancy for AFP'. Most likely this viewpoint exists because ATLMs have limitations on the type of parts they can produce (i.e. relatively flat with mild contours), while AFP machines have few shape limitations. An AFP machine can build any part an ATL can make, but the opposite is not true.

The patent database indicates that, once the basics of AMP equipment operation became common knowledge (e.g. gantry setup, material deposition methods and head movement schemes), equipment vendors attempted to establish competitive

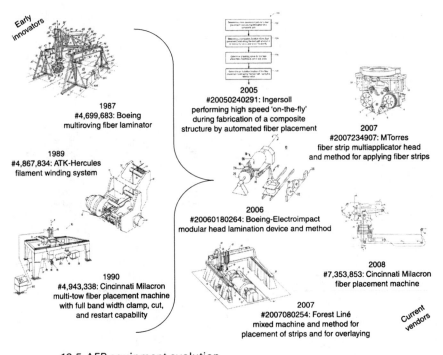

13.5 AFP equipment evolution.

Table 13.4 Early AFP programs and components

Date	Company	Program	Components
1980s–1990s	ATK Boeing Cincinnati Rockwell Northrop Grumman Lockheed McDonnell Douglas	Development programs	Missile cases, tailcones, fuselage sections, inlet ducts, c-spars, cryogenic tanks
1990s	Boeing Bell McDonnell Douglas Lockheed Raytheon Boeing Northrop Grumman Raytheon ATK Vought CASA	V-22 Agusta tiltrotor F-18 E/F F-22, JSF Premier Hawker Sea launch F/A-18, C-17 F-22 Atlas IV V-22 Arian	Fuselage skins, doors, etc. Duct and skins Inlet duct, JSF inlet duct Fuselage Fairing and Delta IV Inlets and cowlings Pivot shafts Rocket casings Miscellaneous Rocket components

advantage by capturing specific machine functionality aspects (e.g. cutter mechanisms, guidance systems, compaction rollers and software-based ply construction generation methods). Since the late 1990s, the quest for AMP machines with unique capabilities has spawned many derivative SP equipment offshoots. A representative list of these equipment branches is provided below:

- Robotic fiber placement (RFP) – a machine that utilizes a specialized AFP head on a highly articulated industrial robotic arm; see Fig. 13.6,
- *In situ* – a machine that places and cures (e.g. electron beam) or consolidates (e.g. compacts thermoplastic material) composite prepreg on-the-fly; see Fig. 13.7,
- Fabric placing (FP) – a machine that places fabric broad goods,
- Tape cassette – a machine that cuts special material shapes and places them on a roll so they can subsequently be laid by an ATL faster than if the ATL shaped the material,
- Multi-head – an ATL or AFP machine that uses more than one head as a means of increasing material lay-down rate; see Fig. 13.8.

Patent assignees identified during the current study are provided below:

- Aerospace companies:

  ○ ATK (also known as Hercules and Alliant Techsystems Inc.),
  ○ Boeing,

1988
#4,750,960: Rensselaer Polytechnic Institute
Robotic winding system and method

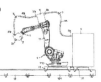

2006
#2008202691: Coriolis Composites
fiber application machine with
fiber supply flexible tubes

2006
#20060180264: Boeing-Electroimpact
modular head lamination device and method

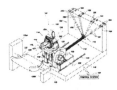

2006
#6,994,324: ATK
fiber redirect system, multi-axis robotic
wrist and fiber placement apparatus
incorporating same and related methods

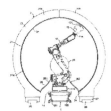

2006
#2008196825: Coriolis Composites
method and apparatus for making structures
of composite material in particular
airplane fuselage section

13.6 Robotic placement equipment evolution.

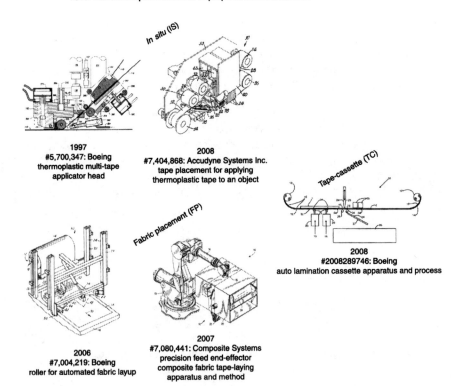

1997
#5,700,347: Boeing
thermoplastic multi-tape
applicator head

2008
#7,404,868: Accudyne Systems Inc.
tape placement for applying
thermoplastic tape to an object

2008
#2008289746: Boeing
auto lamination cassette apparatus and process

2006
#7,004,219: Boeing
roller for automated fabric layup

2007
#7,080,441: Composite Systems
precision feed end-effector
composite fabric tape-laying
apparatus and method

13.7 Miscellaneous special-purpose equipment evolution.

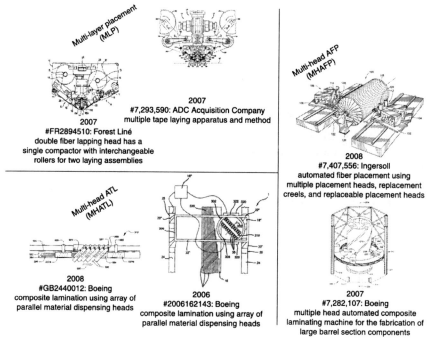

13.8 Multi-head AMP equipment evolution.

- ○ General Dynamics,
- ○ Kawasaki,
- ○ Lockheed,
- ○ LTV (also known as Vought),
- ○ Northrop Grumman,
- ○ Rohr.

- Equipment vendors:

  - ○ Accudyne Systems Inc.,
  - ○ Automated Dynamics (also known as ADC Acquisition Company),
  - ○ Camsco,
  - ○ MAG Cincinnati (also known as Cincinnati Machine and Cincinnati Milacron),
  - ○ Composite Systems Inc.,
  - ○ Coriolis Composites,
  - ○ Electroimpact,
  - ○ Forest-Liné,
  - ○ Goldsworthy,
  - ○ Ingersoll Machine Tools (also known as Ingersoll Milling Machine), and
  - ○ Manual Torres Martinez (also known as MTorres).

- Miscellaneous players (government entities and universities):
  - Air Force, and
  - Rensselaer Polytechnic Institute.

Table 13.5 provides, by decade, a list of AMP patents. This list reiterates the overall industry timeline presented previously:

- ATLMs were invented in the 1970s,
- AFP machines were conceived 10–15 years later, in the 1980s,
- SP equipment is just emerging, and
- All AMP branches are experiencing renewed interest and growth.

Table 13.5 AMP equipment patent history

		1970s	1980s	1990s	2000s	Total
ATL						
	Rohr	2				2
	Goldsworthy	2				2
	LTV		1			1
	Lockheed		1			1
	Northrop Grumman	1	1			1
	Camsco		1			1
	General Dynamics	1	2	1		4
	Ingersoll		3	1		4
	Cincinnati		3	9		12
	Kawasaki				1	1
	MTorres			2	1	3
	Boeing		3	1	2	6
	Forest-Liné		1		3	4
	Total	4	16	14	7	41
AFP						
	Cincinnati		1	1	1	3
	ATK		1	4	3	8
	Forest-Liné				1	1
	Coriolis				2	2
	MTorres				7	7
	Boeing		3	1	13	17
	Ingersoll				13	13
	Total	0	5	6	40	51
SP						
	RPI		1			
	Air Force				1	1
	Accudyne				1	1
	Automated Dynamics				1	1
	Composite Systems				1	1
	ATK				2	1
	Coriolis				3	2
	Boeing			3	7	10
	Total	0	1	3	16	20

In the field of ATLMs, Cincinnati, the industry's leader, has the most patents (primarily from the 1990s). More recent players such as MTorres and Forest-Liné have some of the newest patent suites. ATL patent distribution mirrors previous literature review findings, wherein early equipment development activities involved numerous vendors (i.e. Goldsworthy, Camsco and Ingersoll) and airframers (i.e. General Dynamics, Rohr, LTV, Northrop Grumman and Lockheed), but more recent advancements are being propelled by a smaller group of players (Boeing, Cincinnati, MTorres and Forest-Liné) (Grant, 2006; Sloan, 2008a). In 2005, Ingersoll stopped offering ATL machines and chose to concentrate on their AFP business.

In the field of AFP, two airframers, Boeing and ATK, have numerous patents dating back to the initial evolution of the equipment. Further, they are still active in the field, with 16 patents since 2000. This fact coincides with ATK's push for next-generation launch vehicle work and Boeing's launch of the 787. One of Boeing's latest AFP patents, for a modular head lamination device and method, was co-developed with Electroimpact and has resulted in a new product line. At the time of writing, the first AFP machine using this technology is being installed at Spirit AeroSystems for production of the 787 nose sections (Sloan, 2008a). As with ATLMs, Cincinnati, MTorres and Forest-Liné are major players in the field of AFP equipment. Of these three, MTorres has received twice as many patents since 2000 as the other two companies combined, coinciding with their decision to drop ATLMs and concentrate on AFP equipment. Ingersoll has 13 AFP patents since 2000.

Although there are a couple of special-purpose patents from the 1980s and 1990s (i.e. RPI and Boeing), the bulk of the development in this field is post-2000. Mostly, the companies with patents in this branch of equipment are smaller relatively new companies offering niche machines:

- Accudyne – *in situ* and multi-head,
- Automated Dynamics – *in situ*, right-sized machines, robotics,
- Composite Systems – robotics and fabric placement, and
- Coriolis – robotics.

Still, Boeing is the most active company in the field of SP equipment. They possess half the field's total patent count. Most of their SP patents cover multi-head and fabric placement machines.

Figure 13.9 is a graph of the number of patents for each AMP equipment branch by decade. As originally proposed by Gilfillan (1963), the nature and pace of inventions in a given field often follow common evolutionary processes or life cycles. A technology's life cycle typically depends on a company's ability to field the patent, the pace of commercial adoption and the inevitability that success will breed other patents that may diminish the value of the original patent.

As shown, the number of FW patents appears to have peaked in the 1990s and is currently decaying. This trend could be attributable to the fact that FW has been around for roughly five decades and many of the patents in the field have expired

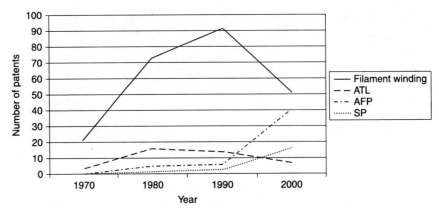

13.9 AMP patents by year.

and become common social knowledge. Also, due to the sheer volume of existing patents, hence machine options, it is conceivable that current equipment is capable of meeting most product size, rate and configuration demands (Entec, 2007).

The total number of ATL patents is about one-tenth of that identified for FW. This can be attributed to the fact that FW equipment is used in a variety of industries (e.g. appliance, construction, consumer goods, marine, transportation and sporting) while AMP equipment is used primarily in the production of aerospace components (Campbell, 2004). As pointed out by Pottisch (2005), about 10% or 181.4 million kg/y of all composite (e.g. carbon, fiberglass and aramid) components are made using FW equipment. Alternatively, Red (2007) placed the volume of composite laid by AMP equipment at about 1.13 million kg/y.

The number of ATL patents seems to have peaked in the 1980s and 1990s and is in decline. This may be attributable to the fact that ATLMs are only used to make a few part families (e.g. skins and doors) and there are only three companies currently making full-scale machines. These vendors may be able to capture business by keeping certain features as trade secrets instead of patenting them. Consequently, they may not publicly disclose innovations until fully matured, patented or obsolete. Alternatively, since the prime ATL equipment vendors are also leaders in the AFP industry, it may be possible that their AFP patents serve dual duty by also covering advancements applicable to ATL.

An AMP equipment tipping point fueled by the launch of new LCA carbon airplanes appears to be noticeable in the number and variety of AFP and SP patents issued since 2000. Of the total 112 AMP patents identified in the current study, 56 or 50% of them fall into this category. This patent count seems to indicate that the AFP and SP branches of the AMP industry will drive transformation of aerospace lay-down methods. During this period of patent escalation it may be common for duplicate, equivalent or simultaneous inventions to arise in response to common problems. For example, several companies appear to be concurrently working on new approaches for *in situ*, fabric and/or multi-head placement.

13.12 AMP equipment family tree

Figure 13.10 was created to show how machine types can be grouped and how certain machine evolved from or were inspired by other machines. As pointed out by Grant (2006), AFP machines resulted from a marriage of FW and ATL machines

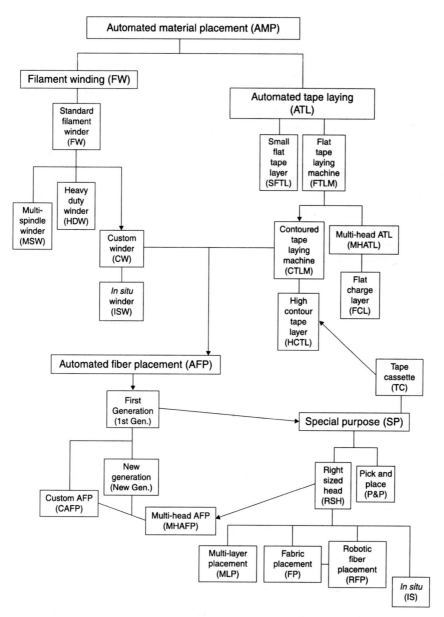

13.10 AMP equipment family tree.

technologies (i.e. placing bodies of revolutions and ply tailoring). Other equipment types were created to meet the needs of particular product types. For example, flat charge laminators (FCL) were developed to place long narrow wing stringer stacks prior to drape forming and pick and place (P&P) machines were made to lay curved multi-leg components such as fuselage frames (Cope, 2007).

13.13 Conclusion

This chapter provided background information on human creativity and innovation as they apply to intellectual capital and patent rights within the aerospace industry. As a means of illustrating these concepts, a historic case study of AMP equipment evolution was presented. The production of multi-ton composite major assemblies for the LCA industry requires the installation and verification of all-new AMP equipment at Boeing, Airbus and partnering facilities. The long-term health of these global supply chains and the LCA industry as a whole depends on the ability of these manufacturers to increase AMP throughput as production delivery demand increases. Although current-generation machines have matured to a point at which they can consistently, albeit slowly, place large wing and fuselage aircraft members, it will be a substantial industry-wide challenge to evolve these machines to a degree where composite fabrication costs/flows are competitive with aluminum panelized manufacturing processes. This chapter provided equipment evolution and functionality information that grounds the current state of the AMP industry.

13.14 Sources of further information

For early pivotal work on human creativity/innovation and the sociology behind invention the reader is referred to Gilfillan (1963) and Rossman (1964). Taylor and Getzels (1975) and Ogle (2007) extend this area of inquiry by discussing timely advances in the field. Stim (2007) contains a general overview of intellectual property and the patent process. For actual patents and filed applications, the reader should visit the European and US on-line searchable patent databases (http://ep.espacenet. com/advancedSearch?locale=en_ep and http://www.uspto.gov/patft /index.html). Newhouse (2007) contains a comprehensive overview of the LCA industry, whereas Campbell (2004) and Miracle and Donaldson (2001) are good sources for generic composite material and processing information. More specific to the current study, Sloan (2008a; 2008b) provides a discussion of ongoing trends within the AMP industry. Finally, for the latest information on fielded AMP equipment, the reader should visit the websites of the equipment vendors listed in this chapter.

13.15 References

Anthony J, Foutris F, Banuelas R and Thomas A (2004). Using Six Sigma. *Manufacturing Engineering*, **132**(2): 10–12.

Beckwith S (2008). Filament winding vs. fiber placement manufacturing technologies. *SAMPE Journal*, **44**(2): 54–55.

Biggs R, Schieliet G and McBain M (2001). Advanced composite manufacturing methods. *SAE Technical Paper Series*, 2001-01-2601.

Blair M (2007). Composites: Success, opportunity and challenge. *The Society of Manufacturing Engineers' Composites Manufacturing Conference 2007*. Salt Lake City, UT.

Campbell F (2004). *Manufacturing Processes for Advanced Composites*. New York: Elsevier Ltd.

Carbaugh R and Olienyk J (2004). Boeing–Airbus subsidy dispute: A sequel. *Global Economy Journal*, **4**(2): 1–9.

Chandy R, Hopstaken B, Narasimhan O and Prabhu J (2006). From invention to innovation: Conversion ability in product development. *Journal of Marketing Research*, **43**: 494–508.

Christensen C and Raynor M (2003). *The Innovator's Solution: Creating and Sustaining Successful Growth*. Boston: Harvard Business School Press.

Cooper R (1999). *Product Leadership: Creating and Launching Superior New Products*. New York: Perseus Books.

Cope R (2007). Increasing throughput with custom, right-sized machines. *The Society of Manufacturing Engineers Composites Manufacturing Conference 2007*, San Diego, CA.

Crawford C and Di Benedetto C (2000). *New Products Management*, 6th ed. New York: Irwin McGraw-Hill.

The Economist (2004). The cost of ideas. *The Economist*, **373**(8401): 1–2.

Entec (2007). *Filament Winding*. Entec Composite Machines Inc. brochure. Available at http://www.entec.com/brochures/filament.pdf.

Espacenet European Patent Office (2008). Available at http://worldwide.espacenet.com/advancedSearch?locale=en_ep.

Evans H (2004). What drives America's great innovators? *Fortune*, **150**(8): 1–3.

Fauconnier G and Turner M (2002). *The Way We Think: Conceptual Blending and the Mind's Hidden Complexities*. New York: Basic Books.

Fixler D (1983). Uncertainty, market, structure and the incentive to invent. *Economica*, **50**, 407–23.

Gallini N and Scotchmer S (2001). Intellectual property: When is it the best incentive system? *Innovation Policy and the Economy*. Available at http://www.dklevine.com/archive/scotchmer-when-is-ip-best.pdf.

Garrett A B (1963). *The Flash of Genius*. Princeton, New Jersey: D. Van Nostrand Company, Inc.

Gilfillan (1963). *The Sociology of Invention*. London: MIT Press. (Original work published 1935.)

Gladwell M (2000). *The Tipping Point: How Little Things Can Make a Big Difference*. New York: Little, Brown and Company.

Grant C (2006). Automated processes for composite aircraft structure. *Industrial Robot*, **33**(2): 117–21.

Henderson S (2002). *The correlates of inventor motivation, creativity and achievement*. PhD dissertation, School of Education, Stanford University, Stanford, CA.

Huber J (1998). Invention and inventivity is a random, Poisson process: A potential guide to analysis of general creativity. *Creativity Research Journal*, **11**(3): 231–41.

Izco L, Isturiz J and Motilva M (2006). High-speed tow placement for complex surfaces with cur/clamp & restart capabilities at 85 m/min. *Aerospace Manufacturing and Automated Fastening Conference and Exhibition, 2006*, Toulouse, France.

Kerton R (2006). Competition policy: Quality matters. *Consumer Policy Review*, **16**(5): 176–81.

Kisch R (2006). Automated fiber placement historical perspective. *The Society of Manufacturing Engineers' Composites Manufacturing Conference 2007*. Seattle, WA.

Kock W (1978). *The Creative Engineer: The Art of Inventing*. New York: Plenum Press.

Lamb C, Hair J and McDaniel C (2000). *Marketing*, 7th ed. Canada: South-Western College Publishing.

LeBlanc A (1994). *The human dynamics of invention: A phenomenological case study of inventors*. PhD dissertation, School of Administration and Management, Walden University, Minneapolis, MN.

Llobet G (2000). *Sequential innovation and patent design*. PhD dissertation, College of Arts and Sciences, University of Rochester, Rochester, NY.

Maini S and Nordbeck B (2003). Critical moments, the creative process and research motivation: A discussion of results based on empirical studies of scientists. *International Social Science Journal*, **25**(1/2): 190–204.

Markiewicz K (2003). University patenting and the rate of knowledge exploitation. *Academy of Management Proceedings, 2003*, St Louis, Missouri.

McConnell V (2005). Composites in the sky with Dreamliner. *The Seattle Times*, 1 May 2005. Available at http://www.aviationtoday.com/am/issue/cover/256.html.

Miracle D and Donaldson S (2001). *ASM Handbook Volume 21: Composites*. New York: ASM International.

Murdock R (2006). Commercial fiber placement. *The Society of Manufacturing Engineers' Composites Manufacturing Conference 2007*. Salt Lake City, UT.

Nelson R, Colyvas J, Crow M, Gelijns A, Mazzoleni R *et al.* (2002). How do university inventions get into practice? *Management Science*, **48**, v–ix.

Newhouse J (2007). *Boeing Versus Airbus: The Inside Story of the Greatest International Competition in Business*. New York: Alfred A. Knoph.

Ogle R (2007). *Smart World*. Boston, Massachusetts: Harvard Business School Press.

Orkin N (1984). Rewarding employee invention: Time for change. *Harvard Business Review*, **62**(1): 56–7.

Palmisano S (2006). The globally integrated enterprise. *Foreign Affairs*, **85**(3): 1–7.

Pottisch N (2005). A filament winder buyer's guide. *Composites World*. Available at http://www.compositesworld.com/ct/issues/2005/August/947.

Prince G (1975). Creativity, self and power. In Taylor I A and Getzels J W (eds) *Perspectives in Creativity*, pp. 249–77. Chicago: Adaline.

Red C (2007). Automated manufacturing the boom in advanced composites. *The Society of Manufacturing Engineers' Composites Manufacturing Conference 2007*. Salt Lake City, UT.

Redelinghuys C (2000). Proposed measures for invention gain in engineering design. *Journal of Engineering Design*, **11**(3): 245–63.

Romano R (1987). A note on market structure and innovation when inventors can enter. *The Journal of Industrial Economics*, **35**(3): 353–8.

Rossman J (1930). The motives of inventors. *Quarterly Journal of Economics*, **45**(3): 522–8.

Rossman J (1964). *Industrial Creativity: The Psychology of the Inventor*. New York: University Books.

Sloan J (2008a). ATL and AFP: Defining the megatrends in composite aerostructures. *Composites World*. Available at http://www.compositesworld.com/articles/atl-and-afp-defining-the-megatrends-in-composite-aerostructures.

Sloan J (2008b). ATL and AFP: Signs of evolution in machine process control. *Composites World.* Available at http://www.compositesworld.com/articles/atl-and-afp-signs-of-evolution-in-machine-process-control.

Stevens T (1999). A modern-day Ben Franklin. *Industry Week,* **248**(5): 20–5.

Stim R (2007). *Patent Copyright & Trademark.* Berkeley, CA: Consolidated Printers.

Stonecipher H (2004). *Outsourcing: The real issue.* A speech given to the Orange County Business Council Annual Meeting & Dinner, Crystal Cove, Irvine, California, June 2.

Strong A (2008). Composites manufacturing materials, methods and applications. *Society of Manufacturing Engineers 2008 Conference,* Dearborn, MI.

Suarez-Villa L (1990). Invention, inventive learning and innovative capacity. *Behavioral Science,* **35**(4): 290–312.

Taylor I (1975a) A retrospective view of creativity investigation. In Taylor I A and Getzels J W (eds) *Perspectives in Creativity,* pp. 1–36. Chicago: Adaline.

Taylor I (1975b). An emerging view of creative actions. In Taylor I A and Getzels J W (eds) *Perspectives in Creativity,* pp. 297–325. Chicago: Adaline.

Taylor I A and Sandler B E (1972). Use of a creative product inventory for evaluating products of chemists. *Proceeding of the 80th Annual Convention of the American Psychological Association,* **7**: 311–12.

US Patent and Trademark Office Electronic Business Center (2008). *Patent full-text and full-page image databases.* Available at http://www.uspto.gov/patft /index.html.

Wallace J (2007). Boeing Dreamliner coming to life. *Seattle Post Intelligencer,* June 27. Available at http://www.seattlepi.com/business/275465_japan27.html.

White R (1990). Inventors, invention and innovation. *Vital Speeches of the Day,* **56**(19): 593–6.

13.16 Appendix: AMP acronym list

The following list summarizes acronyms generated to describe the various types of machines that constitute the AMP equipment industry. Family-level acronyms (e.g. ATL and AFP) are universally used and recognized within the industry; others (e.g. MHAFP and RSH) were created as part of the current study.

1st Gen.	First generation
AFP	Automated fiber placement
AMP	Automated material placement
ATL	Automated tape laying
ATLM	Automated tape lamination machine
CAFP	Custom automated fiber placement
CTLM	Contoured tape laying machine
CW	Custom winder
FCL	Flat charge layer
FTLM	Flat tape laying machine
FW	Filament winding
HCTL	High contour tape layer
HDW	Heavy duty winder
IS	*In situ*
ISW	*In situ* winder
MHAFP	Multi-head AFP
MHATL	Multi-head ATL
MHTLM	Multi-head tape laminating machine

MHCLP	Multi-head composite lamination platform
MLP	Multi-layer placement
MSW	Multi-spindle winder
P&P	Pick and place
PFE	Precision feed end-effecter
RFI	Resin film infusion
RFP	Robotic fiber placement
RSH	Right-sized head
New Gen.	New generation
SFTL	Small flat tape layer
SP	Special purpose
TC	Tape cassette
TTL	Thermoplastic tape layer
VAPPS	Vacuum-assisted ply placement system

14

Cost, time and technical performance risk mitigation in large, complex and innovative aeronautics development projects

T. BROWNING, Texas Christian University, USA

Abstract: Large, complex, innovative development projects, such as many in the aeronautical industry, occur under conditions of consequential uncertainty, i.e. risk. Risk is manifest with respect to project cost, schedule, and technical performance. These risks need to be managed in relation to each other as well as to the project schedule, budget, and technical requirements. This chapter presents an integrated model to plan and monitor project value in terms of all of these areas. The model complements conventional approaches such as earned value management, which does not directly account for technical performance or any kind of risk. The model allows a user to quantify a project's overall value, the portion of that value at risk, and the respective contributions of cost, schedule, and technical performance to that value and risk.

Key words: project management, risk management, project value, uncertainty.

14.1 Introduction

A project is 'a temporary endeavor undertaken to create a unique product, service, or result' (PMI 2008). Unique circumstances, novel requirements, and limited resources all contribute to make projects uncertain and sometimes even ambiguous regarding exactly what work should be done and when (Loch *et al.* 2006). Uncertainty is at its greatest in large, complex, innovative development projects, which are common in the aerospace industry. When uncertainty allows potential outcomes that carry a negative consequence or impact to a project, uncertainty causes risk – risks of cost and schedule overruns, and risks of failing to provide desired levels of quality, functionality, or technical performance.

This chapter shows an integrated, quantitative framework for monitoring and managing project cost, schedule, technical performance, and overall uncertainties, impacts, and risks. The framework (Browning 2011) combines two existing frameworks: one for cost and schedule risks (Browning and Eppinger 2002; Browning 1998) and one for technical performance risks, the risk value method (RVM) (Browning 2006; Browning *et al.* 2002; Browning 1998). This combination into an integrated framework – effectively an extension of the RVM – provides a basis for exploring risk tradeoffs in projects. Risk tradeoffs are critical because reducing risk in one part of a project often just pushes risks into other parts, often inadvertently and unexpectedly. This supposed risk mitigation by risk transfer contributes to negative syndromes such as 'whack-a-mole' management where,

305

like in the popular arcade game, managers try to 'whack' each new problem as it arises, but in doing so merely cause new problems elsewhere. Hence, an integrated framework is needed to support more optimal decisions at the overall project level.

After introducing concepts and definitions pertaining to project cost, schedule, and technical performance uncertainty and risk, this chapter briefly shows the integrated decision-support framework and provides an example of the development of an unmanned combat aerial vehicle (UCAV). The framework and example provide the basis for subsequent discussion of issues pertaining to risk tradeoffs. The chapter concludes with suggestions for further reading and references on this topic.

14.2   Interdependence of development cost, schedule, and technical performance

As a temporary endeavor, a project must produce its result with limited resources, especially time (deadlines or due dates) and money (budgets). While specific types of resources, such as skilled individuals or specialized facilities, may also be constrained, we will simplify the discussion in this chapter to the constraints posed by time and money. (The proposed framework could be extended to account for other specific resources of interest.) Collectively, we can think of all the resources expended for a project as the sacrifice for or investment in it.

The benefits of a project – its desired results – should meet expectations for quality. While time and money are relatively easy to measure, the quality of a result is often less so. Depending on the type of result expected, quality may be considered in terms of numbers of features or functions, numbers of goals met, degrees of capability, absence of 'bugs' or non-conformances to requirements, and/or aesthetic virtues. In this chapter, we will use the term 'technical performance' (or just 'performance') to refer to the degree of capability or 'goodness' of a project's result. Technical performance should be understood in a general and generic sense as the attributes of the benefits or returns provided by a project, against which the costs or sacrifices are weighed.

Joint consideration of benefits for costs or sacrifices leads to the concept of value. A project's value depends on the technical performance provided for the time and money spent. Exactly how these three variables are connected will be discussed below, but their general relationship implies that overall project value will tend to increase as time and money are spent, to a point, past which that value will begin to decrease (Browning 2003). In other words, everyone realizes that any worthwhile result will require some investment of resources, but at some point the marginal benefits cease to justify the marginal costs. Exactly where this point occurs depends on the specifics of the project and the desires of its stakeholders.

Another way to think of overall project value is illustrated in Fig. 14.1 as the volume of a rectangular solid with the three dimensions of project time (i.e. from

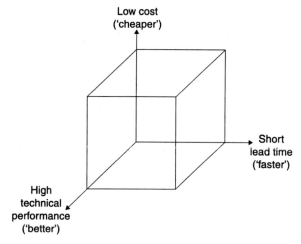

14.1 A stylized view of project value as the volume of a rectangular solid.

the point of view of the customer, lead time), cost, and the technical performance of its result. As the time and cost of a project increase, the volume of the solid decreases, while an increase in technical performance increases the volume. However, an increase in technical performance usually requires an expenditure of time and money, so the three dimensions are not independent. It is precisely these interdependencies that cause time–cost–performance tradeoff problems in engineering design and management.

14.3   The aspect of risk

14.3.1 Uncertainty and the probability model

While definitions and usages vary by context and author, we will use the following definitions, adapted from Schrader *et al.* (1993):

- Uncertainty: the state of having some knowledge of the set of potential outcomes for a variable but not knowing what the outcome will actually be; 'knowing what you don't know.'
- Ambiguity: the state of not even knowing about the existence of a variable or its set of potential outcomes; 'not knowing what you don't know'; unknown unknowns or 'unk-unks.'

In this chapter, we will focus on uncertainty, which addresses most of the issues on all but the most unprecedented of projects. We will presume that the ultimate outcomes of variables such as project cost, duration, and technical performance cannot be known with certainty, especially early in a project, but that, firstly, these variables are known to exist and, secondly, estimates of them can be made. Most project risk-management methodologies and literature deal with aspects of uncertainty. By its very nature, ambiguity is more challenging to address in a formal way, although its presence and potential should never be ignored. (Ambiguity is most likely to affect a project with dynamic goals. That is, if the whole point of the project is changing, then estimates of the time and cost required to achieve a superseded goal become irrelevant.)

A conventional way to model a variable about which we are uncertain is to use a random variable represented by a probability distribution, $\tilde{P}(x)$. Often, such distributions are inferred from only a few data points. For example, suppose we have three estimates of a project's duration: an optimistic or best-case scenario for fast completion, a; a pessimistic or worst case scenario for slow completion, b; and a 'most likely' completion time, m, somewhere between a and b. Traditional project management methods (e.g. Meredith and Mantel 2009) use these three points to fit a beta or a triangle distribution (see Fig. 14.2), from which measures such as the mean, variance, and skewness can be inferred. Since the height of a probability distribution is normalized so that the sum of the area under the distribution equals one, $\tilde{P}(x)$ can be used to address questions concerning the probability of an outcome. For example, in Fig. 14.2, the area under the triangle and to the left of the vertical bar represents the probability of completing the project by the deadline of 160 days (about a 42% chance). Meanwhile, about 58% of the potential outcomes imply a schedule overrun.

Much of the information about our level of certainty in an outcome is represented by the variance (or standard deviation) of a distribution – or, quite simply, its width. A distribution with small variance represents a narrow set of potential

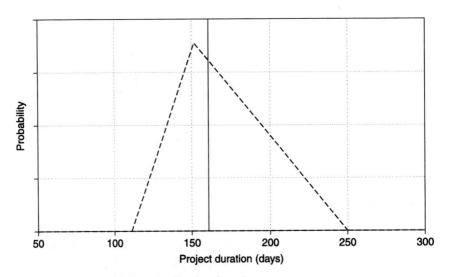

14.2 Probability distribution functions.

outcomes and, therefore, relative precision and confidence in the eventual outcome. A wide distribution conveys a broad range of potential outcomes and a lack of confidence in knowledge of the eventual outcome.

14.3.2 Impact and expected loss (risk)

While probability distributions provide a very useful model of uncertainty, a project manager faces a myriad of uncertainties, and some are more important than others. While more uncertainty in any dimension is usually bad, which uncertainties deserve the most attention? Moreover, in using only cumulative probabilities, we lose information about the outcomes themselves: are all outcomes that miss the deadline, for instance, equally bad? Is it worse to miss the deadline by 50 days than to miss it by one? In a simple probability model, both of these outcomes are equally undesirable. Hence, we supplement our basic model of uncertainty in outcomes with a model of the consequences or impacts of those outcomes.

Figure 14.3 shows an impact function, $I(x)$, a mathematical expression of the consequences of each potential outcome in $\tilde{P}(x)$. In this example of a project's duration relative to a deadline, early completion provides no reward, but being late has a penalty – $100\,000 for the first day plus $10\,000 per additional day. A project manager would prefer to make decisions that would increase confidence in the completion date (narrow the distribution) or move it up (shift the distribution to the left).

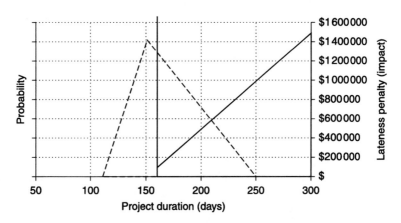

14.3 Probability distribution with impact functions.

Risk is not the same thing as uncertainty; it is uncertainty weighted by its impact. Standard approaches to risk management (e.g. PMI 2008) multiply probability and impact to arrive at a risk index. We can derive a richer risk index by looking at the impact of each potential outcome, weighted by its probability, to find the overall expected penalty or loss inherent in a set of potential outcomes:

$$\mathcal{R} = \int_{G}^{\infty} \tilde{P}(x) I(x) \, dx \qquad\qquad [14.1]$$

where $\mathcal{R}$ is the risk index and G is the goal (e.g. a deadline). Equation 14.1 pertains to a set of outcomes where 'smaller is better' (SIB), such as project duration, so the integration is performed over the region of undesirable outcomes that exceed the goal. For a set of outcomes where 'larger is better' (LIB), we would integrate over $-\infty$ to G. For a set of outcomes where some nominal value is best (NIB), two integrations must be performed, one over each region of adversity. For the example in Fig. 14.3, where

$$I(x) = \begin{cases} 0 & x \le G \\ \$100\,000 + \$10\,000(x - G - 1) & x \ge G + 1, \end{cases} \qquad [14.2]$$

$$\mathcal{R} = \int_{160}^{\infty} \tilde{P}(x) I(x) \, dx = \$228\,546.43 \qquad\qquad [14.3]$$

That is, the risk index conveys the expected (in the sense of statistical expectation) penalty the project will face, given the prevailing uncertainties (i.e. the possibilities of a particular range of undesired outcomes). If the impact function is specified in monetary units, as in this example, then $\mathcal{R}$ is the expected loss or 'value at risk' due to project duration. Impacts may also be expressed in other terms, such as customer utility, sales revenue, or sales volume. In any case, $\mathcal{R}$ conveys the expected (average) penalty in those terms.

Note that $\mathcal{R}$ is a function of G. All else being equal, for a SIB (LIB) attribute, $\mathcal{R}$ decreases (increases) with G. In other words, using the analogy of a high jumper, as the 'bar' is raised, the risk of not clearing it increases. Equation 14.1 accounts for risk due to the height of the bar (G), the capability of the jumper ($\tilde{P}(x)$), and any consequences of failing the jump ($I(x)$).

14.3.3 Cost, schedule, and performance risks

While some project risk management methodologies treat risk as a fourth dimension along with cost, schedule, and performance, it seems that, since the risks threaten each of these three dimensions, they should be accounted for within them. Therefore, each risk is defined as follows (Browning 2011):

- Cost risk ($\mathcal{R}_C$): the expected additional cost (beyond the planned budget) to be incurred given, (a), the uncertainties in the project's ability to develop an acceptable result, (b), within a given budget and, (c), the impacts of any cost overruns.
- Schedule risk ($\mathcal{R}_S$): the expected additional project duration (beyond the planned deadline) to be required given, (a), the uncertainties in the project's

ability to develop an acceptable result, (b), by a given deadline and, (c), the impacts of any schedule overruns.

- Technical performance risk ($\mathcal{R}_p$): the expected loss due to, (a) the uncertainties in the project's ability to develop a result, (b) that will satisfy technical performance goals and, (c), the impacts of any performance shortfalls.

14.3.4 Risk tradeoffs

The previous section mentioned the potential for cost, schedule, and performance tradeoffs in projects. Indeed, time–cost tradeoffs in particular have received much attention in the literature on project scheduling (e.g. Roemer *et al.* 2000). However, most if not all of this literature discusses the tradeoffs in terms of single-point estimates for cost, duration and quality, neglecting the effects on the distributions of potential outcomes. Indeed, some tradeoffs may affect characteristics of the distribution besides the mean or most likely outcome and therefore could be inadequately represented by other models. For example, an investment in information (at a cost), such as performing a test activity, might improve confidence in a dimension of technical performance, thereby reducing the variance in its distribution, but without shifting its mean. This trades an expense (which may increase $\mathcal{R}_C$) for a reduction in $\mathcal{R}_p$. To determine whether this tradeoff is rational, we need an integrated model to support such analyses and decisions.

14.4   An integrated decision-support model – the risk value method (RVM)

The components presented in the previous section combine to form an integrated model with the following inputs:

- Dimensions of technical performance: the J critical-to-quality (CTQ) characteristics or attributes of the project's result,
- Project capabilities ($\tilde{P}$): estimates of the potential outcomes in each dimension of cost, schedule, and the J dimensions of technical performance, consolidated into a vector of length $J + 2$ of probability distributions,

$$\tilde{P} = [\tilde{P}_C(x)\ \tilde{P}_S(x)\ \tilde{P}_1(x)\ \tilde{P}_2(x) \ldots \tilde{P}_J(x)],$$

- Project goals (G): a vector of length $J + 2$ of goals, targets, or requirements for each dimension, including the budget, the deadline, and a level of technical performance in each dimension,
- Impact functions (I): a vector of length $J + 2$ of impact functions, representing the penalties (in similar terms) of outcomes that fail to achieve their goal.

Using these inputs and Eq. 14.1, we can find the vector of length $J + 2$ of risk indices ($\mathcal{R}$). To get a sense of the project's overall risk, we can combine the dimensions of risk in one of the following ways:

$$\mathcal{R}_{\text{Project}} = w_C \mathcal{R}_C + w_S \mathcal{R}_S + w_1 \mathcal{R}_1 + w_2 \mathcal{R}_2 + \ldots + w_J \mathcal{R}_J \qquad [14.4]$$

where all of the w sum to one (an arithmetic mean), or

$$\mathcal{R}_{\text{Project}} = \text{MAX}(\mathcal{R}_C, \mathcal{R}_S, \mathcal{R}_1, \mathcal{R}_2, \ldots, \mathcal{R}_J), \qquad [14.5]$$

where overall project risk depends on the weakest attribute, or a weighted geometric average (formula not shown), or others. Multi-attribute decision analysis is notoriously difficult due to the interdependencies among the dimensions, and each of the potential methods has its benefits and drawbacks. Equation 14.4 has the advantage of simplicity and the disadvantage of potentially obscuring some high risks. Equation 14.5 is also simple and provides a constraint-based approach. Whichever methods a project manager selects, the intent is to keep tabs on the evolving levels of risk in the critical dimensions of a project and to support decisions about risk management and tradeoffs.

In a simplistic way, $\mathcal{R}_{\text{Project}}$ may be thought of as *the portion of the project's overall value that is threatened by the potential occurrence of adverse outcomes.* Augmenting the stylized representation in Fig. 14.1, one can think of $\mathcal{R}_{\text{Project}}$ as the potential reduction in volume down to the smaller, shaded solid in Fig. 14.4. As $\mathcal{R}_{\text{Project}}$ decreases, the shaded solid grows until $\mathcal{R}_{\text{Project}} = 0$, at which point the shaded solid is the same size as the unshaded one.

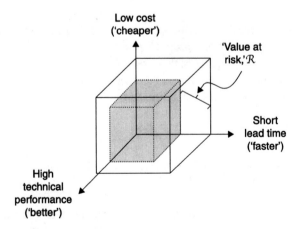

14.4 A stylized view of the portion of a project's value at risk.

This framework, an extension of the risk value method to include project time and cost, maintains the RVM's connection of project progress and added value with the reduction of the portion of a project's value at risk.

At the same time, it is important to note that the framework can also account for the *benefits* of the uncertainties facing projects – namely, the possibility of opportunities residing in the upside potential of each probability distribution (i.e. the portion of the outcomes that exceed the goal). If exceeding a goal has a positive reward, then an impact function can similarly be used to account for such benefits. For further information on opportunities and their relationship to risks, see Browning (2011) and Hillson (2003).

14.5 Example: an unmanned combat aerial vehicle (UCAV) development project

This section uses the example of the preliminary design phase of a UCAV development project to illustrate the application of the risk framework. The example is hypothetical but based on an actual UCAV development project at The Boeing Company (Browning 1998).

14.5.1 Model inputs and initial conditions

We will focus on four attributes of technical performance for the UCAV: maximum payload weight (in pounds), maximum range (in miles), reliability (in terms of mean time before failure (MTBF)), and stealth (in terms of a ratio to the performance level of the previous generation product, such that a measure above 1.0 represents an improvement). Adding project cost and schedule (duration) to these yields six dimensions of project attributes. Consultation with experts in each of these areas led to the initial estimates of project capabilities shown as the probability distributions in Table 14.1. A variety of inputs were distilled into three point values (*a*, *m*, and *b*) from which the triangle distributions were derived:

$$\tilde{P}(x) = \begin{cases} \dfrac{2(x-a)}{(m-a)(b-a)} & a \leq x \leq m \\[2mm] \dfrac{2(b-x)}{(b-m)(b-a)} & m < x \leq b \\[2mm] 0 & \text{otherwise} \end{cases} \qquad [14.6]$$

Other types of distribution could be used instead of triangles. In this example, the distributions for project cost and duration were derived from project simulations (Browning and Eppinger 2002). Table 14.1 also shows the goal for each dimension,

Table 14.1 Initial (t_0) project data and risk calculations

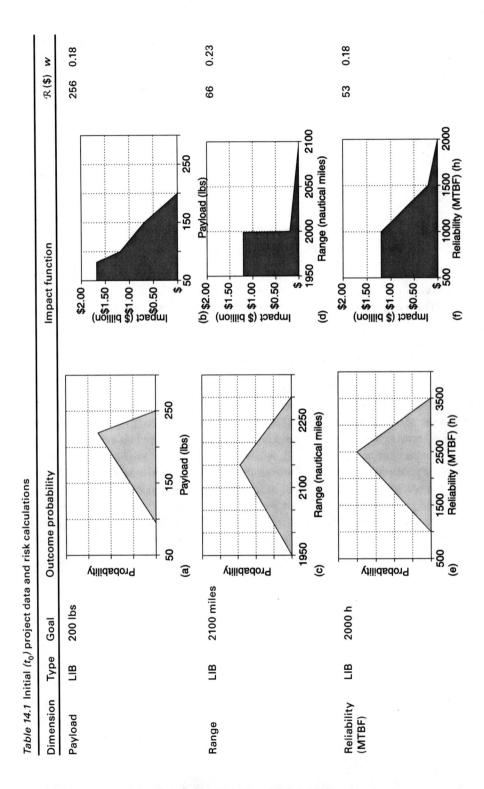

Dimension	Type	Goal	Outcome probability	Impact function	$\mathcal{R}$ ($)	w
Payload	LIB	200 lbs	(a)		256	0.18
Range	LIB	2100 miles	(c)	(b)	66	0.23
Reliability (MTBF)	LIB	2000 h	(e)	(d)	53	0.18
				(f)		

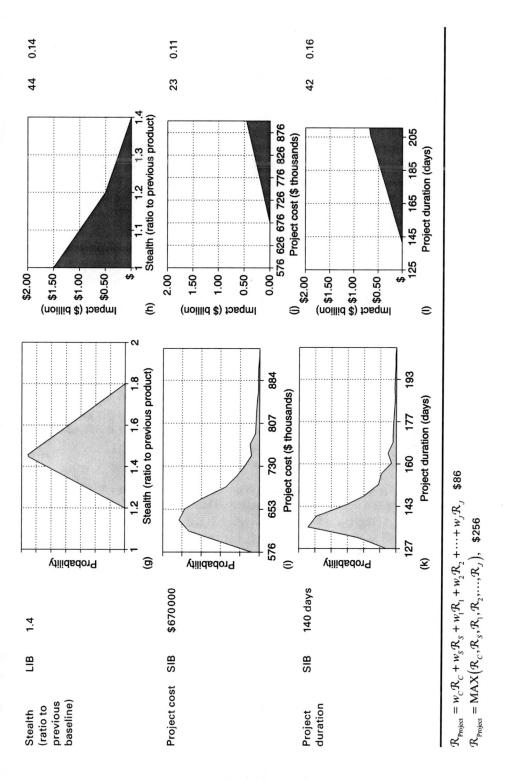

$$\mathcal{R}_{Project} = w_C \mathcal{R}_C + w_S \mathcal{R}_S + w_1 \mathcal{R}_1 + w_2 \mathcal{R}_2 + \cdots + w_J \mathcal{R}_J \quad \$86$$

$$\mathcal{R}_{Project} = \text{MAX}\left(\mathcal{R}_C, \mathcal{R}_S, \mathcal{R}_1, \mathcal{R}_2, \ldots, \mathcal{R}_J\right), \quad \$256$$

the impact function (in terms of the potential for lost revenue), and the implied level of risk, as calculated with a discrete form of Eq. 14.1. The weights shown in the second-to-last column are used in Eq. 14.4 to calculate the overall project risk (shown at the bottom of Table 14.1). The following simplifying assumptions are also used to relate project cost and duration to product unit cost and delivery date. (These assumptions are necessary since the latter attributes are actual customer CTQs, whereas preliminary design cost and duration are not.) Firstly, each dollar of cost overrun in the preliminary design phase project would likely cause a two-dollar increase in the product's unit cost. Secondly, each day of overrun in the project's duration would likely increase the product's final delivery date by two days.

14.5.2 Project progress

As the UCAV preliminary design project proceeded, workers completed activities, which produced new information and knowledge, which enabled updates to the probability distributions. Doing the work consumed time and money, which replaced some cost and schedule uncertainties with actual results, thus narrowing the cost and schedule distributions. (Regardless of which way the mean shifts, the typical trend is for cost and schedule variance to decrease over the course of a project.) Note that we use the term 'variance' in the statistical sense (referring to the width of the distribution), not in the sense of a difference from a planned value, which is how the term is used in earned value management (e.g. PMI 2008). The work resulted in new information about the dimensions of technical performance that tended to narrow their distributions, albeit without any guarantee of the mean shifting in a desirable direction. Figure 14.5 plots the weighted risk indices derived from Eq. 14.4 over project time.

We can observe the following trends as the project proceeded:

- The width of the distributions tends to decrease as ignorance is replaced with knowledge and uncertainty with certainty. This does not imply that a project will always progress in a desirable direction (i.e. the mean of each distribution could shift in an undesirable way), but at least one gains more precise knowledge of what the project's outcomes will actually be.
- While uncertainty decreases over the course of the project, risk may grow. Since the risk values depend on the overall location of the probability distribution, not just its variance, we may find a project moving in the direction of more certainty about a bad outcome, which of course implies increasing risk. For example, over the course of the UCAV project, the weight of the aircraft trended upwards, which implied a reduction in aircraft range (weeks 4–10 in Fig. 14.5). When such a trend is observed, a project manager may choose to reallocate resources to address the issue.
- Attributes can be traded off, as guided by their prevailing level of risk and their relative weighting (in Eq. 14.4). For example, at week 17 the project reduced the estimated range of the aircraft by decreasing the size of its fuel

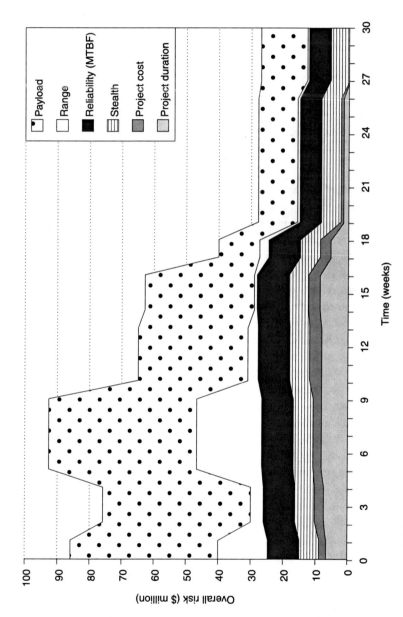

14.5 Evolution of project risk dimensions over time.

Legend:
- Payload
- Range
- Reliability (MTBF)
- Stealth
- Project cost
- Project duration

X-axis: Time (weeks)

Y-axis: Overall risk ($ million)

tanks, which allowed for carrying additional payload. While this slightly increased the risk in the range attribute, it provided a major reduction of the risk in the payload attribute.

14.6   Discussion

14.6.1 The notion of progress in projects

'To solve our basic problem [of improving the product development process], any methodology that is to be developed must be useful in evaluating the partially-developed product at any time during its development life.'

Sidney Sobelman (1958).

The problem of evaluating a partially designed product and thus measuring progress (or the value earned) in a product development project is a long-standing one. Approaches such as earned value management (EVM) (e.g. Fleming and Koppelman 2000, PMI 2008) attempt to measure progress as a function of activities completed and time and money spent in relation to a plan. However, EVM does not account for cost and schedule uncertainty or risk or for any aspects of technical performance. Moreover, it is not clear that true progress results from doing a preconceived slate of activities. As distilled by Steward (2001), progress is what you accomplish, not what you do. The framework presented in this chapter enables progress to be measured in terms of the attributes that determine the overall value of a project to its stakeholders.

Moreover, by accounting for the prevailing uncertainties and their implications (risks), the framework allows value to be added to a project simply by gaining confidence in the outcomes. For example, suppose a person is about to embark on a coast-to-coast road trip across the US. This person plans to use his own personal automobile, which is seven years old, but he is not fully confident about its suitability for such an adventure. So, he takes the car to a mechanic and asks him to check it over. After an hour of work, the mechanic says that the car is indeed good to go – and charges $100. Did this guy just waste $100? Did he receive any value from the transaction? Actually, he was quite happy to verify that his car was in good condition, and he left with a confidence that he did not have before. Similarly, some activities in a product development process (e.g. tests and verifications) add value and contribute to progress merely by producing information that increases confidence in the outcome (by ruling out the possibility of some outcomes). While approaches such as Lean may classify such activities as non-value-adding, this designation seems suspect (Browning and Heath 2009).

14.6.2 Relationship to technical performance measure (TPM) tracking

The probability distributions used to represent a project's capabilities and estimations regarding each dimension of technical performance are related to

TPM tracking, which has long been used in the development of complex systems (e.g. Coleman *et al.* 1996; DSMC 1990; Kulick 1997; NASA 1995; Pisano 1996) to monitor the status of important attributes of an evolving system design. Figure 14.6 shows an example of a basic TPM tracking chart, where the estimated maximum range of an aircraft is updated over the course of a project, albeit in hindsight, as new data become available. Some TPM tracking charts have been known to exhibit definite tendencies and trends that can be anticipated as typical, such as the notorious increase in weight over the course of an aircraft development project. TPM charts can be augmented to display planned trajectories, requirements, and margins. While it is also possible to augment the display at each point with an estimate of the prevailing uncertainty (an *a* and *b* to go with the *m*), this seems to be seldom done in practice. Research and observations of TPM tracking that did include uncertainty estimates (Browning *et al.* 2002; Browning 1998) led to the finding that, while the direction of change in the estimate of the most likely level of the TPM (*m*) was relatively difficult to predict from one period to the next, the uncertainty bounds around that most likely estimate tended to shrink more reliably. Thus, estimates of the uncertainties around any TPM estimate are quite important, and they can be used as a basis for determining $\tilde{P}$, as in Eq. 14.6.

While the set of TPMs worth tracking on a complex development project is larger than the set of high-level dimensions included explicitly in the RVM, the larger set of TPMs plays a supporting role and might even be formally related via a TPM breakdown structure or tree diagram. For example, the TPM 'aircraft weight' is not included in our UCAV example. While it could have been included, one could also argue that many customers do not care directly about a UCAV's

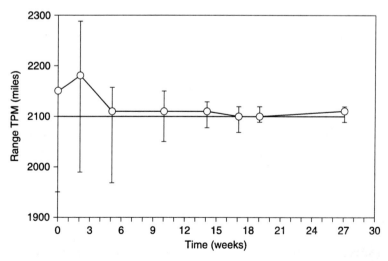

14.6 Example of a TPM tracking chart for aircraft maximum range.

weight; rather, they care about the implications of weight on attributes such as maximum payload, maximum range, endurance, and cost. In such a case, while designers might track the weight TPM explicitly, it would be an input to revisions of $\tilde{P}$, even if it was not an actual member of that set of distributions.

14.6.3 The role of project activities in the RVM

Activities – packages of work that consume resources – have long been the basic unit of analysis for project time and cost models. However, the results of activities – namely, the information they produce that adds value by revising $\tilde{P}$ – have received much less attention. Hence, activities serve as the basic building blocks of an RVM model. For further details on how to associate the results of activities with changes in $\tilde{P}$, see Browning *et al.* (2002). For a simulation that treats activities as the agents in an agent-based model of a development project, see Lévárdy and Browning (2009).

14.6.4 Benefits of a holistic framework

Many projects incorporate the conventional techniques of risk management, including TPM tracking, project scheduling, budgeting, and earned value management. However, especially on large projects, such activities are often undertaken by different individuals or teams, and they may not always communicate frequently or completely. The result is that the various plans and schedules used to direct the project can become disintegrated and unsynchronized. For example, the scheduling group may have one list of project activities in their plan, while the risk management group may have another (one that includes additional activities they have nominated to mitigate risks). Moreover, as mentioned previously, some groups (e.g. the 'managers') tend to plan and monitor the use of resources while others (e.g. the 'systems engineers') focus on the implications of the work done on the system development. What is needed is a framework that integrates these two perspectives and information sets, as well as a third perspective, that of the customer (or, as their proxy, the marketing department), which is represented in the RVM by the goals and impact functions. The RVM provides a common language for discussions among these groups about the sources and implications of project risks. Without a holistic framework, a project's decision makers will operate with either a subset of the various available plans, schedules, and technical data or else conflicting assumptions and recommendations from the various groups attending to each subset.

14.7   Conclusion and future trends

Implementation of the RVM on a large scale would be greatly facilitated by a software tool that automated the input, analysis, and representation of the relevant data. As more projects adopt this way of thinking about and modeling risk, the demand for such a tool will increase.

Practical experience has shown that the first RVM models built by an organization do not come easily, because their construction prompts questions that have not been asked before. While much of the needed information can be pulled together from existing risk management plans, project schedules, etc., estimates of impact functions and project capabilities in each critical dimension of performance tend to require more work. RVM implementation provides a 'forcing function' to gather information that a project should have anyway but usually does not. Once an initial model is built, it is much more easily revised and improved. Once project participants have gotten used to thinking about risk and its implications for project value in the new way, they 'get down the learning curve' and find it possible to provide much better estimates.

14.8   Sources of further information and advice

For further background on technical performance risk management methods, see the primary works upon which this chapter is based (Browning 1998, 2006, 2011; Browning *et al.* 2002). To see relationships of this type of work to decision analysis, see Ullman (2006). For primers on general methods for project risk and opportunity management, see Smith and Merritt (2002) and Hillson (2003). For focus on project cost and schedule risks and time–cost tradeoffs, see Browning and Eppinger (2002) and Roemer *et al.* (2000). Finally, for a simulation that incorporates many of the methods discussed in this chapter to represent decision-making in an adaptive process, see Lévárdy and Browning (2009).

14.9   References

Browning, T.R. (1998). 'Modeling and analyzing cost, schedule, and performance in complex system product development.' PhD Thesis (TMP), Massachusetts Institute of Technology, Cambridge, MA.

Browning, T.R. (2003). 'On customer value and improvement in product development processes.' *Systems Engineering*, **6**(1): 49–61.

Browning, T.R. (2006). 'Technical risk management.' In Hillson, D. (ed.) *The Risk Management Universe: A Guided Tour*, pp. 292–320, London: BSI.

Browning, T.R. (2011). 'A quantitative framework for managing progress, uncertainty, risk, and value in projects.' TCU Neeley School of Business, Working Paper.

Browning, T.R. and Eppinger, S.D. (2002). 'Modeling impacts of process architecture on cost and schedule risk in product development.' *IEEE Transactions on Engineering Management*, **49**(4): 428–42.

Browning, T.R. and Heath, R.D. (2009). 'Reconceptualizing the effects of lean on production costs with evidence from the F-22 program.' *Journal of Operations Management*, **27**(1): 23–44.

Browning, T.R., Deyst, J.J., Eppinger, S.D. and Whitney, D.E. (2002). 'Adding value in product development by creating information and reducing risk.' *IEEE Transactions on Engineering Management*, **49**(4): 443–58.

Coleman, C., Kulick, K. and Pisano, N. (1996). 'Technical performance measurement (TPM) retrospective implementation and concept validation on the T45TS Cockpit-21 program.' Program Executive Office for Air Anti-Submarine Warfare, Assault, and Special Mission Programs, White Paper.

DSMC (1990). *Systems Engineering Management Guide*. Fort Belvoir, VA: Defense Systems Management College.

Fleming, Q.W. and Koppelman, J.M. (2000). *Earned Value Project Management*, 2nd ed. Upper Darby, PA: Project Management Institute.

Hillson, D.A. (2003). *Effective Opportunity Management for Projects: Exploiting Positive Risk*. New York: Marcel Dekker.

Kulick, K.A. (1997). 'Technical performance measurement: A systematic approach to planning, integration, and assessment (3 Parts).' *The Measurable News*.

Lévárdy, V. and Browning, T.R. (2009). 'An adaptive process model to support product development project management.' *IEEE Transactions on Engineering Management*, **56**(4): 600–20.

Loch, C.H., DeMeyer, A. and Pich, M.T. (2006). *Managing the Unknown*. New York: Wiley.

Meredith, J.R. and Mantel, S.J. (2009). *Project Management*, 7th ed. New York: Wiley.

NASA (1995). *NASA Systems Engineering Handbook*. NASA Headquarters, Code FT, SP-6105.

Pisano, N.D. (1996). 'Technical performance measurement, earned value, and risk management: An integrated diagnostic tool for program management.' Program Executive Office for Air ASW, Assault, and Special Mission Programs (PEO(A)), US Air Force, Unpublished white paper.

PMI (2008). *A Guide to the Project Management Body of Knowledge*, 4th ed. Newtown Square, PA: Project Management Institute.

Roemer, T.A., Ahmadi, R. and Wang, R.H. (2000). 'Time-cost trade-offs in overlapped product development.' *Operations Research*, **48**(6): 858–65.

Schrader, S., Riggs, W.M. and Smith, R.P. (1993). 'Choice over uncertainty and ambiguity in technical problem solving.' *Journal of Engineering and Technology Management*, **10**(1): 73–99.

Smith, P.G. and Merritt, G.M. (2002). *Proactive Risk Management: Controlling Uncertainty in Product Development*. New York: Productivity Press.

Sobelman, S. (1958). *A Modern Dynamic Approach to Product Development*. Dover, NJ: Office of Technical Services (OTS).

Steward, D.V. (2001). 'Perceiving DSM as a problem solving method.' *Proceedings of the 3rd MIT Design Structure Matrix Workshop*. 29–30 October, Cambridge, MA.

Ullman, D.G. (2006). *Making Robust Decisions*. Victoria, BC: Trafford.

15
Innovation in aeronautics through Lean Engineering

E. M. MURMAN, MIT, USA

Abstract: The dynamics of innovation theory indicate that, for products as mature as aircraft, process innovation is an important contributor to product success and innovation. Many aerospace companies have adopted Lean Thinking as an enterprise-wide continuous improvement strategy. This chapter extends Lean Thinking to the engineering domain with a Lean Engineering framework based upon observational findings from a decade of research in the aerospace domain, published works on Toyota and Southwest Airlines, and practitioner input. Examples illustrate how the framework maybe be applied. Lean Engineering is not totally new to aerospace, and it continues to evolve. Future challenges are briefly summarized.

Key words: Lean Engineering, Lean Product Development.

15.1 Introduction

A series of MIT studies and publications on industrial productivity has led to this chapter, the objective of which is to present a framework of basic Lean Thinking principles in a format that aerospace engineers can understand and embrace. Engineers are responsible for decisions that dictate most of the lifecycle costs of aerospace products, play a key role in getting products introduced in a timely way, provide the know-how to make the products perform, and consume a significant portion of the development costs. Yet engineers often feel that Lean Thinking does not apply to what they do. It is hoped that, after reading this chapter, engineers will understand the innovation–process improvement linkage and embrace the fundamental principles of Lean Thinking. The result will be not only a stronger aerospace industry, but also happier, more satisfied engineers.

Innovation in aeronautics can take many forms, including new product architectures, new technology, new product subsystems, and new product processes. Scholarly studies of the dynamics of innovation (e.g. Utterback, 1994; Christensen, 1997) reveal that innovation is influenced not only by the evolution of technology, but also by the evolution of the product. This topic is addressed briefly in Section 15.2 to reveal the linkage between product innovation and process innovation, the main topic of this chapter.

In the mid-1980s MIT initiated a series of studies out of concern for lagging US productivity and its implications for future economic growth. In the book *Made in America* (Dertouzous *et al.*, 1989), the MIT Commission on Industrial Productivity identified fundamental weaknesses in American industrial practice, including the

323

commercial aircraft sector. The book concluded that 'without major changes in the ways Americans learn, produce, work with one another, compete internationally, and provide for the future, no amount of macroeconomics fine-tuning will be able to provide a rising standard of living in the long run.' Shortly afterwards, in 1990, the MIT International Motor Vehicle Program published *The Machine That Changed the World* (Womack *et al.*, 1990), a comparative study of the Japanese, North American and European automotive industry.

The authors identified a fundamentally new industrial paradigm they called 'Lean' based on the Toyota Production System. Lean is considered a successor to mass production and differs from it in fundamental ways, as summarized in Section 15.3. The automotive study, together with strategic planning in the MIT Aeronautics and Astronautics Department in 1990–1 and the urging of the United States Air Force in 1992–3, led to the creation of the Lean Aircraft Initiative, or LAI. Subsequently it was renamed the Lean Aerospace Initiative and recently the Lean Advancement Initiative as its scope evolved. That research effort, centered at MIT and involving many US aerospace companies and agencies, resulted in the 2002 publication of *Lean Enterprise Value* (Murman *et al.*, 2002), which translated the underlying Lean principles to an aerospace enterprise context. Section 15.4 provides a short summary of LAI and the current state of Lean implementation in the US aerospace industry and government agencies. The author was a founding member of LAI and much of the work reported in this paper is based on findings from that program.

One of the LAI case studies (Stanke, 2001) involving the F/A-18E/F uncovered a 'best in class' program. Although considered an upgrade to the C/D model, the E/F was 25% larger and had significantly improved technical performance in range, payload, and low observables, together with future growth capabilities and reduced operating costs. By all measures it qualifies as a major product upgrade. Unlike many DoD programs that overrun their development budgets, exceed their development time, and may fall short of technical expectations, the F/A-18E/F was completed within budget and schedule, and exceeded its performance goals. Equally interesting was the observation that the engineers working in the program were happy, energized and fully engaged. Additional insight came from findings of other case studies together with reports from Toyota (Morgan and Liker, 2006), Southwest Airlines (Gittell, 2003), and LAI (McManus, 2005).

With this knowledge, the author working with colleagues has constructed a Lean Engineering framework that embodies observational data in a format that, hopefully, connects with aerospace engineers. As presented in Section 15.5, the framework provides a solid conceptual basis at a high-level actionable form. Practitioners and educators will need a further level of supporting practices for implementation. Engineering is applied to a wide range of tasks from basic research and development to product support. Section 15.6 illustrates the tailoring of Lean Engineering to different applications. A common phrase is 'Lean is a journey, not a state', and Lean Engineering is no exception. Section 15.7 briefly summarizes current challenges in embedding Lean Engineering in aerospace practice.

One caveat is offered on the content of this chapter. It was originally written for the SAE/AIAA 2008 William Littlewood Memorial Lecture, which deals with a 'broad phase of civil air transportation considered of current interest and major importance.' In keeping with this theme, the examples are slanted towards that product domain. However, the framework draws on many military aircraft, avionics, missiles, launcher and satellite systems. Just as Lean Thinking was not restricted to automotive products, the Lean Engineering framework should be broad enough to cover aerospace domains other than civil air transportation.

15.2 Dynamics of innovation

Utterback's classic book *Mastering the Dynamics of Innovation* (Utterback, 1994) established that opportunities for innovation in a given product area vary with phases of the product's overall lifecycle. A plot of the number of companies producing the product by year serves to identify the relevant phase. Figure 15.1 shows such a plot of major US aerospace companies (Murman *et al.*, 2002). With the exception of the great depression and a few other dips, from 1907 to 1957 there were more major companies entering aerospace than exiting. Starting in 1958, mergers and acquisitions have reduced the number of major aerospace companies. Figure 15.1 does not give a detailed or complete picture of aerospace product domains. For example, in 1941 there were 52 US aircraft manufacturers listed in *Jane's All The World's Aircraft* (Jane's, 1941), while Fig. 15.1 shows a total of only 18 major firms. Or, if one looked at a plot of unmanned aerial vehicles (UAVs), it might well show an increasing number of firms towards the end of the twentieth century. However, Fig. 15.1 is representative of the overall trends of the US aircraft industry for the purposes of this chapter.

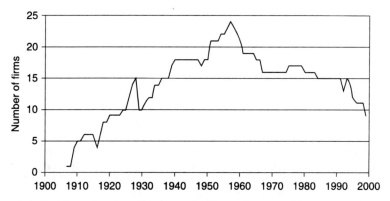

15.1 Number of major US aerospace firms. (Source: Murman *et al.*, 2002.)

Utterback's analysis reveals three important phases of a product's lifecycle:

1 The Fluid Phase represents the early years when the product architecture and features are evolving as new ideas emerge and new companies enter the market. In this phase, more companies enter than exit.
2 During the Transitional Phase, a dominant design emerges that represents the best combination of technologies, operational constraints, and customer desires. This phase surrounds the peak years of the number of companies producing the product.
3 The product then enters the Specific Phase with the dominant design driving industry consolidation, with more companies exiting than entering the market.

Table 15.1 summarizes the significant characteristics of each phase. It can be seen that the opportunities for innovation are quite different in the Fluid and Specific Phases.

Table 15.1 Selected characteristics in the three phases of industrial innovation

Attribute	Fluid phase	Transitional phase	Specific phase
Innovation	Frequent major product changes	Major process changes required by rising demand	Incremental for product and with cumulative improvements in productivity and quality
Source of innovation	Industry pioneers; product users	Manufacturers; users	Often suppliers
Products	Diverse designs, often customized	At least one product design, stable enough to have significant product volume	Mostly undifferentiated, standard products
R and D	Focus unspecified because of high degree of technical uncertainty	Focus on specific product features once dominant design emerges	Focus on incremental product technologies; emphasis on process technology
Competitors	Few, but growing in numbers with widely fluctuating market shares	Many, but declining in numbers after emergence of dominant design	Few; classic oligopoly with stable market shares
Basis of competition	Functional product performance	Product variation; fitness for use	Price
Vulnerabilities of industry leaders	To imitators, and patent challenges; to successful product breakthroughs	To more efficient and higher-quality producers	To technological innovations that present superior product substitutes

(Source: Utterback, 1994.)

In the Fluid Phase, there are frequent major product changes, diverse designs, less directed research and development, and competition based upon product performance. Innovation in the Specific Phase is incremental for the product with an emphasis on quality and productivity, suppliers are often the source of innovation, and there is a greater emphasis on process innovation. Indeed, Utterback argues that there is a shift from product innovation in the Fluid Phase to process innovation in the Specific Phase. One might recognize that commercial and manned military aircraft are in the Specific Phase with dominant designs in each speed range. However, UAVs could still be in the Fluid Phase. Consequently, opportunities for innovation could be different for these product domains.

With the above picture in mind, this chapter addresses opportunities for process innovation for aircraft, specifically for the development phase of the product lifecycle. Principles similar to those presented apply to manufacturing and product support activities. It is important to realize that if continuing innovation is to occur for aircraft with a dominant design, aeronautical engineering needs to be executed in a productive manner to focus limited resources in the most effective way.

15.3   Lean Thinking

Lean Thinking is evolutionary and rooted in actual practice. Its origins reach back nearly 60 years to the post-World War II Japanese automotive industry. Faced with limited resources and no market position, Japan had to find ways to build competitive products with limited resources. The outcome of their quest is well known, particularly to the US automobile companies. It is important to grasp that Lean Thinking is based upon years of best practices, not on some new theory or latest consultants' approach – although both have contributed to the body of knowledge. Not being based upon absolutes such as laws of science, Lean Thinking continues to evolve as more experience and knowledge are accumulated. Lean Engineering fits right in stride with this.

A predecessor to Lean, Total Quality Management (TQM) was based on the thinking of Deming (1982). TQM was adopted by US organizations in the 1980s with many beneficial results (e.g., see the Citation X case in Section 15.6.2). Lean encompasses all the aspects of TQM, but goes beyond to include a fundamental focus on elimination of waste. Another major quality movement, Six Sigma (6σ) (e.g. Mikel and Schroeder, 2000), developed by Motorola in the 1980s, is also an outgrowth of TQM. Six Sigma became very popular after Jack Welch implemented it corporate-wide at GE, starting in 1995. The heart of Six Sigma is the elimination of variation in all process outcomes – from drilling holes to administrative processes to product features that the customer desires. Lean very much relies on an effective quality system such as Six Sigma. The awareness of Lean rapidly followed the publication of *The Machine That Changed The World*

(Womack *et al.*, 1990). Today, aerospace enterprises are implementing combinations of Lean and Six Sigma practices, often including elements of the Theory of Constraints (Goldratt, 1997), which addresses techniques for reducing time of tasks on the critical path.

Beyond Lean is Agile, which pushes flexibility and adaptability to its limits by focusing on very rapid response to customer demands and market opportunities. Becoming Lean might be considered a step to becoming Agile, but not all business opportunities align with an Agile approach. One aerospace area that works well with Agile is software development. Reengineering (Hammer and Champy, 1993) is another improvement model that is outside the TQM, 6σ, Lean, Agile family. It is much more of a top-down radical restructuring approach rather than an evolutionary approach based on fundamental quality principles and worker involvement. Reengineering has had mixed results due to its radical approach.

Table 15.2 compares attributes of Craft, Mass and Lean production systems. It is essential to recognize that Lean Thinking is a 'way of thinking' and not just a set of tools. Labeling an organization with the Mass mentality as 'traditional' and one that exhibits Lean Thinking as 'lean', let us contrast how the attributes in Table 15.1 might be observed in an engineering context. In a traditional organization, an engineer might be thinking about how to enhance the product from a purely technical perspective, while with a Lean mindset the same engineer would also be thinking about how to enhance the product to better meet customer needs and expectations. Customer 'wants' will drive the engineering innovation. Using traditional thinking, an engineer would strive to produce the greatest output, regardless of whether the customer for that output is ready to receive it. A Lean-thinking engineer would strive to produce output just as the overall design/development cycle needs it. The same Lean-thinking engineer would always

Table 15.2 Production system attributes

	Craft	Mass production	Lean Thinking
Focus	Task	Product	Customer
Operation	Single items	Batch and queue	Synchronized flow and pull
Overall aim	Mastery of craft	Reduce cost and increase efficiency	Eliminate waste and add value
Quality	Integration (part of the craft)	Inspection (a second stage after production)	Inclusion (built in by design and methods)
Business strategy	Customization	Economies of scale and automation	Flexibility and adaptability
Improvement	Master-driven continuous improvement	Expert-driven periodic improvement	Worker-driven continuous improvement

(Source: Murman *et al.*, 2002.)

question whether the work is adding value to the overall effort and seek ways to continually eliminate waste in the overall effort (see below). Another engineer coming from a traditional perspective might be trying to increase their output with reduced resources, without knowing whether that output is adding any value. The Lean-thinking engineer would always be designing to meet quality objectives, interpreting quality in the broadest possible sense. A traditional-thinking engineer would be leaving quality concerns to others. The Lean-thinking engineer will be flexible and adaptable, not only to the tasks undertaken, but also to their span of technical competencies. Finally, the Lean-thinking engineer is empowered by management, trained and rewarded to continually make their contribution to enterprise better, and taking delight in doing so. An engineer enmeshed in a traditional organization is waiting for management to make things better, and likely cynical that management is so out of touch with how they do things.

Implementation of Lean Thinking took a major step forward when Womack and Jones (1996) operationalized value creation principles with the five fundamental steps listed in Fig. 15.2. Absolutely central to Lean Thinking is to fully understand what the customer values. That customer may be the one who buys the product (external customer) or the next person along the line (internal customer) who receives the engineer's output. Value is not an attribute many engineers are comfortable with, as it is difficult to quantify or measure. Yet that same engineer makes many value decisions as a customer in allocating their own personal resources. The next fundamental concept to understand is the value stream – the set of actions that transforms raw inputs to finished outputs. In manufacturing, hardware is usually transformed by a series of largely non-iterative actions. But in engineering, information is transformed, and iterations are often needed, but should be planned. At the start of a new design, there are requirements, a project plan, previous designs to draw upon, a base of knowledge, etc. At the end of a design is a set of specifications or Build-To-Package including manufacturing, maintenance, training and other documented information. In between are analyses,

1. Specify *value*: Value is defined by customer in terms of specific products and services

2. Identify the *value stream*: Map out all end-to-end linked actions, processes and functions necessary for transforming inputs to outputs to identify and eliminate waste

3. Make value *flow* continuously: Having eliminated waste, make remaining value-creating steps 'flow'

4. Let customers *pull* value: Customer's 'pull' cascades all the way back to the lowest level supplier, enabling just-in-time production

5. Pursue *perfection*: Pursue continuous process of improvement striving for perfection

15.2 Five Lean Thinking fundamentals. (Source: Womack and Jones, 1996.)

simulations, tests, reviews, and other information-generating activities. 'Tons' of information flows in an engineering value stream. A Value Stream Map – a process map with quantitative data about resources used – needs to be made and analyzed (value stream mapping and analysis (VSMA)) to identify which actions or inactions are adding value and which represent waste.

Identifying value-added and non-value-added, or wasteful, activities is central to Lean Thinking. Figure 15.3 gives the definitions that have emerged from practice. As an engineer, think about an activity taking time or other resources. Does it meet the test of a 'value-added activity'? If not, in a Lean Thinking context it is waste. But some waste is required or necessary. Maybe you spend time in a weekly coordination meeting. Without spending that time, you likely would be doing the wrong job and creating more waste. But the customer has not asked for that meeting. So the meeting falls into the 'necessary waste' category. Finally, maybe you skipped the meeting and spent the next week doing work that was not needed and has to be redone. This is 'pure waste'.

Following the second item in Fig. 15.2, once you identify these activities, you would immediately eliminate pure waste, seeking to continually minimize or eventually eliminate the necessary waste, and spend as many of your resources as possible on the value-added activities. Although this sounds simplistic, it turns out to be extremely important. Figure 15.4 shows quantitative estimates from several sources on the levels of engineering waste in aerospace and automotive domains. The left pie chart shows the estimated levels of waste for an engineer's effort. Only about 30% of their time is allocated to value-added activities. The right pie chart shows the estimated waste for an engineering job package going through the

Value-Added Activity

- Transforms or shapes material, information or people

- And it is done right the first time

- And the customer wants it

Non-Value-Added Activity – Necessary Waste

- No value is created, but cannot be eliminated based on current technology, policy, or thinking

- Examples: project coordination, regulatory constraints, company mandate, law

Non-Value-Added Activity – Pure Waste

- Consumes resources, creates no value in the eyes of the customer, and is not needed to support value-added activities

- Examples: idle/wait time, inventory, rework, excess checkoffs, accidents

15.3 Value-added and non-value-added definitions. (Source: Womack and Jones, 1996.)

Published by Woodhead Publishing Limited, 2012

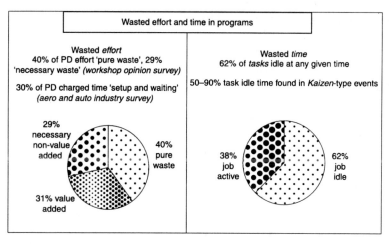

15.4 Estimated engineering waste. (Source: McManus, 2005.)

system. About 62% of the time no work is being done on the job. It is just waiting for someone to add value to it. But, when that someone starts working, only about 30% of their time is devoted to adding value. Although the math is notional, this means that only about 11% of the total time spent on an engineering job package is valued-added activity! Great improvements seem possible and, as we will see later in the paper, these can be realized. Freeing up the wasted engineering resources allows them to be directed to innovation, and the engineers will be happier than when using their talents for wasteful purposes.

Returning to Fig. 15.2, once the waste has been identified and as much eliminated as possible, the next goal is to get the work process to flow as smoothly as possible. This means evening out the load amongst engineers, synchronizing completion of tasks (as much as possible), co-locating teams and people, minimizing handoffs between engineers, managing iterations, and so forth. With the work now smoothed out, backward chaining can be done from the desired end state to initiate work so that the next step in the design process is ready to accept it. That is, work should not be started until the time when its completion is needed by the next activity in the value stream. The final step is to follow cycles of continuous improvement to eliminate waste.

Engineers may argue that their work is not so repetitive and that such principles would interfere with the creative aspects of their work. Although there is certainly some legitimacy to this, by and large engineers repeat the same work steps on different projects, albeit adapting to new challenges and opportunities. This varies with the type of engineering. On one hand, there is product support such as small upgrades and customer adaptations and/or engineering support such as testing, which are mostly repetitive. On the other hand, there is basic research and advanced development, which are much less repetitive. These principles definitely need tailoring, but the basic ideas of customer value and seeking to eliminate waste are equally applicable.

Having shown the data in Fig. 15.4 to engineers in many talks, the author usually sees heads nodding in agreement that much of their efforts are non-value-added. On the other hand, when talking with groups that have implemented principles of Lean thinking, they feel much more satisfied with their work and are by and large happier. Before turning to further elaboration of strategies for implementing Lean Engineering, let us pause to take stock of the current picture of Lean implementation in aerospace engineering.

15.4   Lean Thinking and aerospace

As the Japanese began their Lean journey after World War II, the US aerospace industry started down the same path, but backwards! There is a certain amount of *déjà vu* for aerospace, for at one time it exhibited Lean Thinking. In the engineering area, one only need think of Kelly Johnson's Skunk Works (Rich and Janos, 1994) to see this. Be that as it may, by and large by the 1970s, and even today, the aerospace industry could be characteri zed as a craft industry organized and managed as a traditional mass production system. Figure 15.5 gives a brief summary of the aerospace Lean journey. Following the appearance of *The Machine That Changed The World*, many aerospace companies began implementing Lean Thinking, but only in their manufacturing and supply chain areas, as those consume the most resources. Eventually, they discovered that, without adopting Lean Thinking in their office and engineering areas, they could not achieve the real benefits. Just as GE implemented Six Sigma enterprise-wide, so too does Lean Thinking need enterprise-wide implementation. Significant strides have been made in implementing Lean Thinking in various office areas, but Lean Engineering implementation is lagging behind.

- Aeronautical organizations before about late 1950s exhibited many Lean Thinking behaviors
- By the 1980s, aerospace was a non-lean craft industry with a traditional mass production mentality
- 1980s – beginning of refocus with TQM approaches
- 1990 – *The Machine that Changed the World* coined 'Lean' based on Toyota Production System
- 1993 – Lean Aircraft Initiative started at MIT
- 1990s – Most aerospace companies implemented some form of Lean in manufacturing
- 2000s – Enterprise implementation of Lean, including engineering. Government agencies implement Lean

15.5 Brief history of Lean Thinking in aerospace.

In 1992, Lt General Tom Ferguson of the USAF Aeronautical Systems Center read *The Machine That Changed the World* and contacted Professor Roos at MIT to inquire whether these principles could apply to military aircraft. That led to the start of LAI at MIT, a consortium of aerospace industries, government agencies, and MIT engineering and management faculty. The goal of the consortium was to perform research, develop useful products for implementation, implement Lean practices, and reflect on the outcomes. In essence, it is a learning community at the national level. With the addition of space and commercial aerospace organizations, the name changed to Lean Aerospace Initiative in 1998, and in 2007 further broadening of the stakeholder group led to a new name: Lean Advancement Initiative. Following a 2002 directive from the LAI Executive Board to develop and deploy a university-level Lean curriculum, an Educational Network was formed in 2003, which as of the time of writing has about 60 member colleges and universities on five continents. More information about LAI and the EdNet can be found at http://lean.mit.edu.

By 2007, most aerospace companies and government agencies had an enterprise-wide improvement program incorporating Lean Thinking and Six Sigma. Largely based upon the same fundamentals, each organization tailors the practices and priorities to meet their own culture and location on the Lean journey with a branded name. Some representative names for these programs are:

- Boeing – Lean+,
- United Technologies – ACE,
- Lockheed Martin – LM21,
- Textron – Textron Six Sigma,
- Raytheon – R6σ,
- Rockwell Collin – Lean Electronics,
- USAF – AFSO21,
- NAVAIR – AIRSpeed.

A notable exception on this list is NASA, which in the 1990s initiated its Faster, Better, Cheaper program with mixed outcomes. Unlike the above, it was not based upon the accumulated body of knowledge cited in this paper, but sought breakthrough performance with new approaches.

If you are an engineer in an aerospace organization with an enterprise improvement program, or soon to be entering one, sooner or later you are likely to be challenged to implement Lean Thinking. The following section may give a starting point for that journey.

15.5   Lean Engineering framework

The Lean Engineering framework flows from the value perspectives already introduced, and in particular from the value creation framework introduced in Murman *et al.* (2002) and shown in Fig. 15.6. Conceptually the value creation

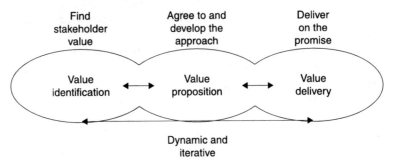

15.6 Value creation framework. (Source: Murman *et al.*, 2002.)

framework embodies three interrelated elements. One is to identify the relevant stakeholders and determine their value expectations: what value they expect to give and to receive from the endeavor. Stakeholders for an engineering-related undertaking might include the customer, end user, program management, other engineers, production, operators, regulators, and others. The second element is to formulate and communicate a value proposition that addresses (but likely not completely satisfies) all the value expectations. This would be embodied in high-level objectives, detailed program plans, requirements, partner agreements, and other documents and agreements. The final element is to deliver on the value proposition. Since the external environment, knowledge base, program progress and other factors are continually evolving, there is a certain amount of dynamic adjustment needed during the lifetime of the endeavor. Making no adjustment leads to obsolescence, while too frequent adjustment leads to instability. The value creation framework is a more sophisticated representation of the familiar adage 'Do the right job and do the job right'. It makes no sense to do the job right before being sure one is doing the right job.

With the value creation framework as a top-level architectural anchor, a Lean Engineering framework consisting of 11 high-level actionable practices is displayed in Fig. 15.7. The remainder of this section carries these practices one level lower, including some illustrative examples. The practitioner and educator will need another level of enabling practices and tools for implementation. This remains to be constructed. The presented framework represents a solid architecture for assembling the enablers. In line with most social–technical frameworks, the Lean Engineering framework is a collection of people-oriented practices, process-oriented practices and tools.

The Lean Engineering framework is empirical, based upon observations of numerous engineering undertakings. A number of the practices come directly or adapted from the book by Morgan and Liker (2006), which presents a similar framework for the Toyota product development system. The practice on perfect coordination is based on Gittell's (2003) relational coordination construct

Capable people

- Engage customer early and often
- Organize for Lean behavior
- Strive for perfect coordination
- Pursue excellence and continuous improvement

Effective and efficient processes

- Implement Integrated Product and Process Development (IPPD)
- Design for lifecycle value
- Assure smooth information flow
- Optimize process flow
- Manage all risks

Adaptable tools

- Utilize integrated engineering tools
- Employ Lean process improvement methods

15.7 Lean Engineering framework.

emerging from research on Southwest Airlines. The author has observed practices reported by Morgan and Liker and Gittell in aerospace organizations. Having them appear in multiple domains gives more solid foundation to their inclusion. A number of practices come from LAI research performed by the author and/or his colleagues. Some of these were embedded in LAI's Lean Enterprise Model (LAI, 1996). Further findings come from a series of student case studies of nine aircraft designs, which illustrated that many Lean Engineering practices have been employed for years in aircraft engineering (Haggerty and Murman, 2006). Two Minta Martin Lectures at MIT provided valuable insight into Lean Engineering for aircraft (Haggerty, 2002) and spacecraft (Leopold, 2004). Members of EdNet and INCOSE's Lean Systems Engineering Working Group have provided practices. Finally, several colleagues with both industrial and academic experience have provided a great deal of wisdom in constructing the framework, and are mentioned in the Acknowledgements. Any shortcoming on the synthesis of this observational data is the responsibility of the author. There is certainly no reason to believe the framework is unique, and other versions might be equally valuable.

An earlier paper by McManus *et al.* (2007) presented a less specific, but very consistent, framework for Lean Engineering. The current framework is broader, encompassing more practices, and draws heavily on this earlier work.

15.5.1 Engage the customer early and often

As displayed in Fig. 15.2, the customer is the focus of Lean Thinking. So a first step is to identify the customer(s) and their value expectations. The engineer may have one or more internal customers (e.g. manufacturing) and the program will have external customers. There are many types of external customers, even within civilian aviation. Beyond that, for aerospace there are military customers, communication companies and others. The customers may not have well-formulated value expectations, and the value expectations may change. All this can be very frustrating to an engineer. Nevertheless, unless the customer ends up satisfied, the engineer's work will be in vain.

Figure 15.8 lists some practices that are observed in Lean Thinking organizations, while Fig. 15.9 illustrates the key elements underpinning customer value. Customers always weigh some combination of what the product will cost them (often over its lifecycle), when it will be available, and its quality, e.g. the features, maintenance, lack of defects. Engineers need to realize that customer value is broader than technical performance. Today's Lean Thinking civilian aircraft companies often have external customer advisory councils that are formed very early in the conceptual design phase, and are frequently involved during the entire development cycle. Council members, or their representatives, participate in design reviews, change boards and various validation events. Representatives

- Identify external and internal customers
- Establish customer value
- *Listen carefully* for customer priorities
- Be willing to challenge customer's assumptions
- Look for unspoken requirements and future needs
- Understand the operational environment
- Involve customer frequently during design and development

15.8 Practices for involving the customer early and often.

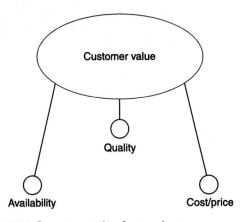

15.9 Customer value for products.

from members' operators, maintainers, and other functional areas work with the engineers, often as part of Integrated Product Teams (see Section 15.5.2). It is very important for the engineer to fully understand how the customer intends to use the product and the operational environment. Effective engagement of the customer is often facilitated by a system representation (Dare, 2003) such as a sketch, mockup, solid model, prototype, user interface, or other visual aid. Such devices, sometimes called boundary objects (Carlile, 2002), greatly improve mutual understanding and communication between the engineer and his or her customers.

15.5.2 Organize for Lean behavior

An organization and organizational mentality aligned with Lean Thinking is essential to achieving Lean Engineering. Since the customer–enterprise–supplier and management–worker relationships are completely different from a traditional organization, this may not be an easy transition. However, no amount of Lean tools or processes can be effective within the wrong organizational structure. Figure 15.10 lists actionable practices characteristic of Lean Engineering organizations. An important one is the role of a chief engineer who has expertise and authority to direct the project *and* make key technical decisions – Kelly Johnson comes to mind in aerospace. Yet this practice is taken (almost verbatim) from Morgan and Liker's study of Toyota, which states: 'The unique role of Toyota's chief engineer is to be the glue that holds the whole PD system together.' This does not mean the chief engineer makes all the decisions. Many are made by Integrated Product Teams (IPTs). But the chief engineer assures the technical integrity of the product development as well as aligning resource expenditures with priorities. Such a person may be called a program manager (e.g. Section 15.6.2) or other title, but the important point is that a single person is responsible for both technical and key resource decisions.

The need to balance functional area expertise with cross-functional integration within a large project is best addressed with IPTs as illustrated in Fig. 15.11. Other

- Have chief engineer integrate development from start to finish
- Identify and involve all relevant stakeholders
- Use co-located IPTs with capable leaders to balance functional expertise with cross-functional integration
- Make decisions at lowest appropriate level with flowdown of responsibility, accountability and authority (RAA)
- Integrate suppliers early into design and development
- Promote teamwork at all levels
- Overcome narrow specialization

15.10 Practices for organizing for lean behavior.

Published by Woodhead Publishing Limited, 2012

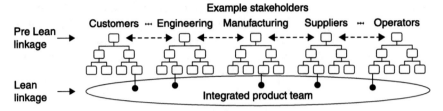

15.11 Stakeholders and integrated product teams. (Source: LAI, 2008.)

names may be used for these teams – the 777 used Design Build Teams and the Citation X used Integrated Design Teams – but the structure and functioning of the teams are largely the same. Team members include the relevant stakeholder groups that affect all aspects of the product design. There will be core team members who devote a major amount of time, and other members who participate on a regular but less frequent basis. Figure 15.11 indicates that the interactions between the functional groups take place at the team level, rather than at the management level. For effective decision making, there needs to be a flow down of responsibility, authority and accountability (RAA) to the teams from the chief engineer. In a large project with many IPTs, integrative mechanisms (Browning, 1997) are needed to keep the teams synchronized. For smaller projects such as the HondaJet (Section 15.6.1), a formal IPT structure is not needed as the whole project can function as one multifunctional team.

Another key Lean Engineering practice is to involve the suppliers early in design and development. An outstanding example of this outside the aircraft domain is the Joint Direct Attack Munition (JDAM), which involved suppliers and the customer as early as the proposal phase. This resulted in the suppliers' knowledge influencing the system architecture, and an impressive reduction in unit costs (Murman *et al.*, 2002).

A Lean Engineering environment is a team environment, and teamwork skills must be emphasized at all levels from hiring to rewarding. Engineers who lack the ability to work effectively in teams are basically unemployable in a Lean Engineering organization. The current emphasis on teaching teamwork skills in college is essential to educating engineers for the future workforce.

Perhaps most the difficult practice for engineers is the need to avoid narrow specialization. This does not mean engineers should be shallow (see Section 15.5.4), but it means they must not have a narrow 'job description'. They should look upon job rotations and learning new technical skills as opportunities for productive careers, rather than deflections from comfort zones.

15.5.3 Strive for perfect coordination

A Lean organization functions as an integrated whole, not a collection of disconnected departments. To achieve this state, it needs to excel in coordination and communication. Engineering is not unique in this respect, as other work areas deal with task interdependence, uncertainty, and time constraints – factors that also affect the departure operations of airlines. In *The Southwest Airlines Way*, Gittell (2003) presents findings from studying the departure operations of four airlines. Data for the airlines displayed in Fig. 15.12 show a high correlation between performance and coordination. Gittell's findings on organizational practices are in near-perfect agreement with observations from aircraft programs studied by Stanke (2001). Key practices for achieving coordination are listed in Fig. 15.13.

Working towards a common objective greatly enables coordination, and may be essential. For Southwest departures, the shared goals are: safety, on-time departure, and customer satisfaction. For the F/18-E/F, it was 'the airplane is the boss', while for the 777-300ER it was 'fly more people farther'. Having all

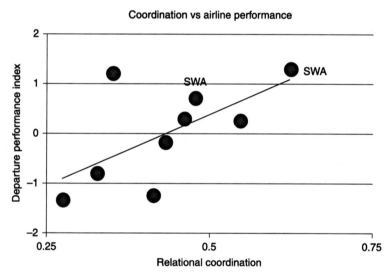

15.12 Departure performance versus relational coordination for four airlines. (Source: Gittell, 2003.)

- Align organization with shared goals
- Develop relationships based on mutual trust and respect
- Make processes and activities visible to all stakeholders
- Practice open, honest, frequent and timely communication
- Use simple, efficient communication techniques
- Hire and train employees for communication and coordination

15.13 Practices for achieving perfect coordination.

employees working towards a common goal overcomes suboptimization at the individual and group level. An absolutely essential practice is to develop mutual trust and respect among all stakeholders – easy to say, but takes time and excellent leadership to achieve. It is hard to be coordinated if individuals do not know what other members of the team are doing. Sharing knowledge, developing process maps, and other information exchange mechanisms can enable this. Most important is effective communication, as listed in the fourth practice in Fig. 15.13. Southwest uses problems as opportunities for communication to develop mutual respect and trust and shared knowledge. Rather than suppressing problems or looking for blame, employees use them to improve.

Gittell groups the first four practices shown in Fig. 15.13 under the name 'Relational Coordination'. Added to these is one from Toyota (Morgan and Liker, 2006) to use simple communication methods, particularly the A3 sheet. At Toyota, information is communicated on one A3-size piece of paper using a standard layout. Other simple techniques such as stop-light charts and wall charts are common in Lean aerospace groups (Stanke, 2001). Communication and coordination – in fact teamwork in general – is so important that it is an important discriminator in hiring and is supported by training (Leopold, 2004).

15.5.4 Pursue excellence and continuous improvement

The final fundamental value principle listed in Fig. 15.2 – Pursue Perfection – reflects the Lean Thinking mindset that any individual or organization is on a continual journey of improvement. Practices shown in Fig. 15.14 reflect the multiple ways that engineers can pursue excellence and continuous improvement. Aerospace success demands technical excellence, and Lean aerospace engineering organizations should expect and support the technical competence of their engineers. Rockwell Collins supports the technical competence of its engineers with a series of internal Communities of Practice that meet regularly to share technical knowledge. Many companies have a technical fellows program to emphasize the importance of technical competence. Selection for a corporate technical fellow position is both rigorous and prestigious. Involvement with

- Develop towering technical competence in all engineers
- Get hands-on experience. Go see for yourself – *Genchi genbutsu*
- Standardize what can be to free resources for innovation
- Share lessons learned across programs
- Challenge all assumptions
- Treat failure as an opportunity for learning
- Continuously improve process and tool capability

15.14 Practices for pursuing excellence and continuous improvement.

professional societies and academic experts, as well as effective technical mentoring, are characteristics of engineers in Lean Thinking organizations.

But excellence is more than purely technical. It also involves engineers getting meaningful exposure to the practice of engineering. This involves engineers 'getting their hands dirty' by actually doing non-technical tasks. Toyota starts its new engineers in sales, working directly with customers. They then spend time in manufacturing, on the factory floor. Finally they are ready for design engineering. Throughout their careers they practice *genchi genbutsu* – going and seeing themselves. Cessna subsidizes flight training for all of its employees so that they may better understand the operational aspects of their products (Denis *et al.*, 2003). A noteworthy example of *genchi genbutsu* is The Boeing Welliver Faculty Fellowship Program (http://www.boeing.com/educationrelations/ facultyfellowship/index.html). Motivated by a concern that engineering faculty lacked any knowledge of actual engineering and accompanying business practices, in the mid-1990s Bert Welliver convened a group of academics to suggest to the Boeing Company ways in which they could help change this situation. The author was fortunate to be among the group that proposed what soon became the Welliver program after Bert passed away. Over 125 faculty have spent eight-week summer internships 'looking over the shoulders' of Boeing engineers, returning to campus better prepared to teach and research in their specialties. Disappointingly, not another aerospace company has followed this exemplary effort, as the return on their investment seems too distant to justify the cost.

Other ways to pursue excellence and continuous improvement are to avoid needless 'reinvention of the wheel' by standardizing engineering tasks whenever possible and learning from the lessons of the past. Engineers have a strong tendency to invent rather than adopt. The Iridium satellite program made rapid progress, partly because they 'developed only what needed developing, invented only what needed inventing, and created miracles only where miracles were necessary' (Leopold, 2004). One of the payoffs of Rockwell Collins' Lean Engineering efforts has been to free up resources so engineers can spend more time on innovation (Jones, 2006). Other practices given in Fig 15.14 represent characteristics of learning organizations and should be self-explanatory.

15.5.5 Implement Integrated Product and Process Development (IPPD)

IPPD has become the norm for Lean Engineering. Built around a series of practices shown in Fig. 15.15, IPPD enables an integrated approach to product development that assures the engineers' output is implementable and will satisfy the customer's needs. It is interrelated and dependent upon other Lean Engineering framework practices – something that can generally be said for all the practices. One example of the impact of IPPD on design is shown in Fig. 15.16, where 'before' and 'after' levels of drawing changes are shown for large aircraft structures for three different

- Adopt systems engineering tailored to program
- Form Integrated Product Teams (IPTs) involving all relevant stakeholders, including suppliers and regulatory representatives
- Assure continuity of IPTs during transitions in project phases
- Use integrated design tools *adapted* to fit project and people
- Exploit rapid and/or virtual prototyping to maximize learning
- Track Key Performance Parameter metrics during design, development, production and sustainment

15.15 Practices for implementing IPPD.

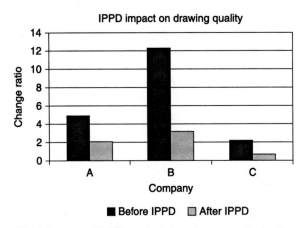

15.16 Impact of IPPD on drawing change ratio for three airframe companies. (Source: Hernandez, 1995.)

manufacturers. Change ratio is the average number of changes made to each drawing after it is released from design. Such changes result in expensive rework and time delays. The waste eliminated by IPPD is clear from these data.

Systems Engineering (Haskins, 2006; Jackson 1997) is the way to translate customer needs into project requirements and engineering specifications, allocate them to subsystems and components, and later verify and validate that the engineers' designs meet requirements and the customer's needs. For large aerospace engineering projects a formal method is required, while for smaller efforts a less formal process is appropriate. An International Council on Systems Engineering (INCOSE) Lean systems engineering working group has recently applied Lean Thinking to systems engineering by developing a set of Lean Enablers for Systems Engineering (Oppenheim *et al.*, 2011).

Adding to the IPT practices introduced in Section 15.5.2 is the need to maintain continuity of team membership during transitions from design to manufacturing to delivery to sustainment. Although leadership will change, loss of key members during these transitions will result in loss of tacit knowledge and result in rework.

Rapid learning can greatly enable implementable solutions that meet the customer's expectations. Engineers tend to want to work problems to perfection before releasing the results. The trouble is that sometimes they are working the wrong problem to perfection and wasting resources, and meanwhile they may be delaying other important tasks. Opportunities to try prototypes or other mockups should be exploited for rapid learning. Both physical and virtual mediums are needed, each offering different advantages. Finally, engineers need to be focused on the key performance metrics, and their acceptable ranges, to assure their design meets the project specifications.

15.5.6 Design for lifecycle value

Enabled by IPPD and IPTs, designing for lifecycle value assures that the engineer's creations will meet both enterprise and customer value expectations. Aircraft have long lifecycles, and lifecycle costs are largely realized in later phases than engineering. Yet engineers are responsible for making decisions that by and large determine those costs. They do this through the choice of materials, subsystems, reuse of existing technology, commonality with other systems, ease of production and maintenance, and other factors. It is generally agreed that 80% of the lifecycle costs are determined by the end of detailed design (Fabrycky and Blanchard, 1991). Therefore early decisions by the engineer have a large impact. Figure 15.17 lists Lean practices for designing for lifecycle value.

Investment in engineering during the early design phase allows trade studies when maximum opportunity exists for selection of the product design. Toyota uses a process called Set Based Concurrent Engineering (Morgan and Liker, 2006) to keep the design options open as far as possible into the development cycle. Multiple design solutions are carried in parallel, one or more of which are known to work. Others may have uncertainty, which needs further work to resolve. Aerospace has a heritage of going for 'breakthrough' technology, but in a Lean Engineering paradigm its development should be taken off the critical path of

- Front load the process while maximum flexibility exists
- Use system engineering requirements flowdown to Design for X (DFX), where X includes safety, quality, environment, production, testing, reliability, maintainability, human factors, operability, support, disposal and lifecycle costs (LCC)
- Exploit commonality and design reuse
- Design in supply chain strategy
- Build in flexibility for upgrade and change during product lifecycle

15.17 Practices for designing for lifecycle value.

Published by Woodhead Publishing Limited, 2012

product development to be funded and organized as advanced research and development (see Section 15.5.9).

Effective DFX requires getting experienced people from the various X areas involved in the design through IPTs. Any one engineer simply does not have the knowledge of so many areas. Table 15.3 illustrates the benefits of designing for production. A rather new area in aerospace is including supply chain architecture in the design of the product, what Fine (1998) calls '3D IPPD'. As more and more of the design is outsourced, this becomes more and more important. With such long lifecycles, designing for future upgrades and/or derivatives also needs to be factored in.

15.5.7 Assure smooth flow of information

Engineering is very much about processing information – lots of it. Therefore, smooth flow of information is essential to efficient engineering. Often this is not the case and the engineer is bombarded with, or creates, too much information, or might be starved for information. Getting the right amount of information at the right time is needed for Lean Engineering.

Figure 15.18 shows the results of a survey of 25 companies that were asked 'what was the most frequent source of information waste in their organization for the product development process' (Slack, 1998). Almost half of the replies fell into two categories: waiting for information and over-processing information. On the one hand, information was not available when it was needed and waste was generated as people waited to complete their tasks. On the other hand, the generator of the information was spending too much time over-processing – getting more accuracy than needed or non-valued-added formatting or whatever.

Figure 15.19 lists five practices to enable the smooth flow of information. These largely self-explanatory practices do not fully address *how* to reduce the information wastes shown in Fig. 15.18. As with other Lean Engineering practices,

Table 15.3 Example of benefits of designing for reducing manufacturing variability – floor beams for commercial aircraft

	B-747	B-777
Assembly strategy	Tooling	Toolless
Hard tools	28	0
Soft tools	2/part #	1/part #
Major assembly steps	10	5
Assembly hours	100%	47%
Process capability	$C_{pk} < 1$ (3.0 σ)	$C_{pk} > 1.5$ (4.5 σ)
Number of shims	18	0

(Source: Koonmen, 1994; Hopps, 1995.)

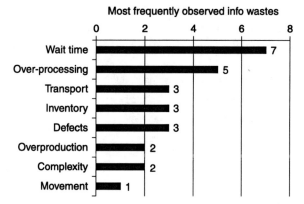

15.18 Most frequent sources of product development information waste for 25 organizations. (Source: Slack, 1998.)

- Align organization with simple visual communication
- Use intranets for ubiquitous access to data and documents
- Employ common or interoperable databases
- Minimize documentation: assuring traceability and visibility
- Co-locate physically or virtually

15.19 Practices for assuring smooth flow of information.

they are interdependent with others; in this case, for example, with optimizing process flow, employing Lean process improvement methods, and striving for perfect coordination.

15.5.8 Optimize process flow

Just as smooth flow of information enables Lean Engineering, so does smooth flow of tasks or activities. As noted above, these are quite interrelated. Figure 15.20 lists practices to help achieve this, all of which flow from the five Lean Thinking fundamentals in Fig. 15.2. The objective is to eliminate as many non-valued added tasks as possible using value stream mapping and analysis techniques, and then synchronize the valued added tasks by eliminating wait time, poor communication, unnecessary iterations, and so forth. For a small engineering project, the team leader might informally handle flow optimization. But for a large project, involving many teams and external partners, this is far from an easy task. Tools (see Section 15.5.11) are needed. Oppenheim (2004) illustrates how to achieve this for an engineering undertaking that is well developed, stable and proven.

- Use customer-defined value to separate value-added from waste
- Eliminate unnecessary tasks/streamline remaining tasks
- Synchronize flow with integration events (stand up meetings, virtual reality reviews, design reviews, etc.)
- Minimize handoffs to shorten cycle time and avoid rework
- Maximize horizontal (value stream) flow, minimize vertical flow
- Iterate early – minimize 'downstream' iterations

15.20 Practices for optimizing process flow.

15.5.9 Manage all risks

Risk reduction through active management is very important to Lean Engineering, and Fig. 15.21 shows practices to accomplish this. Not taking unnecessary risks is a good place to start. One of the most important is to take unproven technology off the critical path of product development and manage it as research and development. By and large, this is well done in commercial aircraft development, and often poorly done in military procurements, leading to program stretch out (e.g. F-22 Raptor) or cancellation (e.g. A-12 Avenger II). Another is to keep things simple, in particular looking for simple solutions. Aerospace engineers are often enamored with the elegant. While personally satisfying, elegance can sometimes have unintended negative consequences.

Lean engineering programs consistently (i.e. across the whole program and for its duration) take a risk management approach that involves: identification of risks, assessment of their estimated likelihood and consequences, plans for mitigation, and tracking to assure resolution. Figure 15.22 illustrates a way to manage risks that is frequently seen in large aircraft programs (Stanke, 2001). Although engineers might like a more quantitative method, this relatively simple approach enables good visibility and tracking of hundreds of risks. The approach is good for handling 'known unknowns', or those risk areas that can be identified. Also needed is a way to handle risks from 'unknown unknowns', or those risk

- Take unproven technology off the critical path – separate research from design and development
- Look for simple solutions – avoid unnecessary elegance
- Consistently use a tailored risk management approach for identification, assessment, mitigation and tracking
- Keep management reserves at program level
- Balance stability and adaptability

15.21 Practices for managing all risks.

Published by Woodhead Publishing Limited, 2012

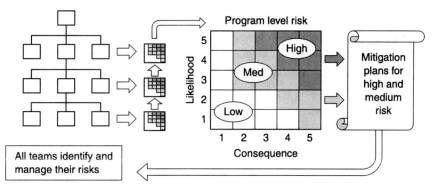

15.22 Illustration of risk management approach. (Source: Haggerty, 2004.)

areas that foresight does not identify. The best approach is to retain management reserves (funds, performance margins, human or test resources) at the program management level (Leopold, 2004).

Another area that can introduce risk is not being flexible enough, or being too flexible, to changing environmental factors. Experience is needed to make the right decisions here – one reason for having a chief engineer (Section 15.5.2).

The final two tool-oriented Lean Engineering practices draw directly on a number of the supporting practices listed above. For example, interoperable databases in Section 15.5.7 are a Lean practice that applies to integrated engineering tools. The practice introduced above to look for simple solutions and avoid unnecessary elegance applies to all the tools in Section 15.5.10 and 15.5.11. With this observation, the final two Lean Engineering practices are supported by lists of tools rather than lists of additional practices.

15.5.10 Utilize integrated engineering tools

Sophisticated design tools based upon solid science and mathematics are a hallmark of aerospace engineering, and their evolution from basic research to application is a marvelous success story of the international aerospace community. Figure 15.23 lists some of the tools found in Lean Engineering organizations. The list is likely to be incomplete, but representative of current practice. Many of these tools may be familiar to aerospace engineers and only a brief summary of each is offered.

- Preliminary design methods such as multi-attribute trade space exploration and integrated concurrent engineering allow rapid exploration of possible designs and frequent interaction with the customer and other stakeholders (McManus *et al.*, 2007).

- Preliminary design methods supporting uncertainty in specifications
- Design for Six Sigma methods, including variability reduction and key characteristics
- Integrated software/hardware design tools
- Multidisciplinary design and optimization for fluids, structures, controls, dynamics, etc.
- Design for manufacturing and assembly (DFMA)
- Product lifecycle management (PLM) tools
- Production and system simulations
- Verification and validation mockups, prototypes, tests

15.23 Engineering tools for lean engineering.

- Design for Six Sigma methods encompasses a set of tools and strategies to design for quality, starting with Voice of the Customer methods through to detailed design (Egbert *et al.*, 2008).
- Software design tools with automatic code generation, e-cad tools, model-based engineering approaches and others are needed for the increasingly information-intensive aerospace products (Lempia, 2008).
- Computational design and simulation of fluids, structures, controls and other disciplines in a coupled, optimizing approach.
- Design for manufacture and assembly (DFMA) tools to reduce part count, manage tolerance build-up, etc.
- PLM for configuration control and management.
- Digital pre-assembly has replaced physical engineering mockups to ensure that parts fit and can be assembled (Haggerty, 2002).
- Rapid and/or partial physical mock ups and prototypes help engineers learn and avoid downstream iterations and rework (Ulrich and Eppinger, 2008). Testing for risk reduction and verification is needed.

A Lean Engineering organization requires more than a chest of stand-alone sophisticated tools. These tools need to be integrated into the organization, into the time constraints of the design cycle, and into the information technology infrastructure. Since the Lean Engineering organization is closely engaged with partners, the tools have to support both local and virtual collaboration. This means common databases, user interfaces, and versions. Beyond integration comes standardization, already mentioned. Unrealistic resources will be needed to maintain and support users unless standardization is continually pursued. Another Lean practice is continuous improvement of tool capability (Section 15.5.4). An excellent example of this is the development of aircraft modeling tools by SAAB (Fredricksson, 1994). An aircraft simulation model is continuously updated and improved as new test and operational information is obtained.

Figure 15.24 illustrates the impact DFMA can have on downstream production and lifecycle costs through the reduction of part count. Each part represents an

F/A-18 E/F is 25%
larger but has 42%
fewer parts than C/D
due to DFMA

Total parts	
C/D	E/F
14104	8099

15.24 Impact of design for manufacture and assembly on F/A-E/F part count. (Source: LAI, 2008.)

accumulated cost to design, manufacture, maintain and stock spares, etc. The major reduction in part count for the F/A-18E/F contributed towards meeting its development and operational cost goals.

15.5.11 Employ Lean process improvement methods

Hopefully the reader grasps that Lean is a way of thinking, not just a set of tools. Nevertheless, there are Lean tools for process improvement. Many of them are applicable to engineering. Figure 15.25 lists some that have been observed in case studies. Since these may be less familiar to the reader than engineering tools, they are briefly described.

- Value stream mapping and analysis is a quantitative visual tool combining process mapping with process data. It is used to capture an existing process, identify sources of waste, analyze opportunities for improvement, and develop a future state process map. Engineering of aerospace products can involve hundreds or thousands of people and cut across multiple organizations. Often an engineer will simply not understand the entire process of which their activity is but one part. Developing a VSM is one way to make this visible (McManus, 2005).
- Rapid process improvement (RPI) events are focused action workouts of one to five days where workers tackle a process improvement opportunity in a data-driven approach to develop, and maybe even partially implement, process improvements. They are an instance of *Kaizen*, the Japanese word for continuous improvement (Imai, 1986).
- Sort, straighten, shine, standardize, sustain (5S) is a process to organize the workplace to avoid time wasted in locating documents, tools, and other materials needed to complete a job. Simple in concept, it is often the starting point for a Lean implementation and can have rapid payoffs (Murman *et al.*, 2002).
- Root cause analysis/corrective action is the strategy for making sustained improvements. A number of tools and techniques are used, including cause and effect diagrams and the 'five whys' – asking 'why' five times before being satisfied the root cause has been found (LAI, 2008).

- Value stream mapping and analysis
- Rapid process improvement events
- 5S
- Root cause analysis/corrective action
- Pareto and PICK charts
- Visual displays in a 'big room' – *Obeya*
- Fever charts

15.25 Lean tools for process improvement.

- Pareto charts (e.g. Fig 15.18) are one tool used for root cause analysis and other purposes. PICK are four-quadrant charts with axes of Difficulty and Impact, on which candidate improvements are located by an improvement team to decide which will give the most payoffs (LAI, 2008). These simple ways of displaying data help focus on the most important things to address.
- Visual displays of value stream maps, Pareto charts, schedules, and other material in a room where everyone can easily see the status of ongoing events is an effective way to communicate (Morgan and Liker, 2006).
- Fever charts come from the Theory of Constraints (Goldratt, 1997) and show progress relative to plans for critical path activities. Those that get behind schedule enter the red zone indicating the 'temperature' is rising.

There are many other Lean tools, some of which may be very useful for implementing Lean Engineering. The above list is by no means exhaustive.

An example of Lean Engineering implementation using such tools is shown in Table 15.4 for process improvement in Class II Engineering Change Proposals in the F-16 program (Goodman, 2000). Support of existing product lines consumes considerable engineering resources, and process improvements can free up those resources for more productive uses. In this situation, a relatively straightforward application of Lean tools led to dramatic reduction in the time needed to process the engineering change.

15.5.12 Lean Engineering framework wrap up

The above framework is a starting point for codifying the principles and practices of Lean Engineering in an aerospace context. More detailed supporting practices need to be listed for practitioner use, and testing the framework would undoubtedly lead to further improvements. Much of the content is not new or revolutionary. Indeed, comparison of practices like that with nine aircraft programs ranging from the B-52 to the B-777 revealed many of these practices are tried and true (Haggerty and Murman, 2006). Yet few programs apply them consistently and therefore most miss achieving their full potential. One beneficial use of the framework is to provide a checklist for benchmarking internal practices. Since the practices are

Table 15.4 Process improvements in Class II engineering change proposals in the F-16 program after implementing Lean Thinking. Results are for processing 849 build-to-packages

Category	Reduction (%)
Cycle time	75
Process steps	40
Number of handoffs	75
Travel distance	90

(Source: Goodman, 2000.)

interdependent, all need to be considered to some degree. This topic is addressed in the next section.

15.6   Tailoring Lean Engineering

A starting point for implementing the Lean Engineering framework is to apply Lean Thinking principles to identify which value-added practices of the framework to adopt for a particular application. Certainly one size does not fill all needs. Each organization has its current 'as is' starting point and its own culture. Beyond that, not all engineering undertakings are the same. Therefore, the Lean Engineering framework needs to be tailored to fit the engineering activity and group at hand. This section addresses several different engineering applications and illustrates the application of Lean Engineering by an example for each.

One might think of categorizing aerospace engineering as shown in the left-hand column of Table 15.5. The right-hand column suggests some Lean Engineering tailoring strategies for each. At one end of the engineering spectrum is an advanced research and development undertaking where teams are relatively small, technology is relatively immature and risk is high, but payoff is equally high. Processes may or may not be well established, and people may or may not have years of experience. Such programs are best executed by small, hand-picked groups in an environment protected from a management structure designed for large programs – also known as a Skunk Works® type organization (Rich and Janos, 1994). The HondaJet is chosen as an example of how the Lean Engineering framework applies to this category.

At the other end of the spectrum lie much more stable, repetitive engineering activities in product and engineering support, or incremental product upgrades. Engineering activity flow can often be non-iterative, quite standardized, and low-risk. Lean Thinking techniques familiar in manufacturing are more applicable here. One example is already given in Table 15.4. Below we consider ground testing of engines at Arnold Engineering Development Center (AEDC).

Published by Woodhead Publishing Limited, 2012

Table 15.5 Examples of tailoring Lean Engineering

Program type	Possible Lean Engineering approach
Advanced research and development X-vehicle Prototype	• Small, focused, co-located team in 'protected' environment – also known as a Skunk Works® type organization • Rapid design-build–fly cycles for learning, risk reduction, tool calibration, and lifecycle experience
New product development Major upgrade Derivative	• Direct involvement of customer throughout design • Strong focus on lifecycle value and IPPD with integrated digital and Product Lifecycle Management tools • Utilize 'lessons learned' from past programs • Avoid unneeded reinvention, risky technology, unproven tools and unnecessarily elegant solutions
Engineering testing Product support Small upgrades	• More standardized tasks and lower engineering risk allows direct adoption of many lean practices and tools used in manufacturing • Continuous improvement through value stream mapping and analysis

Skunk Works® is a registered trademark of the Lockheed Martin Corporation.

Between these two extremes we find major new product development or upgrades or derivatives. These are characterized by large teams expending large resources, and in the commercial area, often 'betting the company'. Programs like the B787 and A380 come to mind, but are too early in their lifecycle to use as examples. The Citation X program is used to illustrate a mapping of the Lean Engineering framework.

15.6.1 HondaJet – Lean Engineering applied to advanced research and development

The HondaJet is classified as an advanced Very Light Jet, being at the high end for speed and cabin volume of the just-emerging product lines for this category. Developed by the Honda Aircraft Company arm of the Japanese automaker, the approach is rooted in a culture of Lean Thinking. The proof-of-concept program got its official launch in December 1997 after a market study and customer engagement spanning over a year. Those studies identified a candidate configuration with several innovative features: natural laminar flow (NLF) wing, NLF fuselage, and an over-the-wing engine mount configuration. The first two years of the proof-of-concept program were devoted to technology risk reduction with simulations, ground tests, and further design. Go-ahead was given in late 1999 to build a prototype for the four/six-passenger jet. First flight occurred in December 2003 (Fujino, 2004).

Table 15.6 lists Lean Engineering-related practices used at Honda Aircraft Company (Warwick, 2007; personal communication 30 November 2007). Honda used a strong chief engineer model in the spirit of Kelly Johnson. Michimasa Fujino directed all aspects of the program, from pre-launch customer interaction and market assessment through to the current product development phase. He was responsible for technical integration and many technical developments and decisions. A relatively small hand-picked team of engineers and technicians did all the work, from early engineering analysis and risk reduction through prototype fabrication and flight testing. They operated as a single team reporting to the chief engineer. Communication was open, point-to-point and there were virtually no meetings. Timely decisions were made and communicated directly to whoever needed to know. They all shared a large room with no dividers, enabling direct communication.

Customer needs drove the engineering design, and only needed technology was developed by the team. Paramount was the desire for a large cabin and good fuel efficiency and range. These drove the unconventional engine placement and Natural Laminar Flow (NLF) surfaces. Advanced engineering design tools were used, but chosen to match the state of development. For example, Computational Fluid Dynamics (CFD) was used for final design, but more parametric approaches were used for initial design.

Formal Lean tools were not used, but Lean Thinking was intuitive to the team. Mr Fujino insisted on Three Reality Principles: go to the real spot, understand the real situation, make the decision based on the reality – what the Japanese call *genchi genbutsu*.

Table 15.6 Lean Engineering practices and tools used during HondaJet proof-of-concept phase

People	Processes	Tools
• Customer engaged one year before program launch • Chief engineer driven • Small co-located team: 25 engineers, ten technicians • Flat organization • Engineers did design, production, testing • *Genchi genbutsu* • Vision aligned the team	• Customer needs drove technical design • Design for 'X' (DFX) for choices aircraft 'will live with' • IPPD • New technology only where needed • Co-location, no walls • Rapid communication and decisions, no meetings • Two-year risk reduction study	• Rapid simulation tools for early studies • Rapid prototyping wind tunnel models • Simple/partial mockups for engineer learning • State of art computational simulation and testing • *Obeya* – big room

(Source: Warwick, 2007, personal communication 30 November 2007.)

Published by Woodhead Publishing Limited, 2012

The successful proof-of-concept phase led to a product launch announced in July 2006 at the Oshkosh show. At $3.65 million the HondaJet is priced above other Very Light Jets, but below business jets with the same cabin volume (Warwick, 2007). Over 100 orders have been placed for this innovative new aircraft product.

15.6.2 Citation X – Lean Engineering applied to new product development

The fastest civilian jet currently in operation, the Cessna Citation X was developed in the early 1990s at the same time that Boeing was developing the 777. Both programs utilized Total Quality Management (TQM) principles that have now been subsumed under the broader Lean Thinking title. The Citation X design incorporates 5 to1 bypass turbofan engines with low noise and fuel burn, a highly swept 37° wing, and glass cockpit. Falling in the medium-to-large business jet category, it can carry 8–12 passengers transcontinental in the US and make some city pair connections across the North Atlantic.

Table 15.7 displays some of the Lean Engineering practices utilized in the Citation X program (Denis *et al.*, 2003; personal communication 26 December 2007). New for Cessna was the formation and active involvement of the Customer Advisory Council comprising key customer and supplier representatives. Their input to and review of program decisions and designs had a large impact on the engineering effort. Much like the HondaJet, a single program manager directed all top-level program and technical decisions and coordination.

Table 15.7 Lean Engineering practices for the Citation X program

People	Processes	Tools

• Proactive customer advisory council
• Program manager as chief engineer
  – On the floor during build
• Integrated Design Teams
  – Co-located near hardware
  – Flowdown of RAA
• *Genchi genbutsu*
  – Engineers can install their parts on prototype
• New employee mentoring
	• Designed for:
	  – Safety
	  – Quality
	  – Manufacture and Assembly
	  – Maintenance hours
	• Open, frequent meetings
	• Suppliers as partners
	• Risk reduction:
	  – Prototype
	  – Design for simplicity
	  – Test for verification
	  – Limit percentage of new employees
		• CATIA (see http://www.3ds.com/products/catia) and CFD
		• Prototype for learning
		• Common databases with good IT support
		• Lessons learned widely available
		• Simple process based management tools
		• Schedule used to synchronize flow

(Source: Denis *et al.*, 2003, personal communication 26 December 2007.)

The program manager reported to the Cessna senior management. Differing from the HondaJet in scale with 500 engineers, Integrated Design Teams (IDTs) were formed and given full decision responsibility, accountability and authority. Briefings to senior management were given by team members, not by program management. Both the program management and IDTs practiced *genchi genbutsu*. During the prototype build phase, the program manager's desk was located on the shop floor and occupied for all three shifts by him or his direct designates. Program engineers were expected to spend significant time on the floor, and often oversaw or installed their designed parts. All engineering employees were given 40 hours of TQM training through Wichita State University. All of this led to a 'hands-on' culture.

Design for lifecycle value placed major emphasis on safety, quality, manufacturing and assembly, and reliability. An indicator of mutual trust and respect was that any engineer was welcome at any meeting, whether invited or not. Key suppliers were treated as partners and expected to have direct contact with the customers. A formal risk reduction program was not practiced, but risk reduction was very much a part of the program plan, as noted in Table 15.7. One interesting feature was to limit the number of new engineering employees on the program to 25%, and each was aligned with a senior, experienced mentor. Engineering tools that were used represented the state of the art. No Lean tools *per se* were adopted in engineering, although they were in manufacturing.

The Citation X product has been very successful, with over 260 units delivered and one million flying hours achieved without a serious accident. In March 2007, Cessna announced a 65% increase in production to meet the market demand in 2006–2010. Another important outcome of the program was a cadre of satisfied and capable engineers ready to take on the next aircraft development project. The Citation X won the 1997 Collier Trophy.

15.6.3 Engine testing at AEDC – Lean Engineering applied to engineering support

The last example of tailoring Lean Engineering is in the testing area, an engineering activity that consumes considerable resources that, if freed up, could be invested in new product development. As part of the LAI effort, AEDC undertook the application of Lean Enterprise principles to improve the testing of full-scale engines in its Aeropropulsion Systems Test Facility (ASTF) (Smith and Kraft, 2005). After considering several areas for a Lean improvement effort, the first target was the ASTF Cell-2 installation procedure for large commercial turbofans, motivated by pressing demands to test the Rolls Royce Trent 900 and Engine Alliance GP7200 engines for the A380. The as-is installation was projected for 74 days. At a cost of $10–20 million per month, there were opportunities for significant savings as well as meeting pressing testing schedules.

Lean Engineering practices and tools used in the effort are displayed in Fig. 15.26. In line with Table 15.5, there is a heavy emphasis on Lean tools and value

- Voice of the customer
- Integrated product team
- Rapid process improvement events
- Value stream mapping and analysis
- Pareto and PICK charts
- Critical path analysis
- *Obeya* – visual displays
- Standard work

15.26 Lean Engineering practices used for engine testing lean improvement. (Source: Smith and Kraft, 2005.)

stream mapping. Payoff from the Lean approach was significant. The engine installation time was reduced from 74 to 36 days. With an investment of $10 000 to achieve this, and monthly costs of about $10–20 million for an engine development, the payoff is significant.

An important finding from the AEDC effort was the challenge of looking across multiple organizations – from USAF Testing and Evaluation labs, to depots, to industry. No one organization controls the funding. This finding leads into the next section of the chapter, on Lean Engineering challenges.

15.7   Lean Engineering challenges

Looking towards the future, there are a number of challenges to reach the full potential that Lean Engineering can offer the aerospace industry. Three are briefly summarized.

15.7.1 Implementation

The principles of Lean Engineering are not complicated. However, its implementation is neither simple nor quick. It requires a different culture from that found in aerospace in the 1970s and 1980s. Add to that the engineer's skepticism that any kind of Lean Thinking could apply to their areas, and one can see the challenges that lie ahead. Beyond that, engineering efforts can span enterprises from suppliers to customers. It takes committed leadership, training and education, and demonstration of successful examples to promulgate changes of this magnitude. Much of this is what engineers call the 'soft stuff'. Chris Cool of Northrop Grumman told an LAI conference, 'The soft stuff is the hard stuff' (Murman *et al.*, 2002). Toyota has taken years to develop its Product Development System and it continues to evolve (Morgan and Liker, 2006). As the examples in this chapter show – and there are many others – it can work, and companies are implementing Lean thinking in engineering.

15.7.2 Body of Knowledge (BoK)

Lean Engineering is based upon observational findings from ongoing implementations, and this BoK is evolving over time. It is a different BoK than engineers are used to generating through research, and absorbing through experience. It is based upon principles rather than laws of science, and expressed as best practices compared with formulae or graphs. Furthermore, it requires holistic thinking rather than the reductionist approach that is more familiar to engineers.

15.7.3 Education

Topics covered in this paper are infrequently, and surely incompletely, addressed in engineering education. Most faculty lack exposure or experience to these topics and are reluctant to replace existing constrained curriculum with topics of Lean Engineering. Yet graduating engineers would be better prepared for their careers with at least an exposure to this material. They will likely face it soon upon employment.

This latter topic is one of current interest to the author and other members of the LAI Educational Network. Collectively they have developed an introductory level LAI Lean Academy® course (LAI, 2008) and taught it to over 1000 students in industry and on campus. The experience has been that engineers relate to and grasp the material, but want to see more engineering-related examples and references. Many EdNet faculty members have integrated parts of this curriculum into their courses in ways that enhance rather than displace existing material (e.g. Eastlake and Attia, 2007).

15.8   Summary

Lean Engineering has emerged in the aerospace industry over the last two decades, starting with the introduction of Total Quality Management in the 1980s. It is not some kind of new-fangled engineering that ignores the laws of nature and mathematics. In fact, in many respects it is a rediscovery of past practice. Nor is Lean Engineering necessarily less engineering. It is the right amount of engineering at the right time to meet the right objectives. Lean Engineering is a smart combination of science-based engineering and process-based engineering built on the foundations of Lean Thinking, which emerged from the Japanese automotive industry. Examples in this paper, and other examples not presented, show that, when Lean Thinking concepts are applied thoughtfully and consistently to engineering, superior product development and happier engineers can be the result. Lean aerospace engineering is excellent aerospace engineering.

15.9   Acknowledgments

This chapter was originally presented as the SAE/AIAA 2008 William Littlewood Memorial Lecture and published as AIAA Paper 2008-4, January 2008. Permission by the AIAA to publish this revised addition is acknowledged and appreciated.

The author has greatly benefited from the association and input of many colleagues over the past ten years and during the preparation of this paper. Allen Haggerty, retired VP of Engineering and Manufacturing with McDonnell Douglas and the Boeing Company, and Senior Lecturer at MIT, has been an enormous inspiration and source of wisdom. Professors Stan Weiss, currently at Stanford, and Bo Oppenheim, of Loyola Marymount University, have provided insight, stimulation, and critical feedback. Drs Alexis Artery of Eclipse Aerospace Inc., Hugh McManus of Metis Design, and Eric Rebenstisch of LAI have been wonderful research colleagues over many years. Ray Leopold from the Iridium Program gave a valuable perspective from space systems. Fred Stahl was patient enough to thoroughly critique the manuscript. Members of the LAI Educational Network, LAI Product Development research community and INCOSE Lean Systems Engineering Working Group have all provided many interesting discussions and thoughts.

15.10   References

Browning TR (1997), 'Exploring integrative mechanisms with a view of design for integration.' *Advances in Concurrent Engineering – CE97*, 83–90.

Carlile PR (2002), 'A pragmatic view of knowledge and boundaries.' *Organization Science*, 13(4): 442–55.

Christensen C (1997). *The Innovator's Dilemma*. Boston, Harvard Business School Press.

Dare RE (2003). 'Stakeholder collaboration in air force acquisition: Adaptive design using system representations.' PhD Thesis in Technology, Management and Policy, MIT, Cambridge, MA.

Deming WE (1982). *Out of Crisis*. Cambridge, MIT Center for Advanced Engineering Services.

Denis A, Freuler PN, Robinson T, Serrano M and Vatz ME (2003). 'Citation X: A case study.' MIT 16.885 Aircraft Systems Engineering Final Report.

Dertouzous ML, Lester RK and Solow RM (1989). *Made in America*. Cambridge, The MIT Press.

Eastlake C and Attia M (2007). 'Employing Lean Engineering principles as a student exercise to modify the design content of traditional aircraft and propulsion courses.' ASEE Paper AC 2007-268, *2007 ASEE Annual Conference*. Honolulu, HI.

Egbert N, McCoy P, Schwerin D, Jones J and Karl A (2008). 'Design for process excellence – Ensuring timely and cost-effective solutions.' *Proceedings of the 26th International Congress of the Aeronautical Sciences*. Anchorage, AK. Available at http://www.icas. org/Papers-from-previous-congresses.html.

Fabrycky W and Blanchard B (1991). *Life-cycle Cost and Economic Analysis*. Upper Saddle River, Prentice-Hall.

Fine C (1998). *Clockspeed: Winning Industry Control in the Age of Temporary Advantage*. Reading, Perseus Books.

Fredriksson B (1994). 'Holistic system engineering in product development.' *The SAAB-SCANIA GRIFFIN*, November, 23–31.

Fujino M (2004). 'Development of the HondaJet.' *Proceedings of the 2004 International Congress of Aeronautical Sciences*. Yokohama, Japan. Available at http://www.icas.org/Papers-from-previous-congresses.html.

Gittell JH (2003). *The Southwest Airlines Way*. New York, McGraw Hill.

Goldratt EM (1997). *Critical Chain*. Great Barrington, North River Press.

Goodman G (2000). *F-16 build-T- Package Support Center Process*. LAI Product Development Team Presentation, MIT, Cambridge, MA.

Haggerty A (2002). *Lean Engineering has Come of Age*. Minta Martin Lecture, MIT Department of Aeronautics and Astronautics, Cambridge, MA. Available at http://www.icas.org/Papers-from-previous-congresses.html.

Haggerty A (2004). *Risk Management*. 16.885 Aircraft Systems Engineering Lecture, MIT, Cambridge, MA.

Haggerty A and Murman E (2006). 'Evidence of Lean Engineering in aircraft programs.' *Proceedings of the 2006 International Congress on Aeronautical Sciences*. Hamburg, Germany.

Hammer M and Champy J (1993). *Reengineering the Corporation: A Manifesto For Business Revolution*. New York, Harper Business.

Haskins C (ed.) (2006). *Systems Engineering Handbook*, INCOSE-TP-2003-002-03.

Hernandez C (1995). 'Challenges and benefits to the implementation of IPTs on large military procurements.' Master's Thesis in Management, MIT, Cambridge, MA.

Hopps JC (1995). 'Lean Manufacturing practices in the defense aircraft industry.' Master's Thesis, MIT, Cambridge, MA.

Imai M (1986). *Kaizen: The Key to Japan's Competitive Success*. New York, McGraw-Hill.

Jackson S (1997). *Systems Engineering for Commercial Aircraft*. Hants, Ashgate.

Jane's Information Group (1941). *Jane's All The World's Aircraft*. London, Sampson Low, Marston & Co.

Jones CM (2006), *Leading Rockwell Collins Lean transformation*. LAI Keynote Talk, http://mitworld.mit.edu/video/378/.

Koonmen JP (1994). 'Implementing precision assembly techniques in the commercial aircraft industry.' Master's thesis, MIT, Cambridge.

LAI (1996). *Lean enterprise model, Lean Advancement initiative*. Cambridge, MA, MIT.

LAI (2008). *LAI Lean Academy® Curriculum*, Vol. 6. MIT LAI Educational Network. Available at http://ocw.mit.edu/OcwWeb/Aeronautics-and-Astronautics/16-660January--IAP--2008/CourseHome/index.htm.

Lempia D (2008). 'Using Lean Principles and MBE in design and development of avionics equipment at Rockwell Collins.' *Proceedings of the 26th International Congress of the Aeronautical Sciences*, Anchorage, AK. Available at http://www.icas.org/Papers-from-previous-congresses.html.

Leopold R (2004). *The Iridium Story: An Engineer's Eclectic Journey*. Minta Martin Lecture, MIT Department of Aeronautics and Astronautics, Cambridge, MA.

McManus HL (2005). *Product Development Value Stream Mapping Manual*. LAI, Version 1.0. Available at http://lean.mit.edu/.

McManus H, Haggerty A and Murman E (2007). 'Lean Engineering: A framework for doing the right thing right.' *The Aeronautical Journal*, **111**(1116): 104–14.

Mikel H and Schroeder R (2000). *Six Sigma*. New York, Doubleday.

Morgan JM and Liker JK (2006). *The Toyota Product Development System*. New York, Productivity Press.

Murman E, Allen T, Bozdogan K, Cutcher-Gershenfeld J, McManus H *et al.* (2002). *Lean Enterprise Value: Insights from MIT's Lean Aerospace Initiative*. Hampshire, Palgrave.

Oppenheim B (2004). 'Lean product development flow.' *INCOSE J of Systems Engineering*, **7**(4): 352–76.

Oppenheim B, Murman E and Secor D (2011). 'Lean enablers for systems engineering.' *J. of Systems Engineering*, **13**:1.

Rich B and Janos L (1994). *Skunk Works*. Boston, Little, Brown & Co.

Slack RA (1998). 'Application of Lean principles to the military aerospace product development process.' Master's Thesis in Engineering and Management, MIT, Cambridge, MA.

Smith VK and Kraft E (2005). 'A Lean enterprise approach to test and evaluation in turbine engine development and sustainment.' Arnold Engineering and Development Center. Unpublished.

Stanke AK (2001). 'A framework for achieving lifecycle value in product development.' Master's Thesis in Aeronautics and Astronautics, MIT, Cambridge, MA.

Ulrich KT and Eppinger SD (2008). *Product Design and Development*, 4th ed. New York, McGraw-Hill.

Utterback J (1994). *Mastering the Dynamics of Innovation*. Boston, Harvard Business School Press.

Warwick G (2007). 'Opening doors: Carmaker Honda's aircraft research and development facility gears up for the HondaJet.' *Flight International*, December 2007.

Womack J, Jones D and Roos D (1990). *The Machine That Changed the World*. New York, McMillan Publishing Co.

Womack J and Jones D (1996). *Lean Thinking*. New York, Simon and Schuster.

Part IV
Conclusion

16

Conclusion: innovations in aeronautics

T.M. YOUNG, University of Limerick, Ireland, and M. HIRST,
Independent consultant, UK

Abstract: In this chapter the editors review contributor content, and provide an overview of the way that individual contributions integrate. The objective is to provide a perspective over developments that are occurring within the established industry, and the influences that are apparent, emerging, or unrecognised, and stem from developments outside the scope of traditional concerns. The prognosis is that there is plenty of scope for new developments, innovation will be essential to driving forward new solutions, and the old problems, whilst they might be ameliorated, are unlikely to go away. The scope of the solutions expected will simply become more complex with time, and it is important for industry to keep abreast of ways of managing this challenge.

Key words: Innovation and risk, technical readiness, disruptive technologies, computer-assisted engineering, education and innovation.

16.1 Introduction

Our strategy for this book was a simple one: we placed the topic of innovation in aeronautics at the centre of the project and then, with the help of a diverse group of experts, looked at the topic from many different angles. Throughout the months of compilation, the addition of each individual contribution brought new perspectives on, and often profound glimpses of, the matter at the core of the book's topic. It has been a privilege for us to witness this volume coalesce into the unified whole you see before you. We sought the assistance of our chapter authors, as we believed that they had a strong message to present. Many of the contributors have published their own books, and all have presented associated material at international conferences or in technical publications. It is not just in terms of the quality of what they have presented, but equally it is the confidence we have in their personal qualities, that ensures this book holds our confidence.

That is not to say that it has been a steady course. We have reached our goal, almost as an aircrew will reach their destination, by tracking selected waypoints. We could have extended our journey to include other exciting views – as originally planned. It is in this respect that our actions are tempered with regrets. We held onto several scripts submitted long before the publication date (wanting to get them into print without delay), whilst we sought alternative or replacement authorities on other subjects that we wished to explore. We prevaricated, facing a dilemma: should we do it quickly and be happy with the outcome, or take genuine care, delay it, and be even more content with the outcome? Our apologies have been proffered to those we delayed and our immense gratitude extended to those

363

who stepped in at the last minute and submitted contributions within crippling timescales.

16.2   Innovation and risk

Embedded within every chapter there is evidence of tremendous efforts to sustain a high level of interest and, throughout the pages, the authors guide us through topical and sometimes controversial issues. Occasionally there is a startling revelation or observation or a statement that makes one sit back and reflect on an unconventional concept – such as the biologically inspired designs of Bar-Cohen (Chapter 1), the morphing wing shapes described by Guiler and Huebsch (Chapter 2) or the single-crew flight deck suggested for an airliner by Harris (Chapter 7). It is evident that, whilst each author considered the same general topic, each had a different viewpoint, leading naturally to discussions of varying approaches, conclusions and, sometimes, almost irreconcilable differences. This is natural, of course; it also reflects our expressed desire that the authors could contribute freely without an imposed bias or prejudice and that they be allowed to draw their own conclusions.

There is a clear message at the centre of this book: that the aerospace industry values innovation. It would seem, however, that the true value of innovative thinkers is often not appreciated. The dichotomy is that, as they open doors, they introduce risks, and they pose uncomfortable challenges to managers and decision-makers, yet they are key to the development of novel, innovative products. Nurturing innovation in the design and manufacture of safety-critical products, as required for the aerospace industry, is clearly a difficult proposition. A theme that plays out in several chapters is the inevitable link between innovation and risk. Browning (Chapter 14) expresses it as follows:

> 'Uncertainty is at its greatest in large, complex, innovative development projects, such as are common in the aerospace industry. When uncertainty allows potential outcomes that carry a negative consequence or impact to a project, uncertainty causes risk – risks of cost and schedule overruns, and risks of failing to provide desired levels of quality, functionality, or technical performance.'

The adoption by commercial organisations of appropriate techniques for monitoring and managing project costs, schedules, technical performance, uncertainties and risks is thus vital to the process that enables engineers to develop fledgling, innovative ideas into successful products. Ultimately, the risks associated with not meeting project costs or timescales or technical performance targets can be expressed as a commercial risk – that is, a risk that the project will not deliver the envisaged financial return on investment. Engineers are sometimes accused of failing to appreciate the aspect of commercial risk. This is a platitude that has to be dismissed. The absurdity is readily dispelled if one considers that engineers depend on their employer to be successful; and to relegate commercial risk to realms where it can be regarded as the concern of others is failing to act in a responsible manner.

Henke (Chapter 10) develops this link between the incorporation of new, potentially risky, technologies into a product (e.g. an airliner) and the associated financial consequences to the project in an intriguing direction. He explains that standard financial measures of internal rate of return (IRR) or net present value (NPV) can be used as a means to assess the outcome of trade studies that evaluate alternative technical scenarios. This technique of technology assessment has been shown to provide valuable inputs into the decision-making process in the evaluation of alternative design choices during the development of a new aircraft. Trade studies can be exceedingly complex, and the identification and quantification of the repercussion that will arise from a decision to include an innovative technology (such as hybrid laminar flow control, for example) in a new aircraft are particularly difficult. Without appropriate tools for such evaluations, it is apparent that conservative decisions will inevitably dominate, and, consequently, innovative technologies will remain unused.

16.3   Technology readiness levels (TRLs)

Henshaw (Chapter 9) notes that 'many organisations that deal with complex systems – such as aircraft – have adopted the concept of technology readiness levels (TRLs) to characterise the maturity of the innovative system under development'. The TRL framework, which was pioneered by NASA in the late 1980s and 1990s, provides a ranking system for assessing the maturity of technologies – typically, between TRL 1 (basic principle observed) and TRL 9 (actual system flight proven).[1] As the maturity of diverse technologies are evaluated using a common metric, project managers – who, in most cases, would not have the technical expertise to fully appreciate the technical risks – are able to make global project decisions, such as whether or not to retain the particular technology in the project. TRLs thus provide a framework for risk management; however, the technique is not without its limitations – for example, it fails to account for the possible obsolescing of technologies. This can also introduce a project risk of a different type: the risk of backing a mature technology when there are newer technologies on the horizon that could 'leapfrog' it, rendering it obsolete, and in the process giving a possible competitive advantage to another manufacturer. Henshaw warns that 'TRLs are sometimes equated to the innovation process, but they are really related to the product development process and the management of risk, and this does not map exactly to innovation'.

The TRL framework has reinforced the idea that research and development can – or even should – be stratified, with categories ranging from 'blue sky' to applied research. The cerebral breakthroughs are the most significant in terms of innovation and might be regarded, therefore, as blue sky. The process of developing mature applications that implement novel ideas is applied research. 'De-risking' of projects is achieved by moving the technically challenging aspects out of the project, usually by insisting that research and development precedes the project. Indeed, it is not uncommon for project managers only to consider technologies at TRL 6 for inclusion in new projects. Although this stratification provides a useful means to describe research and development activities in an organisation, it can – if poorly implemented – impede the adoption of novel technologies. This can happen when a stratified research and development model is imposed on an organisational structure, for example, and innovators are required – often with reservations – to pass the project, with their fledgling ideas, to people operating at the next level in the development process. Several of the contributors to this book describe the almost universal desire of innovators to see their ideas succeed – consequently, company policies that do not accommodate innovators fail to recognise (and exploit) this aspect of human nature.

16.4   Capturing innovation and disruptive technologies

Research and development generates intellectual property, which can be held as a trade secret or patented. McCarville (Chapter 13) describes how the patent process can be used as a means of 'capturing' innovation to provide a competitive advantage for the inventor. He states that 'the ability to innovate and capture competitive advantage by establishing intellectual property rights often drives the pace of technological adoption'. His excellent case study of automated material placement (AMP) equipment, suitable for the manufacture of composite material airframe structures, provides an insight into how the patenting process can act to accelerate the introduction of new ideas (the patent holder usually requires this protection in order to invest in the invention).

McCarville also introduces the idea of disruptive technologies, noting that revolutionary or disruptive inventions are generally made by individuals who are outside the specialist field. Disruptive technologies result in substantial leaps in technological performance or capability of the product. Singh *et al.* (Chapter 4) describe it as follows:

> 'These are innovations that change the rules of the game, introducing a new value proposition. They do not have to outperform the technologies they seek to replace in every sense, but they must offer some new valuable feature that will justify their implementation, at least at a small scale. If enough momentum can be gathered, the new disruptive technology will evolve to replace its predecessor.'

The jet engine is a prime example of a disruptive technology. Frank Whittle saw that the jet engine could address the foremost aeronautical challenges at the time – that is, to fly higher, faster and farther than what was possible with piston engines. Progressive refinements of his design, through thousands of incremental improvements, have produced what is often described as one of the true marvels of twentieth-century engineering: the modern gas turbine engine. During this time, jet engine performance – as measured by specific fuel consumption, thrust-to-weight ratio, noise and exhaust emissions – has continuously improved. But today, we face new challenges related to the environmental damage caused by our global transportation system and the significant dependence of this system on fossil fuels. Singh *et al.* concludes:

> 'While evolutionary improvements to current gas turbines, biofuels, and alternative engine and aircraft configurations are set to lead the way into a new, cleaner paradigm for air travel, evolutionary improvements cannot sustain aviation indefinitely, and we must not lose sight of our long-term aim . . . At current passenger growth rates, a reduction of even 80% in block fuel consumption would still lead to a 20-fold increase in aviation fuel consumption by the end of the century. It is therefore clear that alternative fuels such as hydrogen could one day play a central role in achieving sustainability.'

16.5   Key design drivers

The key design drivers – or challenges – for aeronautics have never been constant: they have evolved as society's needs have evolved. What we see today is an urgent need to address the long-term environmental damage that will be caused by aviation. In Chapter 12, co-editor Hirst looks beyond the 'evolutionary developments of the current air transport system, by responding to defined forces for change that are evident today – such as climate change, hydrocarbon fuel shortages, communication systems in society, and large-scale demographic shifts'. In the process, he explores several revolutionary ideas for the future of air transport.

Poll (Chapter 6) takes a holistic view of this paradigm shift in design priorities, developing techniques that can be used to assess concepts designed to mitigate the environmental damage of air transport. In conclusion, he writes: 'The inescapable requirement to transform aviation into a safe, clean and highly efficient transportation system growing, without constraint, to supply the needs of a sustainable global economy is the greatest challenge yet faced by the aerospace and aviation communities'.

Smith (Chapter 8) considers the case for future supersonic passenger transport aircraft and concludes, like Poll and Hirst, that environmental issues are of paramount importance. Smith notes that,

> '[A reduction of] the sonic boom signature is vital if the [supersonic] vehicle is to be permitted to operate over land and hence have a good chance of being economically viable . . . This single design constraint permeates all aspects of the design, from planform and fuselage geometry, empennage and internal layout, through propulsion integration, fuel and other systems.'

16.6   Moving from concept to implementation

Many authors present a desire to move concepts closer to implementation, believing that the innovators will be able to take impetus from the more certain knowledge that what they contribute will be worthwhile. It is reasonable too to assume that they will appreciate at first hand the constraints that resources, whether time, manpower or money, impose at implementation. This is not easy to achieve once commercial horizons have been set, and the 'best practice' with regard to making any move in this direction is not yet recognised.

Two different approaches do seem to be emerging. One is to 'plan' in depth. Instead of planning with only project constraints as decisive motivators, the breadth of consideration is widened. This is a 'systems engineering' approach, and it has become a more common practice. The focus of each innovative activity tends to be labelled as one or more examples of an 'attribute' or 'property' of the product. Industry has been doing this for a long time. Reliability, maintainability and testability (RM and T) are attributes that have held centre stage for several decades, and are still relevant. The message has been subsumed, however, swallowed and hidden in the attribute matrices that often are treated as the most important management decision-making tools. But innovation is not a matter that fits easily into this jigsaw.

A notably different approach is to enable individuals within a team to take multifaceted perspectives in their stride, thus allowing them to 'manage' the responsibility of technical trade-off more coherently within an application. Whilst well liked, the approach does tend to introduce research into a regime where the uncertainties can lead to much more time being devoted to achieving goals that, whilst they might improve the qualities of the product overall, do not offer enough certainty of a successful outcome to carry confidence overall. The potential that a genuine but false assumption will have consequences that will cascade almost exponentially as the 'complexity' of a programme increases holds too much fear for many managers. Their reticence can be summarised as being critical, but not entirely dismissive, of an approach that has a good orientation, but no guiding light.

16.7   Computer-assisted engineering and design

Much hope emanates from the steady enhancement of computer-aided engineering (CAE) tools. The first computer-aided design (CAD) systems, which began to supplant the drawing board as the preferred means of defining component geometries more than two decades ago, have given way to sophisticated, integrated computer-based tools. These tools enable engineers to readily simulate stress states within structural components, to visualise fluid flow fields around a vehicle and to perform the more mundane tasks of calculating mass, centre of gravity and moments of inertia. In many cases, the associated design databases, which retain the definitive dimensional information and component specifications, can be accessed by sub-contractors and suppliers so that they can contribute to or share data. The ability of a design team to rapidly evaluate the suitability of alternative materials and design solutions using CAE tools is an essential part of modern engineering. And, there is an expectation of increased functionality – and increased connectively between different applications – with each new release of software. This expectation, of course, is rooted in our confidence that the unrelenting increase in computing power, which can be quantified by Moore's Law, will continue unabated. (A popular formulation of Moore's Law – attributed to Gordon Moore, co-founder of Intel – is that the number of transistors on integrated circuits, which is a rough measure of computer processing power, doubles every 18 months.) Indeed, many researchers view high-fidelity simulation as one of the key technology areas that are likely to drive future aeronautics innovation.[2]

CAE tools have captivated the design community's imagination to the extent that a large proportion feel that project innovation can now be expressed and analysed with equivalent cognitive and scrupulous vigour in computer-based systems. It does not necessarily work like that, of course. Henshaw (Chapter 9) provides a note of caution. There is an emerging concern among aerospace companies that their engineers are relying too heavily on these software tools. The essence of the problem is that the tools – while they have speeded up the design process – have also distanced engineers from the underlying scientific principles, mathematical formulae and empirical databases, which are at the core of the engineering design process. This concern was previously noted by McMasters and Cummings,[3] who question whether being able to 'go pretty much from daydream to simulation to some sort of flight-test validation of predictions with the computer and its massive database as the core element of the process' has actually improved the design process. They wrote: 'Have we not actually started merely to codify our biases and assumptions and thus essentially stifle creativity and new configuration explorations by relying on the computer to perform most of the routine mechanical work?'

Looking to the future, however, McMasters and Cummings suggested that the full benefits of computer-based synthesis have yet to be realised and envisage an opportunity to 'use computers to revolutionise the design process, as a complement to, rather than just a copy or extension of, the thought processes of human designers'.[3] Innovation is not just another property of a design; it is an overarching influence. It is liable to change the design paradigm so significantly that the issues that need to be in the adaptive reasoning cannot be placed in the logic tables that are used. Innovation is an aspect of human logic that can be as impenetrable to common wisdom as the apparently inherited mastery of tasks that require lengthy training in most individuals, but seems to be second nature to savants.

16.8   The innovation process

This assertion now begs the question: is there an innovation process and, if so, what does it look like? Henshaw (Chapter 9) provides an insightful evaluation of this question. Drawing on evidence from innovative companies – and individuals – he explains that innovation has a strong cultural attribute. The environment in which people work is the key – employees must have freedom to think, to express their ideas and to be wrong. He cites Steve Jobs, formerly chairman and CEO of Apple Inc., who, when asked 'How do you systemise innovation?', replied: 'You don't . . . You hire good people who will challenge each other every day to make the best products possible'. A workplace environment in which innovation flourishes accommodates the factors that encourage innovators – and, surprisingly, this is not just about financial reward. Innovators, for example, are often driven by a simple desire to succeed in a technical challenge. Success, it appears, can be its own reward.

The development of stealth aircraft is a fascinating example of innovation in action. Virtually every implementation to date has profoundly re-aligned the design balances between aerodynamics, propulsion and structures. It took a dedicated US-based team, working virtually in isolation, to turn concept to reality. In stages, the reality has been improved upon too. Gradually, the concepts are being adopted, refined perhaps, and are now becoming a part of conventional wisdom. There could be other ways of achieving the same goal, but the projects that have matured illustrate what is possible within the commercial and technical capabilities of modern industry. If you doubt that there are other ways, consider the roles of covert aircraft, and consider, amongst the topics introduced in this book, the implications of biologically inspired designs (Chapter 1). Does one large eavesdropping aircraft, vulnerable, but protected, have the same capacity, even capability, as a flock of mechanical pigeons?

At an almost opposite extremity of aeronautical endeavour, one editor (Hirst) recalls a lesson in innovation from human-powered flight. In one week in 1977, he was required to present the technical news (in an international aviation magazine) and was confronted with the claim that an individual, of whom he had never heard, had completed the Kremer Prize challenge (to be awarded for the first human-powered flight covering a one-mile figure-of-eight course). Many had tried since the award was established in 1959 by Henry Kremer, but no one had succeeded. Could it be a scam? A call to a trusted editor solicited an emphatic reply: the man was known to him, he was just the kind of person who, if he set his mind to it, would achieve the goal. The evidence, however, was not convincing; pictures of the *Gossamer Condor* revealed a design with little of the finesse of other designs. The pictures were 'scaled' that evening, and the sheer size of the vehicle raised further doubts about whether this was genuine or a hoax. The evidence from drag calculations was that this was, indeed, an aircraft that could fly using less power than anything else ever designed. There remained a concern, however: the aircraft cruised at barely one-third to one-quarter of the speed of every other known experimenter's aircraft. The following morning, the time of the flight was checked, and the estimated eight minutes (which seemed impossibly long at first sight) proved to be spot on. The *Gossamer Condor* is an excellent example of how a motivated team can 'de-risk' technical problems when an innovative concept is adopted. The team has to be able to analyse the facts, though, through perspectives that hold personal confidence. Innovation is difficult to commit to without being a well-versed practitioner to assess the evidence or without having the necessary level of confidence in the intuitive capacity of the innovator.[4]

The lack of credible analysis techniques and knowledge bases is a dilemma at the core of what so many of our contributors have addressed. They speak in favour of change and welcome the challenge of new pastures being explored, but recognise that the risks involved may not just be difficult to accept, but may be greater than what commercial organisations are willing to accept. Innovation seldom follows a logical course. It frequently relies on novel concepts, which – by their very nature – are reliant on underdeveloped methodologies and knowledge bases. Novel concepts often employ technologies that are at the boundaries of scientific understanding. Inherently, innovative projects can be branded too risky because they are so difficult to analyse. In the human-powered aircraft case, the analysis had required nothing beyond what the team could manage; but, had the aircraft been dependent on something far more esoteric, the right information might not have been available.

The Kremer Prize stimulated a great innovator, Paul MacCready, to compete, and it should not be a surprise to discover that the prize itself was the legacy of an individual who simply sought to stimulate others to address a challenge, the attempt of which, alone, was believed to hold merit. The prize money was neither excessive nor extravagant. Indicative of the success is that the same prize-winning team that had developed the *Gossamer Condor* later developed a human-powered aircraft that crossed the English Channel (winning the second Kremer Prize).

This concept of identifying a goal (or goals) and offering an award was re-stated in more recent time with the X Prize. This competition sought design concepts for a manned aircraft to achieve speed and altitude objectives that were much more challenging, virtually inviting an individual to consider new ways of reaching to the edge of space. Again, the competition drew solutions, many destined never to be funded. In accordance with the rules, only after it had been demonstrated, the prize was awarded to a design from an acknowledged aviation innovator, Burt Rutan. He re-engineered a well-rehearsed concept, using a two-stage approach, with jet and rocket stages in sequence used to exceed the altitude and speed targets associated with the prize. The press interest in more novel ideas was tremendous, and none of these materialised. It is questionable whether Rutan's design was really successful from the perspective of innovation, in that it re-stated an existing, rather than a brand new, paradigm. Prizes do open doors for new ideas, and it is intriguing that recognition, through an award or medal, can be more appealing, it seems, than monetary incentives.[5]

16.9   Developing a culture of innovation

James Albaugh, Executive Vice President of The Boeing Company, has described the aerospace industry as 'risk-averse', where 'long-term visions succumb to short-term profits' and where 'large companies purchase small companies for their innovation rather than innovate on their own'.[6] The challenge for large aerospace companies, therefore, is to develop a corporate culture that values and rewards innovation. Competitions, as described earlier, can feature in such a programme to promote innovation, as they have been shown to be an effective incentive to innovators. This is not so mundane an issue as a company suggestion box that, whilst satisfactory for day-to-day matters, does not set out to develop world-beating inventors. Another aspect of this general problem is that many companies, as they develop into large corporations, especially through acquisitions, lose sight of those people who were vital in the early days of a small business. The team that nurtured the business is often buried beneath the mass of the organisation that grows above them. That team becomes an entity that has a commercial remit that, sadly, too often far outstrips any technical commitment.

Adherents of natural analogy can expound this as an understandable matter. They reason that the situation is no different from a seed being nurtured to grow into a plant, and that the original seed is lost to the world, but by then becomes the source of other seeds. In fact, this is an analogy that comes through in much of what is represented by the authors who have contributed to ideas of 'change', and who have enumerated 'challenges' (as they have been classified in this book). The 'concept' authors are the innovators; they have the ideas.

Murman (Chapter 15) stresses the value of nurturing the desires of individual employees to innovate, and that is probably what industry has to learn most of all. The natural analogy solution would be to identify such individuals and to fund them as a self-sustaining enterprise (they are the saplings, protected by the forest). Murman cites the Boeing Technical Fellow (TF) and Welliver Faculty Fellowship programmes as excellent examples of how a company can firstly nurture valuable employees, and secondly foster a greater understanding of the complexities of the industry within educational institutions (which are training future employees). There are many companies with the equivalent of the Technical Fellow (TF) scheme, but in aerospace there are few that have carried it so far as Boeing. Anyone in academia who has a desire to see deeper into industry welcomes the opportunity to participate in the Welliver programme, and the programme is open to individuals worldwide. The programme provides a way of capturing the hearts and minds of potential innovators who, in this case, are not tied to the company's payroll. It does cost money. It takes time and resources and guarantees nothing.

However, this has not stopped Boeing from offering it. Unfortunately, far too few companies have embraced such thinking.

Innovation has been described as 'a team sport, not a solo sport'. James (Jim) McNerney, Chairman and Chief Executive Officer, The Boeing Company, expressed it as follows: 'It depends on a culture of technical sharing and openness to others. It takes people working together across different groups and organizational lines to make it happen.'[7] The creation of design-build teams, in which specialists from different engineering disciplines (e.g. structures, aerodynamics, propulsion, systems, control and manufacturing) are brought together, is a well-established engineering practice that promotes innovation and reduces development times. Similarly, it has been suggested that the creation of teams that include people from different cultural, ethnic or educational backgrounds can create a dynamic environment in which novel viewpoints may be expressed, and fresh, novel ideas developed.[8]

This team concept is central to the unique management approach of the Defense Advanced Research Projects Agency (DARPA) of the USA, which is described by Wilson in Chapter 11. He explains that there are no permanent research employees and that directors and project managers do not remain at DARPA for longer than four years. Specialists are recruited and given free rein with the clear goal of shortening the line between concept and implementation and, consequently, between instigator and implementer. In some cases, only one individual may be involved. This approach creates a vibrant urgency to develop innovative ideas into prototypes. An environment that encourages innovation is created; an environment which, through structured, managed processes, encourages risk-taking. Wilson quotes Stephen Welby, a former director of DARPA's Tactical Technology Office: 'We start a lot of things, but we also ruthlessly kill them. It's acceptable to fail as long as we have learnt from it.' Ambitious targets are set for new projects. Innovative concepts are conceived – often based on ideas that are at odds with conventional thinking, but without a clear path on how the engineering problems will be solved – and packaged as a challenge to prospective research organisations and companies.

16.10 Innovation 'agendas'

The DARPA approach is very different from that adopted by most state organisations (including the European Union) that fund aerospace research, where the recipients of research funding generally propose their own research projects, albeit to address defined topics. The latter approach encourages funding applications that have a high probability of success (as measured by reaching the defined project targets); but, unfortunately, it discourages proposals that include truly innovative, but inherently riskier, concepts. Evaluators (who are tasked with deciding which proposals get funded) are often sceptical of unconventional thinking. The evaluators can also be constrained by proposal-evaluation guidelines that stipulate that marks be allocated to the perceived probability of success of the project. The net result is that such research funding schemes favour projects that will deliver incremental, evolutionary improvements, above those that have the potential to produce radical, step-change improvements.

In Europe, the European Commission (EC) has played an important role in promoting and guiding aeronautics research. Philippe Busquin, former European Commissioner for Research, tasked a group of eminent aerospace personalities (*ca.* 2000) to produce 'a vision for aeronautics [in Europe] in the year 2020'; a vision that would 'arrive at a safe, efficient and environmentally-friendly air transport system'. The result, *European Aeronautics – A Vision for 2020*, was both imaginative and ambitious.[9] These 2020 targets formed the basis of the ACARE Strategic Research Agendas.[10,11] It is apparent that these 'agendas', in turn, have influenced the direction of aerospace research and innovation in Europe (a sort of grand challenge), not only by guiding the preparation and selection of collaborative EC-funded Framework projects, but also – in a more subtle way – by aligning the objectives of state-funded research and university projects throughout the European Union member states.

There is, however, a duality expressed in these agendas – that is, the requirement to address the needs of society and simultaneously to strengthen the competitiveness of European industries. This duality impacts the outcome of the process: it tends to favour short- to medium-term research and development, where there are clearly defined opportunities for the industrial exploitation of research outcomes, over more radical, high-risk research. The concern is that this process is not sufficiently bold to address the societal issues related to aircraft emissions, environmental damage and the reliance on fossil fuel for commercial aviation – as described by Poll in Chapter 6. Faced with the enormity of these challenges, innovators will need greater freedom to explore unconventional technologies that could hold the key to these problems, and which could deliver quantum changes, rather than incremental improvements. The axiom 'the greater the risk, the greater the reward' clearly does not always apply; nonetheless, the intelligent linking of risk to reward (or technical achievement) needs to be considered in this context. An acceptance of a higher risk of failure by innovators reaching too far needs to be adopted by government agencies and other state organisations that support and fund aeronautics research that addresses these societal challenges.

It would be beneficial to have a forum in which innovators could express their wishes and consider the options. Such a focal point might serve the requirement, clearly outstanding, to stimulate the emergence of new innovation paradigms, but few professional societies have laid a claim to do this. The best residence for such a development would be an all-encompassing, multi-disciplinary organisation. The sightline will gravitate, naturally, towards aviation's learned bodies, but the temptation is to look at a new – by definition fresher – society, for which the International Council of Systems Engineers (INCOSE), founded in 1990, is perhaps the best primogenitor and guardian. INCOSE is, itself, a new seed, and can be argued to be the clear choice if natural analogy is a sensible criterion for guidance in the search. It will bring people from different industries closer to the aeronautical innovators, increasing their likelihood of developing an unfettered view of the horizons they can scan. It is not a definite conclusion, however. The society's own definition of what it addresses is not openly contemplative of innovation. It defines the discipline of systems engineering (always acknowledged to be difficult) as follows:

> Systems Engineering is an interdisciplinary approach and means to enable the realization of successful systems. It focuses on defining customer needs and required functionality early in the development cycle, documenting requirements, then proceeding with design synthesis and system validation while considering the complete problem:
>
> - Operations,
> - Performance,
> - Test,
> - Manufacturing,
> - Cost and schedule,
> - Training and support,
> - Disposal.
>
> Systems Engineering integrates all the disciplines and specialty groups into a team effort forming a structured development process that proceeds from concept to production to operation. Systems Engineering considers both the business and the technical needs of all customers with the goal of providing a quality product that meets the user needs.

INCOSE[12]

The business of systems engineering has adherents and critics, and a loudly expressed criticism is the speed at which its supporters develop block diagrams, espouse 'process' and embrace the formalisation of complex system development programmes. Critics rarely have an answer to the challenge of exponents whose chorus has to be that the process approach works. Henshaw (Chapter 9) argues the case for an innovation process and presents a view of innovation that is formalised, but not hide-bound. In Chapter 15, prepared by Murman, the role of the individual is expressed as a core issue to embrace, and this is a similar call. There is a clear belief that we still have a long way to go before we can embrace innovative concepts, and trade risk with the many easier-to-define properties of a system.

There is a challenge, therefore, for the established fora in organisations such as the American Institute of Aeronautics and Astronautics and the Royal Aeronautical Society, in the USA and UK respectively, to evolve debate from their more focussed specialist teams that will lead towards a blueprint for the better integration of innovation in modern aeronautics.

16.11 Education and innovation

There is also a profound challenge set at the doors of our educational establishments. This concerns the education and training of young engineers, who must become career innovators, not only proficient in the engineering sciences, but also in possession of a deep understanding of the complex, multi-disciplinary nature of the aerospace industry. The difficult question that needs to be addressed is how universities, companies and research organisations, collectively, can produce engineers who will, firstly, challenge conventions with radical thinking, but will still retain the knowledge, skills and aptitude to evaluate these radical ideas through the engineering process; secondly, be visionaries, risk-takers, capable of individual, creative thought, but also team-players who can develop their ideas into useful engineering products; and, thirdly, have integrity and a social responsibility to face the challenges presented by the long-term un-sustainability of our current transportation systems.

These are some of the ideas and conclusions that stem from reading the contributed material and from the mind of the compilation team. It is not unreasonable to expect that others will read these chapters and draw different conclusions, adding strength to arguments that could move innovative developments one way or the other. Meanwhile, the evidence on these pages is that people willing to consider burning issues, and to consider innovative solutions, are still as plentiful as ever. We are convinced that the entries in this book will spawn fruitful debate, which will aid in the development of strategies for the promotion of innovation in aeronautics.

16.12 References

1. Mankins, J.C. (1995). *Technology Readiness Levels: A White Paper*. 6 April 1995, Office of Space Access and Technology, Advanced Concepts Office, NASA.
2. Kroo, I. (2004). Innovations in Aeronautics. AIAA Dryden Lecture, AIAA Paper 2004-0001. *42nd AIAA Aerospace Sciences Meeting*, pp. 1–11. 5–8 January 2004, Reno, NV.
3. McMasters, J.H. and Cummings, R.M. (2002). Airplane design – Past, present, and future. *Journal of Aircraft*, **39**: 10–17.
4. MacCready, P.B. (1978). Flight on 0.33 horsepower – The Gossamer Condor. AIAA-1978-308. *American Institute of Aeronautics and Astronautics, 14th Annual Meeting and Technical Display*. 7–9 February 1978, Washington, DC.
5. McKinsey & Company (2009). *And the winner is . . .: Capturing the promise of philanthropic prizes*. Company report. Available at www.mckinsey.com.
6. Albaugh, J. (2005). *Embracing Risk: A Vision for Aerospace in the 21st Century*. Frank Whittle Lecture, 19 January 2005, Royal Aeronautical Society, London. Available at http:www.boeing.com/news/speeches/.
7. McNerney, J. (2006). *Innovation and the Global Economy*, Distinguished Lecture Series in International Business, 16 May 2006, St Louis University. Available at http:www.boeing.com/news/speeches/.
8. Young, T.M. (2007). Aircraft design innovation: Creating an environment for creativity. *Proc Inst Mech Eng Part G*, **221**: 165–74.

9. Group of Personalities (2001). *European Aeronautics: A Vision for 2020*. Available at http://www.acare4europe.org/html/documentation.asp.
10. Advisory Council for Aeronautics Research In Europe (ACARE) (2002). *Strategic Research Agenda*. Available at http://www.acare4europe.org/html/documentation.asp.
11. Advisory Council for Aeronautics Research In Europe (ACARE) (2004). *Strategic Research Agenda 2*. Available at http://www.acare4europe.org/html/documentation.asp.
12. International Council of Systems Engineers (INCOSE) (2004). *Definition of Systems Engineering*. Available at http://www.incose.org/practice/whatissystemseng.aspx. Accessed 4 February 2011.

Glossary

This glossary concentrates on abbreviations and acronyms. It does not include coefficients used in many technical specialist applications, as these tend to be defined, with the nomenclature appropriate to the application, in individual chapters.

AAC   Active aeroelastic wing
AFP   Automated fiber placement
AFRL   Air Force Research Laboratory
AFTI   Advanced fighter technology integration
AI   Artificial intelligence
AIAA   American Institute of Aeronautics and Astronautics
AMC   Acceptable means of compliance (EASA)
AMP   Automated material placement
APU   Auxiliary power unit
AQP   Advanced Qualification Programme
ARAP   Aeronautical Associates of Princeton
Arinc   Aeronautical Radio Inc.
ASTF   Aeropropulsion systems test facility
ATA   Air Transport Alliance
ATC   Air traffic control
ATL   Automated tape laying
ATLM   Automated tape lamination machine
ATPL   Air transport pilot's licence

BCAR   British Civil Airworthiness Requirements
BoK   Book of knowledge
BPR   By-pass ratio
BWB   Blended wing body

CAFP   Custom automated fiber placement
CFD   Computational fluid dynamics
CFIT   Controlled flight into terrain

CFRP	Carbon fibre reinforced plastic
CHAP	Compact hybrid actuation program
CRM	Crew resource management
CRS	Computerised reservation system
CRT	Cathode ray tube
CS	Certification specification (European Aviation Safety Agency)
CS	Certification Specifications
CTC	Criticality to customer
CTF	Conventional turbo fan
CTLM	Contoured tape laying machine
CTQ	Criticality to quality
CW	Custom winder
DARPA	Defense Advanced Research Projects Agency
DMC	Direct maintenance cost
DOC	Direct operating cost
EAP	Electro active polymers
EASA	European Aviation Safety Agency
ECS	Environmental control system
EFCS	Electronic flight control system
EGPWS	Enhanced ground proximity warning systems
EHM	Engine health monitoring
EI	Emission index (mass of emission/mass of fuel burned)
ETRW	Ratio of energy liberated to revenue work done
EVMS	Earned value management system
F^3	Form fit and function
FAA	Federal Aviation Administration
FAR	Federal Aviation Regulations
FBW	Fly-by-wire
FCL	Flat charge layer
FCS	Flight control system
FFP	Frequent-flier programme
FMS	Flight management system
FPO	Future project office
FTLM	Flat tape laying machine
FW	Filament winding
g	Acceleration due to gravity (9.81 m/s^2 or 32.2 ft/s^2)
GPWS	Ground proximity warning systems
HCTL	High contour tape layer

HDW	Heavy duty winder
HiSAC	Environmentally high speed aircraft
HLFC	Hybrid laminar flow control
HSCT	High speed civil transport
HSR	High speed research
IFSD	In-flight engine shutdown
INCOSE	International Council of Systems Engineering
IP	Intellectual property
IPPD	Integrated product and process development
IPT	Integrated product team
IRR	Internal rate of return
IS	*In situ*
ISW	*In situ* winder
IT	Information technology
JSM	Joint Services munition
LAI	Lean aircraft initiative
LCA	Large commercial aircraft
LCC	Life cycle cost
LCD	Liquid crystal display
LCV	Lower calorific value of fuel ($\approx 43 \times 10^6$ J/kg for kerosene)
L/D	Aircraft lift to drag ratio
LED	Light-emitting diode
LEV	Leading edge vortex
LF	Load factor (actual payload mass/maximum payload mass)
LFC	Cargo load factor (actual cargo mass/maximum possible cargo mass)
LFP	Passenger load factor (actual passenger number/maximum possible)
LIB	Larger is better
LOFT	Line orientated flight training
LOSA	Line operations safety audits
MAV	Micro air vehicle
MEL	Minimum equipment list
MEMS	Micro electro-mechanical systems
MF_{NC}	Mass of fuel carried on journey, but not consumed (reserve + tankered)
MHAFP	Multi-head AFP
MHATL	Multi-head ATL
MHCLP	Multi-head composite lamination platform

MHTLM	Multi-head tape laminating machine
MIT	Massachusetts Institute of Technology
ML	Mass of aircraft on landing
MLP	Multi-layer placement
MMALV	Morphing micro air and land vehicle
MMF	Mass of the mission fuel (fuel actually burned on trip)
MMP	Maximum payload mass (passengers + cargo)
MMTO	Maximum permitted take-off mass
MMZF	Maximum zero fuel mass (maximum mass of aircraft + payload)
MNE	Mixed nozzle ejector
MOE	Aircraft operational empty mass (no payload and no fuel)
MP	Payload mass (passengers + cargo)
MRO	Maintenance repair and overhaul
MSW	Multi spindle winder
MTBF	Mean time between failures
MTF	Mid tandem fan
MZF	Zero fuel mass (mass of aircraft + payload only)
NAL	National Aerospace Laboratory (Japan)
NASA	National Aeronautics and Space Administration
New Gen.	New generation
NIB	Nominal value is better
NLF	Natural laminar flow
NPV	Net present value
NRC	Non-recurring cost
OAD	Overall aircraft design
OEW	Operational weight empty
OPR	Overall pressure ratio
PC	Personal computer
PFE	Precision feed end-effecter
PNF	Pilot not flying
POC	Proof of concept
P&P	Pick and place
psf	Pounds per square foot
QSP	Quiet supersonic platform
QSST	Quiet supersonic transport
R	Great circle distance between departure point and destination (km)
RAINBOW	Reduced and internally-biased oxide wafer
RC	Recurring cost

RDT&E	Research, design, test and evaluation
RF	Radioactive forcing
RFI	Resin film infusion
RFP	Robotic fiber placement
ROM	Rough order of magnitude
RPI	Rapid process impact
RSH	Right sized head
R&T	Research and technology
RVM	Risk-value management
RVM	Risk-value matrix
SAE	Society of Automotive Engineers
SAI	Supersonic Aerospace International
SBJ	Supersonic business jet
SBVG	Sub-boundary vortex generator
SCAR	Supersonic cruise aircraft research
SFTL	Small flat tape layer
SGTDP	Small gas turbine distributed propulsion
SLFC	Supersonic laminar flow control
SMA	Shape memory alloys
SMS	Safety management system
SOBER	Sonic boom European research
SP	Special purpose
SSBD	Shaped sonic boom demonstrator
SSBJ	Supersonic business jet
SST	Supersonic transport
SVS	Synthetic vision system
TAWS	Terrain awareness warning system
TC	Tape cassette
TET	Turbine entry temperature
THUNDER	Thin layer composite uni-morph ferroelastic driver
TQM	Total quality management
TSS	Supersonic transport (France)
TTL	Thermoplastic tape layer
UAV	Unmanned air vehicle (or uninhabited aerial vehicle)
UCAV	Unmanned combat air vehicle
UEET	Ultra efficient engine technology
UIUC	University Of Illinois At Urbana-Champaign
USAF	United States Air Force
V_1	Decision speed

V_S	Stall speed
VAPPS	Vacuum assisted ply placement system
VLCC	Very large container carrier
VSM	Value stream map
VSMA	Value stream map analysis

X	Non-dimensional range ($R.g/(\mathrm{LCV}.\eta_o.L/D)$)
XWB	Extra wide-body

α	Mission fuel mass/take-off mass
β	Mass of fuel carried, but not consumed/take-off mass
ε	Fuel burned minus fuel needed for optimum cruise over same distance
η	Efficiency

Index

CPSIA information can be obtained at www.ICGtesting.com
Printed in the USA
BVOW03*0639240214

345676BV00007BC/108/P